Environmental Science and Engineering

The ultimate goal of this series is to contribute to the protection of our environment, which calls for both profound research and the ongoing development of solutions and measurements by experts in the field. Accordingly, the series promotes not only a deeper understanding of environmental processes and the evaluation of management strategies, but also design and technology aimed at improving environmental quality. Books focusing on the former are published in the subseries Environmental Science, those focusing on the latter in the subseries Environmental Engineering.

Haoqing Xu
Editor

Proceedings of the 8th International Symposium on Water Resource and Environmental Management

Springer

Editor
Haoqing Xu
Department of Geotechnical Engineering
School of Architecture and Civil Engineering
Jiangsu University of Science and Technology
Zhenjiang, China

ISSN 1863-5520 ISSN 1863-5539 (electronic)
Environmental Science and Engineering
ISBN 978-3-032-19087-1 ISBN 978-3-032-19088-8 (eBook)
https://doi.org/10.1007/978-3-032-19088-8

This Springer imprint is published by the registered company Springer Nature Switzerland AG
The registered company address is: Gewerbestrasse 11, 6330 Cham, Switzerland

Preface

This volume brings together 21 selected papers from the 8th International Symposium on Water Resources and Environmental Management (WREM 2025). The contributions reflect the breadth and depth of current research addressing pressing challenges in water resources management, environmental protection, and sustainable development.

The papers cover a wide range of topics, including surface and groundwater systems, water quality and pollution control, ecological restoration, hydrological processes, and emerging water treatment technologies. Several studies focus on groundwater contamination and vulnerability assessment, while others address river basin management, ecological flow determination, and the impacts of large-scale water diversion and ecological engineering projects.

Innovative and interdisciplinary approaches are highlighted throughout the volume, such as remote sensing and UAV monitoring, advanced modeling and optimization techniques, photocatalytic and UV-based water treatment technologies, and nature-based systems-thinking frameworks. Collectively, these contributions illustrate the increasing integration of water, energy, ecology, and carbon considerations in addressing complex environmental challenges.

This volume serves as a valuable reference for researchers, engineers, environmental managers, and policymakers in the fields of water resources, environmental engineering, and sustainability science. It is hoped that the findings presented will stimulate further research, support evidence-based decision-making, and contribute to the development of resilient and sustainable water–environment systems.

Haoqing Xu, Ph.D.
Master Supervisor, Associate Professor, Deputy Director of Department of Geotechnical Engineering, School of Architecture and Civil Engineering
Jiangsu University of Science and Technology
Zhenjiang, Jiangsu, China

Contents

Fostering Marine Sustainability and Coastal Economies with Systems Thinking in Milkfish Aquaculture

Sheila Mae Carungay, Maria Celia Mateo-Fernando, Gretchen Enarle, Sheena Marie Marjolino, Jerome Adriano, Girlie Rueda, Ian Junelord Invierno, Kim Irvin Protacio, and Elaida Fiegalan

Abstract Milkfish aquaculture plays an important role in the Philippines, as it provides food security, rural income sources, and coastal economy boosters. However, the industry is encountering several sustainability issues related to habitat destruction, pollution, resource wastage, and social injustices. This paper seeks to explore how the systems thinking approach can respond to such intertwined challenges and facilitate marine sustainability and economic resilience. Using a documentary analysis, this study analyzes environmental, economic, and social aspects of milkfish aquaculture by examining national statistics, policy papers, and scholarly literature. Based on the concepts of systems thinking and circular economy, the research traces the value chain of the industry and identifies the leverage points through which it could be intervened. The results indicate that it is necessary to combine waste valorization solutions with technological advances, which should include biofloc systems, cooperation with stakeholders, and alignment of policies

---

S. M. Carungay (✉) · M. C. Mateo-Fernando · G. Enarle · S. M. Marjolino · J. Adriano · G. Rueda · I. J. Invierno
Faculty of Management and Development Studies, University of the Philippines Open University, Los Baños, Philippines
e-mail: sheila.mae.carungay@adamson.edu.ph

S. M. Carungay
Industrial Engineering Department, Adamson University, Manila, Philippines

M. C. Mateo-Fernando
Socio-Economic Research and Data Analytics Center, Central Luzon State University, Nueva Ecija, Philippines

S. M. Marjolino
Laguna State Polytechnic University, Santa Cruz, Philippines

K. I. Protacio
Food and Nutrition Research Institute, Department of Science and Technology, Taguig, Philippines

E. Fiegalan
Central Luzon State University Distance, Open, and Transnational University, Nueva Ecija, Philippines

H. Xu (ed.), *Proceedings of the 8th International Symposium on Water Resource and Environmental Management*, Environmental Science and Engineering,
https://doi.org/10.1007/978-3-032-19088-8_1

with international standards. One of the proposed circular economy models shows how sustainability results can be enhanced through the production-to-consumption chain. This study concludes that applying systems thinking in milkfish aquaculture enables the preservation of marine ecosystems, promotes inclusive growth, and supports a sustainable blue economy in the Philippines.

Keywords Milkfish aquaculture · Marine sustainability · Coastal economies · Systems thinking · Circular economy

1 Introduction

The Philippines stands among the top fish-producing nations globally, with milkfish (Chanos chanos), recognized as the national fish, playing a critical role in ensuring national food security, supporting rural livelihoods, and contributing significantly to the country's coastal economies (Bagarinao 1998; Bureau of Fisheries and Aquatic Resources (BFAR): Philippine Fisheries Profile 2023). Per capita consumption of milkfish averages 2.46 kg, accounting for a substantial portion of the 25 kg of fresh fish consumed annually by Filipinos (Philippine Statistics Authority (PSA): Fisheries Situation Report 2019).

However, the industry faces considerable challenges that threaten its sustainability. These include habitat loss from mangrove conversion, nutrient pollution leading to eutrophication and fish kills, high dependence on imported fry, rising production costs, and vulnerability to climate change impacts such as rising sea levels and extreme weather events (Holmer et al. 2002; Macusi et al. 2023; Tahiluddin et al. 2025). Socio-economic issues further complicate the landscape, with inequitable profit distribution, inadequate market access, and potential public health risks from antibiotic misuse (Dunham 2023; Garlock et al. 2024).

Addressing these interconnected challenges requires a holistic, multi-dimensional approach. This paper aims to apply systems thinking to understand the complex relationships shaping milkfish production and consumption in the Philippines, analyze sustainability challenges across environmental, economic, and social dimensions, and propose a circular economy framework that emphasizes waste valorization, technological innovation, and stakeholder collaboration. Ultimately, this study seeks to identify pathways toward a resilient, competitive, and sustainable milkfish industry that supports marine ecosystem health, enhances coastal livelihoods, and aligns with national and global sustainability goals.

To aid in understanding the various acronyms used throughout this paper, a nomenclature is provided in Table 1.

Table 1 Nomenclature

Abbreviation	Full term	Description
BFAR	Bureau of fisheries and aquatic resources	Philippine government agency overseeing fisheries and aquaculture
Codex	Codex Alimentarius	International food code jointly developed by FAO and WHO for food safety and trade
FAO	Food and agriculture organization	UN agency guiding global standards in food and agriculture
IMTA	Integrated multi-trophic aquaculture	Aquaculture system combining species (e.g., milkfish, seaweed, shellfish) for environmental balance
KPI	Key performance indicator	Metrics used to evaluate sustainability outcomes (economic, environmental, social)
SEAFDEC	Southeast Asian fisheries development center	Regional organization promoting sustainable fisheries in Southeast Asia

2 Overview of Milkfish Production and Consumption

According to BFAR, milkfish contributed 37.42% of total aquaculture production value in 2022, valued at PHP 46.41 billion. The industry underpins the livelihoods of thousands of fishers, processors, and small-scale farmers (Bureau of Fisheries and Aquatic Resources (BFAR): Philippine Fisheries Profile 2023). Region I (Ilocos Region) continues to lead milkfish production, with over 143,000 metric tons recorded in 2022 and modest growth in 2023 (Philippine Statistics Authority (PSA): Fisheries Situation Report 2019).

Despite its importance, the sector remains dependent on fry imports from Indonesia and Taiwan, as domestic hatcheries can only supply roughly half of the annual demand (Southeast Asian Fisheries Development Center/Aquaculture Department (SEAFDEC/AQD): Milkfish Fry Sufficiency 2024). Furthermore, shifting market trends, such as the expansion of shrimp farming and price volatility, have contributed to fluctuations in production volumes (Macusi et al. 2022). The growing impacts of climate change, such as extreme weather events and sea level rise, pose additional threats to infrastructure and harvest stability in the aquaculture sector.

3 Sustainability Challenges in Milkfish Industry

Milkfish production in the Philippines faces numerous sustainability challenges spanning environmental, economic, and social dimensions. Environmentally, extensive areas of mangrove forests have been converted to fishponds, reducing biodiversity and undermining crucial coastal ecosystem services (Bagarinao 1998; International Union for Conservation of Nature and Natural Resources (IUCN): Mangroves for

Our Future 2017;). Excessive feed use leads to nutrient loading, causing eutrophication and occasional fish kills (Holmer et al. 2002; San Diego-McGlone et al. 2024). The industry also depends heavily on fossil fuels across hatchery, grow-out, and processing stages, impacting profitability and contributing to greenhouse gas emissions (The Milkfish Technical Committee: The Philippines Recommends for Milkfish 2016).

Economically, high feed costs—representing over 50% of production expenses—along with limited access to capital, technology, and technical support, hinder small-scale farmers' competitiveness (Southeast Asian Fisheries Development Center (SEAFDEC): SEAFDEC Annual Report 2022; Sumalpong and Embornas 2024). Socially, while milkfish aquaculture generates employment and supports rural economies (Shalli et al. 2024), issues like inequitable profit distribution, potential displacement from large-scale farm expansion, and antibiotic misuse raise concerns for public health and social equity (Garlock et al. 2024).

Recent literature reinforces these observed challenges by highlighting specific environmental impacts, socio-economic risks, and potential innovations in Philippine milkfish aquaculture. Table 2 summarizes key studies that support the multi-dimensional sustainability issues discussed, emphasizing both system weaknesses and emerging solutions explored in national and regional contexts.

Table 2 Comparative literature review

Author(s)	Focus	Key findings
Holmer et al. (2002)	Environmental impacts	Milkfish farms in Bolinao contributed to sediment and nutrient loading; waste management needed
Macusi et al. (2023)	Farm typologies, socio-economics	Identified production models and socio-economic risks including antibiotic misuse and inequality
SEAFDEC (2022)	Technological trials	Demonstrated potential of IMTA and biofloc systems for sustainable nutrient use and profitability
Garlock et al. (2024)	Sustainability indicators and benchmarking	Emphasized triple-bottom-line KPIs and benchmarking for aquaculture performance
Zapata et al. (2023)	Value chain and stakeholder analysis	Mapped bottlenecks in production and policy coordination in Region I
Sampathkumar et al. (2023)	Waste valorization	Explored conversion of aquaculture by-products into feed and fertilizer to support circular systems

4 Methodology

This study adopts a qualitative research design using documentary analysis of secondary sources, grounded in systems thinking and circular economy frameworks, to examine the sustainability challenges and opportunities in Philippine milkfish aquaculture. The analysis draws exclusively on secondary data, including government reports, industry roadmaps, peer-reviewed literature, and existing statistical records. Key sources include the National Milkfish Industry Roadmap 2021–2040 (Bureau of Fisheries and Aquatic Resources (BFAR): National Milkfish Industry Roadmap 2021–2040 2022), fisheries production profiles (Bureau of Fisheries and Aquatic Resources (BFAR): Philippine Fisheries Profile 2023), and academic research on environmental, economic, and social aspects of milkfish aquaculture (Macusi et al. 2023; San Diego-McGlone et al. 2024).

A systems thinking perspective guided the mapping of the milkfish value chain, helping to identify stakeholders, feedback loops, and critical leverage points where interventions could improve sustainability outcomes (Iacovidou et al. 2021). The study further benchmarked sustainability indicators and circular economy practices against international standards, such as FAO guidelines on responsible aquaculture and Codex standards for food safety and trade compliance (Food and Agriculture Organization of the United Nations (FAO): The State of World Fisheries and Aquaculture 2018: Meeting the Sustainable Development Goals 2018; Food and Agriculture Organization of the United Nations (FAO) and World Health Organization (WHO): Code of Practice for Fish and Fishery Products 2020).

Insights were synthesized thematically to highlight environmental impacts, economic constraints, and social equity issues, incorporating globally recognized performance indicators and best practices (Garlock et al. 2022; Valenti et al. 2018). This approach enabled the identification of integrated strategies and policy recommendations to support a more resilient and sustainable milkfish industry.

5 Systems Analysis and Sustainable Pathways for Milkfish Aquaculture

Applying systems thinking to milkfish aquaculture provides a holistic lens for examining the industry's interconnected environmental, economic, and social challenges. Mapping the value chain—from fry production to grow-out, processing, and distribution, and consumption— reveals feedback loops where resource inefficiencies, ecological pressures, and socio-economic inequities reinforce each other (Zapata et al. 2023). This systems view clarifies where strategic interventions can foster long-term sustainability (Iacovidou et al. 2021).

As illustrated in Fig. 1, the value chain of milkfish aquaculture in the Philippines reveals interconnected stages and feedback loops, from fry sourcing to final distribution. Data drawn from BFAR statistics and the National Milkfish Roadmap

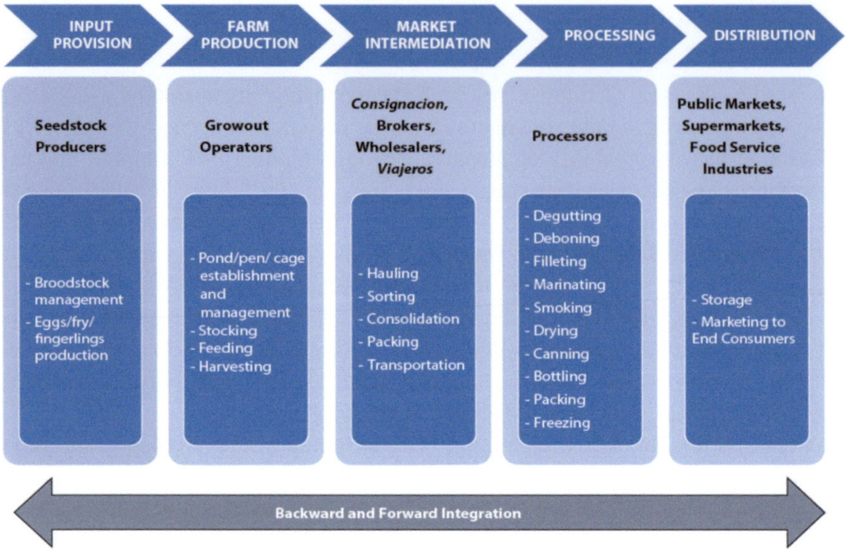

Fig. 1 Value chain mapping of the Philippine milkfish industry, highlighting key production stages and systemic vulnerabilities (Bureau of Fisheries and Aquatic Resources (BFAR): Philippine Fisheries Profile 2023; Macusi et al. 2023; Bureau of Fisheries and Aquatic Resources (BFAR): National Milkfish Industry Roadmap 2021; Zapata et al. 2023)

indicate vulnerabilities such as dependence on imported fry and rising feed costs, particularly in Region I and parts of Western Visayas (Bureau of Fisheries and Aquatic Resources (BFAR): Philippine Fisheries Profile 2023; Bureau of Fisheries and Aquatic Resources (BFAR): National Milkfish Industry Roadmap 2021–2040 2022). Macusi et al. further identify how these issues cascade into socio-economic risks, including profit inequity and antibiotic misuse (Macusi et al. 2023).

5.1 Circular Economy Approaches

A circular economy approach creates pathways to reduce waste, close nutrient loops, and strengthen economic resilience. In milkfish aquaculture, this includes valorizing waste into biofertilizers, animal feed, bioplastics, and nutraceuticals, thereby reducing marine pollution and enhancing coastal livelihoods (Sampathkumar et al. 2023; Venugopal 2021). Enabling policies, market-based incentives, and capacity building are essential to mainstream these practices (Fadeeva and Berkel 2023; Alemu and Azadi 2018).

Figure 2 presents a circular economy model grounded in documented practices in Philippine aquaculture. For example, studies in Bolinao, Pangasinan, have shown how uneaten feed and fish waste contribute to eutrophication, highlighting the need for waste conversion strategies (Holmer et al. 2002; Chambers et al. 2024). In

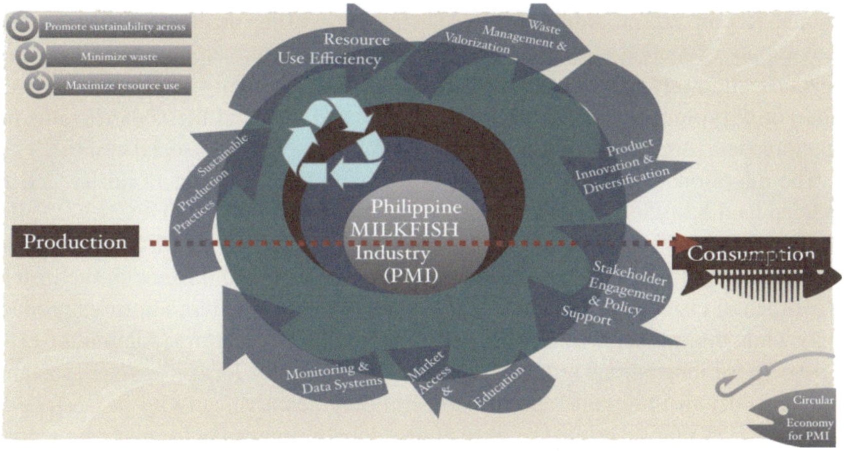

Fig. 2 Circular economy framework for sustainable milkfish aquaculture in the Philippines, integrating real-world waste valorization, feed innovation, and renewable energy initiatives (Holmer et al. 2002; Southeast Asian Fisheries Development Center (SEAFDEC): SEAFDEC Annual Report 2022; ; Sampathkumar et al. 2023; Altamirano et al. 2019)

response, SEAFDEC and other research agencies have piloted the conversion of fish waste into biofertilizers and supplemental feeds using microbial fermentation and solar drying technologies (Sampathkumar et al. 2023). Additionally, IMTA trials integrating milkfish, seaweeds, and mussels in Ilocos Norte have demonstrated reduced nutrient discharge and improved water quality, offering replicable blue economy models (Altamirano et al. 2019).

5.2 *Technological and Management Innovations*

Technological and management innovations are essential in driving the sustainability and resilience of the milkfish aquaculture industry. Biofloc technology, which leverages microbial communities to recycle nutrients and maintain water quality, has demonstrated potential in reducing feed costs and mitigating water pollution, thereby improving farm profitability and reducing ecological impacts (Chambers et al. 2024). Similarly, Integrated Multi-Trophic Aquaculture (IMTA) systems—where milkfish are cultured alongside seaweeds and shellfish—create ecological synergies that reduce waste discharge, stabilize water quality, and diversify income streams for coastal communities (Altamirano et al. 2019).

Feed innovation is another critical area, with the shift toward balanced pelletized diets incorporating plant-based proteins, probiotics, algal paste, and insect meal offering promising solutions to reduce dependence on fishmeal and improve fish health (Southeast Asian Fisheries Development Center (SEAFDEC): SEAFDEC

Annual Report 2022). Additionally, selective harvesting practices ensure continuous supply of market-sized fish while minimizing pressure on marine ecosystems. Beyond technology adoption, innovations in farm management—including improved biosecurity protocols, climate risk assessments, and financial literacy programs for smallholders—support long-term industry competitiveness and social equity.

The adoption of biofloc technology by small-scale farms in Iloilo has led to reduced feed costs and improved water reuse, as reported in SEAFDEC's 2022 pilot studies (Southeast Asian Fisheries Development Center (SEAFDEC): SEAFDEC Annual Report 2022). Likewise, integrated systems pairing milkfish with shellfish in experimental cages in Ilocos Norte demonstrate how polyculture improves productivity while minimizing waste—a key pillar in circular aquaculture (Altamirano et al. 2019). These innovations reflect the growing potential for locally adapted, sustainable practices in Philippine aquaculture. Building adaptive capacity in the face of climate variability is a critical priority under global frameworks for marine resource utilization and sustainable coastal development.

5.3 Complex Network System Analysis

A complex network system perspective offers valuable insights into the multifaceted interactions among stakeholders, resources, and processes in the Philippine milkfish industry. Figure 3 visualizes the stakeholder network supporting milkfish production and trade, showing connections among hatchery operators, processors, traders, and regulators. Value chain assessments in Region I revealed that poor coordination among these actors often leads to market inefficiencies and redundant regulation (Zapata et al. 2023). In Bolinao, data also show how weak enforcement mechanisms allow unsustainable practices to persist, emphasizing the need for better institutional alignment (Holmer et al. 2002). By mapping these relationships, the model highlights where interventions—such as policy harmonization or digital monitoring—can trigger system-wide improvements.

This network demonstrates that challenges such as fry scarcity, environmental pollution, antibiotic misuse, and socio-economic inequality are not isolated problems but emerge from dynamic interactions across the industry (Macusi et al. 2023; Tahiluddin et al. 2025). The diagram highlights critical nodes where interventions—such as circular economy practices, biosecurity enhancements, and sustainability policies—can create positive ripple effects throughout the system (Sampathkumar et al. 2023; Venugopal 2021).

For example, links with research institutions and government agencies indicate the potential for strengthening waste valorization initiatives, digital monitoring systems, and adaptive management strategies (Garlock et al. 2024). Connections with consumer groups and local markets point to opportunities for improving traceability, food safety, and community-based education campaigns (Alemu and Azadi 2018). The network also reveals feedback loops, where improved environmental

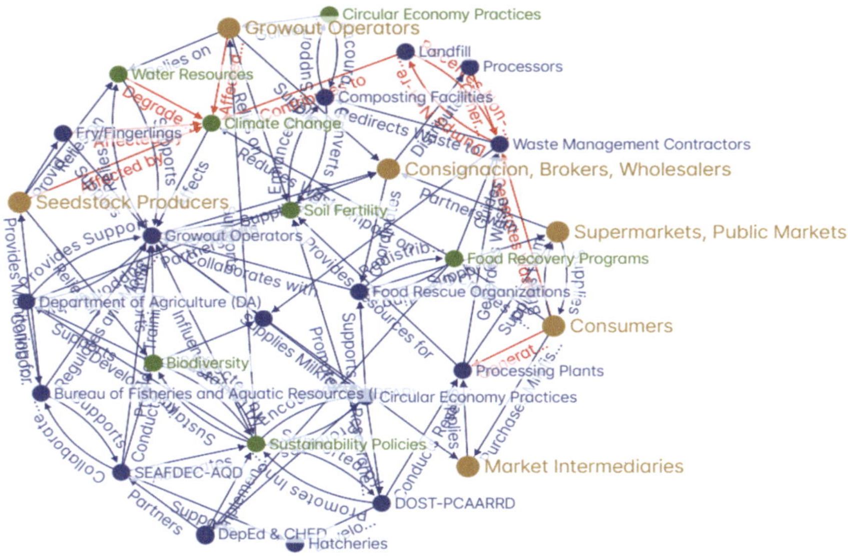

Fig. 3 Complex stakeholder network involved in Philippine milkfish production and consumption, showing interactions between hatcheries, farmers, processors, traders, regulators, and research institutions (Macusi et al. 2023; Bureau of Fisheries and Aquatic Resources (BFAR): National Milkfish Industry Roadmap 2021–2040 2022; ; Iacovidou et al. 2021; Zapata et al. 2023)

practices reduce resource waste, thereby enhancing economic efficiency and building stakeholder trust (Valenti et al. 2018).

Adopting a complex network system perspective aligns with systems thinking by moving beyond linear cause-effect models and recognizing the interdependencies inherent in marine resource utilization, coastal economies, and sustainability transitions (Iacovidou et al. 2021). This approach underscores the importance of collaborative governance, multi-sectoral partnerships, and continuous learning to address the industry's challenges while advancing marine ecosystem health and socio-economic equity.

6 Global Frameworks and Performance Evaluation

Global standards, particularly those set by FAO and Codex, offer guidance for responsible aquaculture, emphasizing food safety, biosecurity, ecosystem protection, and fair trade (Food and Agriculture Organization of the United Nations (FAO): The State of World Fisheries and Aquaculture 2018: Meeting the Sustainable Development Goals 2018; Food and Agriculture Organization of the United Nations (FAO) and World Health Organization (WHO): Code of Practice for Fish and Fishery Products

2020). Although Philippine policies, such as the Fisheries Code, integrate environmental safeguards and spatial planning, enforcement remains inconsistent (Teh and Pauly 2018). Aligning local practices with international frameworks strengthens industry competitiveness and resilience against climate change, consistent with recommendations in the Milkfish Industry Roadmap 2021–2040 (Bureau of Fisheries and Aquatic Resources (BFAR): National Milkfish Industry Roadmap 2021–2040 2022). Integrating these global perspectives supports the transition toward sustainable marine management, reduction of marine pollution, and the advancement of coastal economies.

Effective monitoring and evaluation (M&E) frameworks are vital for tracking progress across sustainability pillars. Economic KPIs should include profitability and capital efficiency; environmental KPIs should cover feed conversion ratios, Greenhouse Gas emissions, and pollution reduction; social KPIs should address equity, gender inclusion, and community well-being (Garlock et al. 2024; Valenti et al. 2018). Technologies like Artificial Intelligence, Internet of Things sensors, and Geographic Information System improve real-time data collection on fish health, water quality, and climate variability (Mandal and Ghosh 2024). These systems enable adaptive management, transparency, and data-driven policymaking aligned with marine ecosystem health and climate change adaptation goals.

7 Synthesis of Findings and Recommendations

This study applied systems thinking to understand the sustainability challenges in Philippine milkfish aquaculture, addressing the first research objective by mapping the complex interdependencies across environmental, economic, and social dimensions. A circular economy framework was developed, answering the second objective by demonstrating how waste valorization, technological innovation, and stakeholder collaboration can build resilience, reduce pollution, and enhance livelihoods. Finally, aligning local practices with international standards, improving monitoring systems, and promoting inclusive policy reforms address the third objective of identifying pathways toward sustainable, competitive, and equitable milkfish aquaculture.

Building on the findings of this study, several strategic actions are recommended to advance the sustainability of milkfish aquaculture, protect marine ecosystem health, and support coastal economies. First, it is essential to harmonize local aquaculture regulations with Codex and FAO standards to strengthen food safety, sustainability practices, and trade competitiveness. Expanding the adoption of IMTA, biofloc technology, and feed innovations will help improve environmental outcomes while enhancing farm profitability. Additionally, government-led capacity building and targeted incentives should support smallholders and producer cooperatives to encourage inclusive participation in the transition toward sustainable practices. Finally, strengthening participatory policymaking and fostering collaboration among stakeholders in planning and regulation processes are crucial steps to ensure long-term industry resilience and equity.

Future studies should quantify the environmental and socio-economic impacts of circular economy practices, assess digital technologies' role in real-time monitoring and risk management, and develop climate adaptation models tailored to milkfish aquaculture systems.

References

Alemu, A.E., Azadi, H.: Fish value chain and its impact on rural households? income: lessons learned from Northern Ethiopia. Sustainability **10**(10), 3759 (2018)

Altamirano, J.P., Nambu, R., Salayo, N.D., Kodama, M. (eds.): Understanding current challenges and future prospects in Integrated Multi-Trophic Aquaculture (IMTA) research. In: Proceedings of the JIRCAS-SEAFDEC/AQD Joint Workshop on IMTA Research, SEAFDmEC/AQD, Tigbauan Main Station, Iloilo, Philippines, 6?8 August 2019. SEAFDEC/AQD, Tigbauan (2022).

Bagarinao, T.: Historical and current trends in milkfish farming in the Philippines. Tropical Maricul. 381?422 (1998)

Bureau of Fisheries and Aquatic Resources (BFAR): National Milkfish Industry Roadmap 2021?2040, release 2022. https://www.bfar.da.gov.ph, Last accessed 29 June 2025

Bureau of Fisheries and Aquatic Resources (BFAR): Philippine Fisheries Profile 2023, release July 2024. https://www.bfar.da.gov.ph, Last accessed 29 June 2025

Chambers, M., Coogan, M., Doherty, M., Howell, H.: Integrated multi-trophic aquaculture of steelhead trout, blue mussel and sugar kelp from a floating ocean platform. Aquaculture **582**, 740540 (2024)

Dunham, R.: In: Aquaculture and Fisheries Biotechnology: Genetic Approaches. CABI (2023)

Fadeeva, Z., Van Berkel, R.: Towards circular economy of food systems: an explorative appraisal of opportunities in fish, seafood value chains. In: Sustainable Food Value Chain Development: Perspectives from Developing and Emerging Economies, pp. 61?86. Springer, Singapore (2023)

Food and Agriculture Organization of the United Nations (FAO) and World Health Organization (WHO): Code of Practice for Fish and Fishery Products. Rome (2020)

Food and Agriculture Organization of the United Nations (FAO): The State of World Fisheries and Aquaculture 2018: Meeting the Sustainable Development Goals.: FAO, Rome (2018)

Garlock, T., Asche, F., Anderson, J., Ceballos-Concha, A., Love, D.C., Osmundsen, T.C., Pincinato, R.B.M.: Aquaculture: the missing contributor in the food security agenda. Glob. Food Sec. **32**, 100620 (2022)

Garlock, T.M., Asche, F., Anderson, J.L., Eggert, H., Anderson, T.M., Che, B., Chávez, C.A., Chu, J., Chukwuone, N., Dey, M.M., Fitzsimmons, K., Flores, J., Guillen, J., Kumar, G., Liu, L., Llorente, I., Nguyen, L., Nielsen, R., Pincinato, R.B.M., Sudhakaran, P.O., Tibesigwa, B., Tveteras, R.: Environmental, economic, and social sustainability in aquaculture: the aquaculture performance indicators. Nat. Commun. **15**(1), 5274 (2024)

Holmer, M., Marbá, N., Terrados, J., Duarte, C.M., Fortes, M.D.: Impacts of milkfish (Chanos chanos) aquaculture on carbon and nutrient fluxes in the Bolinao area, Philippines. Marine Pollut. Bullet. **44**(7), 685?696 (2002)

Iacovidou, E., Hahladakis, J.N., Purnell, P.: A systems thinking approach to understanding the challenges of achieving the circular economy. Environ. Sci. Pollut. Res. **28**, 24785?24806 (2021)

International Union for Conservation of Nature and Natural Resources (IUCN): Mangroves for our future, release September 2017. https://iucn.org, Last accessed 29 June 2025

Macusi, E.D., Estor, D.E.P., Borazon, E.Q., Clapano, M.B., Santos, M.D.: Environmental and socioeconomic impacts of shrimp farming in the Philippines: a critical analysis using PRISMA. Sustainability **14**(5), 2977 (2022)

Macusi, E.D., Macusi, E.S., Bongas, H.P., Cayacay, M.A.: Typology of milkfish (Chanos chanos) farms: their operations, socio-economic viability, and challenges. Preprints 2023120507 (2023)
Mandal, A., Ghosh, A.R.: Role of artificial intelligence (AI) in fish growth and health status monitoring: a review on sustainable aquaculture. Aquacult. Int. **32**(3), 2791?2820 (2024)
Philippine Statistics Authority (PSA): 2023 Fisheries Situation in Ilocos Region, release July 2024. https://rsso07.psa.gov.ph, Last accessed 29 June 2025
Philippine Statistics Authority (PSA): Fisheries Situation Report 2019, release December 2020. https://rsso07.psa.gov.ph, Last accessed 29 June 2025
Sampathkumar, K., Yu, H., Loo, S.C.J.: Valorisation of industrial food waste into sustainable aquaculture feeds. Future Foods **7**, 100240 (2023)
San Diego-McGlone, M.L., Yñiguez, A.T., Benico, G., Lum, W.M., Hii, K.S., Leong, S.C.Y., Leaw, C.P., Iwataki, M., Lim, P.T.: Fish kills related to harmful algal bloom events in Southeast Asia. Sustainability **16**(23), 10521 (2024)
Shalli, M.S., Mmochi, A.J., Rubekie, A.P., Yona, G.K., Shoko, A.P., Limbu, S.M., Mapenzi, L.L.: The contribution of milkfish (Chanos chanos) pond farming to socio-economics and coastal community livelihoods for a sustainable blue economy in Tanzania. Aquacult. Int. **32**(4), 4915?4931 (2024)
Southeast Asian Fisheries Development Center (SEAFDEC): SEAFDEC Annual Report 2022.: Bangkok, Thailand (2022)
Southeast Asian Fisheries Development Center/Aquaculture Department (SEAFDEC/AQD): Milkfish Fry Sufficiency, release 2024. https://www.seafdec.org.ph, Last accessed 29 June 2025
Sumalpong, R., Embornas, A.: Blue economy for small-scale fisheries: challenges from a coastal community in the Southern Philippines. Langkit J. Soc. Sci. Humanit. **13**, 63?82 (2024)
Tahiluddin, A.B., Bornales, J.C., Limbaro, G.R.A., Paudac, M.A.U., Amarille, R.K., Sirad, N.R., Kabirun, M.C., Ujing, R.A., Gonzaga-Torino, F.M., Sabdani, M.H., Bacla-an, R.E., Hairal, M.A.S., Magcanta-Mortos, M.L.M., Esguerra, J.P.: Environmental impacts of aquaculture in the Philippines. The Israeli J. Aquacul.?Bamidgeh **77** (2025)
Teh, L.C.L., Pauly, D.: Who brings in the fish? The relative contribution of small-scale and industrial fisheries to food security in Southeast Asia. Front. Mar. Sci. **5**, 44 (2018)
The Milkfish Technical Committee: The Philippines Recommends for Milkfish.: Philippine Council for Agriculture, Aquatic and Natural Resources Research and Development of the Department of Science and Technology (DOST-PCAARRD), Los Baños (2016)
Valenti, W.C., Kimpara, J.M., Preto, B.L., Moraes-Valenti, P.: Indicators of sustainability to assess aquaculture systems. Ecol. Ind. **88**, 402?413 (2018)
Venugopal, V.: Valorization of seafood processing discards: bioconversion and bio-refinery approaches. Front. Sustain. Food Syst. **5**, 611835 (2021)
Zapata, N.R., Sarmiento, J.A.C., Cruz, M.B.: Development of smart food value chain intervention models for the milkfish industry in Region 1, Philippines. Asian J. Agricul. Developm. **20**(1), 87?108 (2023)

Macrozoobenthos as a Mirror of Karst Springs

Melisa Curić, Tamara Ivetić, Nemanja Pankov, and Branko Miljanović

Abstract Pester Plateau is a unique karst region in Serbia, notable for its ephemeral and permanent springs. Its rich biodiversity, combined with distinctive geomorphological, geological, hydrological, and climatic features, shapes the fauna of benthic organisms inhabiting these springs. The porous soil and low filtration capacity make the springs highly vulnerable to pollution due to strong underground connections. A two-year study analyzed physical and chemical parameters, assessed water quality using macrozoobenthos, examined faunal diversity and community structure, identified new taxa, and proposed conservation measures and reference site designation for the karst springs of Pester Plateau, highlighting macrozoobenthos as a mirror of karst ecosystems. The dominant species across samples was *Gammarus balcanicus* (Schaferna 1922). The most diverse macroinvertebrate group included aquatic insects, such as Trichoptera (45 species), Plecoptera (19), Ephemeroptera (14), and Coleoptera (9 taxa). The rare circumpolar species *Arcynopteryx dichroa* (McLachlan 1872) was also recorded. Several species listed as strictly protected ("Official Gazette of RS," No. 5/2010) and endangered in Serbia were found, including *Dinocras megacephala*, *Nemoura cambrica*, *Protonemura meyeri*, *Protonemura praecox*, *Baetis pavidus*, *Allogamus auricollis*, and *Thremma anomalum*. To assess ecological status and environmental pressures, macrozoobenthos communities were evaluated using the saprobic index, diversity index, BMWP score, and EPT index.

Keywords Pešter Plateau · Karst springs · Macrozoobenthos · Biodiversity

M. Curić (✉) · N. Pankov · B. Miljanović
Department of Biology and Ecology, Faculty of Sciences, University of Novi Sad, Novi Sad, Serbia
e-mail: melisa.numanovic@gmail.com

T. Ivetić
Department of Physics, Faculty of Sciences, University of Novi Sad, Novi Sad, Serbia

H. Xu (ed.), *Proceedings of the 8th International Symposium on Water Resource and Environmental Management*, Environmental Science and Engineering,
https://doi.org/10.1007/978-3-032-19088-8_2

1 Introduction

A karst landscape is defined as a terrain primarily composed of limestone and dolomite, where the topography is shaped by the dissolution of limestone rocks (Field 2002). It is characterized by the presence of sinkholes, sinking rivers, caves, underground lakes, and subterranean drainage systems. What is typical of karst areas is the porosity of the soil, characterized by well-developed networks of underground water and lakes that are connected to surface springs. Due to the high porosity of the surface layers, especially in rocky areas, the amount of surface water is much lower compared to non-karst regions. The presence of groundwater in these areas contributes to the dissolution of limestone rocks. In karst regions, the soil is porous, so water constantly drains into the subterranean water network, which may re-emerge in other parts of the karst field—for example, in caves or springs of various types—only to disappear again after a certain course (Ford and Williams 2007).

Karst systems can represent diverse freshwater ecosystems that provide insight into both underground and surface waters (Smith et al. 2003). The influence of underground water on downstream changes largely depends on the flow rate and the distance between the upstream and downstream sections.

Alterations caused by orographic, hydrological, and other environmental factors, including pollution, can lead to the adaptation of organisms to existing conditions (Hellawell 2012; Kurbanov et al. 2021). Considering that not all organisms have the same capacity to adapt to changes in ecological factors, it is important to note that some are more sensitive. These sensitive species, known as bioindicators (Niemi and McDonald 2004; Stojanović 2017), are typically the first to disappear from altered aquatic ecosystems.

It is particularly important to emphasize that bioindicators may respond differently to changes in aquatic environments, which is usually reflected in alterations in their physiology, behavior, morphology, and, in some cases, survival rates (Stojanović 2017; Pander and Geist 2013). Based on the responses of bioindicators, we can conclude the stability of aquatic ecosystems or the presence of potential pollutants (Stojanović 2017; Knoben and Roos 1995; Dale and Beyeler 2001; Dziock et al. 2006).

One of the striking features of karst areas is the presence of temporary lotic ecosystems (Meyer et al. 2003), where aquatic macroinvertebrates represent a key component in the system's dynamics. These organisms play a crucial role in energy flow and nutrient cycling. Moreover, they are widely used in the biomonitoring of aquatic ecosystems (Sandin and Hering 2004). Given the unique characteristics of karst landscapes and the sensitivity of watercourses within them, it is essential to highlight the importance of biomonitoring in karst waters, especially in karst springs. Among the various abiotic factors that influence aquatic ecosystems, karst springs hold a special place. These are often intermittent (ephemeral) springs that respond dynamically to periods of rainfall and drought (Zollhöfer et al. 2000).

It has been established that the composition of benthic fauna in these springs largely depends on the type of substrate. Earlier studies (Williams 1991; Zalewski et al. 1997; Beierkuhnlein and Gollan 1999) have addressed not only the spring fauna itself but also its immediate surroundings. These studies examined the benthic fauna within the spring as well as changes in community composition along the flow path. However, it is important to emphasize that, overall, there is still a limited number of studies addressing this specific issue (Cantonati et al. 2006). A large number of studies have investigated the structure of benthic fauna in various springs, depending on the substrate type characterized by its granulometric composition (Marković 1998; Fumetti et al. 2006; Sambugar et al. 2006; Dumnicka et al. 2007; Numanović et al. 2021). It has been established that the composition of benthic fauna strongly depends on the substrate type.

Research has shown that in karst springs, physicochemical parameters have very limited influence on the composition of macrozoobenthic communities. In contrast, substrate structure parameters—such as silt, leaf litter, clay, and rock slabs—have a significant impact on community composition (Marković 1998; Fumetti et al. 2006; Sambugar et al. 2006; Dumnicka et al. 2007; Numanović et al. 2021, 2023).

Springs can be viewed as distinct ecological units characterized by the stability of several environmental factors, such as water temperature, specific chemical parameters, and substrate structure. The degree of isolation varies depending on the density of spring distribution and the extent of changes in these parameters along the flow path.

In such environments, taxa with varying ecological requirements have been recorded, including crenobionts, crenophilic, rheophilic, and semi-aquatic species, as well as species with broad ecological valence (Cantonati et al. 2006). The frequency and dominance of particular groups are determined by the characteristics of each spring, ranging from its morphometry to its chemical properties.

The aim of the study was to investigate the macrozoobenthos communities of karst springs on the Pešter Plateau (Southwestern Serbia), with a focus on determining biotic indices and identifying ecological factors influencing the structure of these communities.

2 Materials and Methods

2.1 Research Area

The Pešter Plateau is one of the last well-preserved and largest karst fields in Serbia. This area is characterized by unique geomorphological, geological, hydrogeological, hydrological, and climatic phenomena, as well as distinctive landscapes (Fig. 1).

Pešter is particulary known for its specific microclimate, especially its severe winters. The lowest officially recorded temperature in Serbia, − 39.5 °C, was measured here, which earned this plateou the nickname "Balkan Siberia." The heterogeneous karst landscape of the Pešter Plateau creates a variety of unique habitats and supports numerous endangered species.

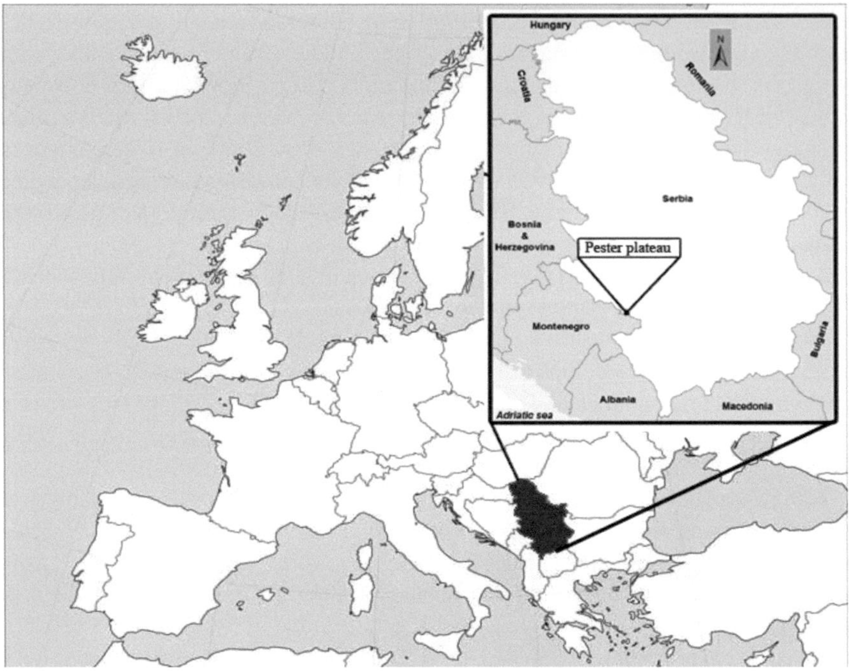

Fig. 1 Geographical position of the Pešter Plateau within Europe

2.2 *Sampling Methods*

In the spring, summer and autumn of 2016 (June to December), we sampled the fauna of the springs of the Pešter Plateau in their upper and middle reaches, as well as in the sinkhole. Benthic macroinvertebrate samples were collected using a Surber sampler (area 500 cm^2 area, mesh size 250 μm) with four replicates.

Since the structure and composition of macrozoobenthos are influenced by the morphological characteristics of the substrate, sampling was carried out in different habitats. Initially, the distribution of substrate type was analyzed by visual assessment, following the previously described methods.

The substrate composition was classified into five categories: fine substrates, sludge (<0.125 mm), sand (0.125–2 mm), gravel (2–64 mm), stone (64–256 mm), and rocks (>256 mm).

A portion of each sample was analyzed in its native state, while the remainder was preserved in 75% ethyl alcohol. Identification of macrozoobenthos was carried out using taxonomic keys.

2.3 *Physical and Chemical Parameters of Water*

The following physical and chemical parameters of water were measured in situ: pH, electrical conductivity, and water temperature using a HI 98,130 HANNA COMBO Tester for pH/EC/TDS/T (USA). Dissolved oxygen concentration and oxygen saturation were measured using a HANNA HI 9146–04/10.

Biochemical oxygen demand over 5 days (BOD_5) was determined according to the Royal Commission on River Pollution methodology, while chemical oxygen demand (COD) was measured using the Kubel-Tiemann method (titration with potassium permanganate in an acidic solution).

Alkalinity and carbonate hardness (expressed in German degrees, °dH) were calculated by multiplying the alkalinity ($CaCO_3$ mg/L) by a factor of 2.8. Concentrations of ammonium, nitrite, nitrate, orthophosphate, iron, copper, silicon, chlorine, and fluoride were determined using an AQUA-CHECK 2 photometer (Söll, Germany).

3 Results

The taxonomic analysis of benthic invertebrates from the springs of the Pešter Plateau revealed the presence of 108 species, belonging to 72 genera and 45 families. The highest diversity was recorded in the orders Trichoptera (45 species), Plecoptera (19 species), Ephemeroptera (14 species), and Coleoptera (9 taxa). Non-insect groups of macrozoobenthos were also present, although represented by only one or two species; however, where they occurred, they were dominant in the samples.

The metric characteristics of macrozoobenthos communities in the Pešter Plateau springs were monitored by analyses of the saprobic index, diversity index, BMWP score, and EPT index. No ecological status assessment was carried out based on the obtained indices, because the examined watercourses on the Pešter Plateau are not defined by the regulations for determining ecological status. It is important to note that the saprobic index, diversity index, BMWP score, and EPT taxa are not consistent accross most of the investigated sites. For example, based on the saprobic index, some sites could be classified as ecological status class I, whereas according to the diversity index and BMWP score they would fall into class III or IV. These results are due to abrupt changes in discharge and water flow velocity, which led to the presence of only those species that have adapted to the aforementioned factors in the studied watercourses. Consequently, these species become the dominant members of the studied macrozoobenthos community (Table 1).

The results of the ANOVA analysis indicate varying levels of statistically significant differences in the values of ecological indices at the investigated karst springs of the Pešter Plateau.

Table 1 Parameters for assessing the ecological status of springs on the Pešter Plateau

Biological Parameters for Assessing Ecological Status						
Name of sampling site	Saprobic Index (method by Zelinka & Marvan)	Diversity Index (Shannon–Weaver method)	BMWP	EPT Index	Geographic coordinates	Altitude (m)
Đerekar spring	1.17	1.59	52.17	22.07	42° 58′ 51″ Nφ 20° 07′ 13″ Eλ	1.297
Peskovito spring	1.28	1.22	31.00	25.64	42° 59′ 00″ Nφ 20° 07′ 25″ Eλ	1.263
Kladenac	1.70	1.15	34.50	11.80	42° 59′ 37″ Nφ 20° 08′ 07″ Eλ	1.219
Česmica	1.04	0.96	22.67	0.35	42° 59′ 37″ Nφ 20° 08′ 07″ Eλ	1.219
Kurtovsko spring	1.74	2.10	50.33	36.99	43° 00′ 56″ Nφ 20° 10′ 11″ Eλ	1.188
Noga	1.30	1.94	60.00	60.16	43° 00′ 37″ Nφ 20° 07′ 57″ Eλ	1.379
Đurđevica	2.86	1.19	24.17	0.77	43° 06′ 04″ Nφ 20° 08′ 52″ Eλ	1.174
Lokve Karajukiće Bunare	2	1.82	15.25	31.28	43° 05′ 39″ Nφ 20° 06′ 15″ Eλ	1.152
Vidrenjak I	2.24	1.31	29.8	40.31	42° 00′ 35″ Nφ 20° 09′ 35″ Eλ	1.191
Vidrenjak II	1.84	1.18	40.83	17.83	42° 00′ 35″ Nφ 20° 09′ 35″ Eλ	1.191

(continued)

Table 1 (continued)

Biological Parameters for Assessing Ecological Status						
Name of sampling site	Saprobic Index (method by Zelinka & Marvan)	Diversity Index (Shannon–Weaver method)	BMWP	EPT Index	Geographic coordinates	Altitude (m)
Vidrenjak III	1.59	1.18	34.8	9.41	42° 00′ 35″ Nφ 20° 09′ 35″ Eλ	1.191
Čukotsko spring	1.21	0.59	34.67	7.71	43° 04′ 40″ Nφ 20° 16′ 32″ Eλ	1.090
Melajsko spring	127	1.57	40.17	10.00	43° 05′ 52″ Nφ 20° 14′ 56″ Eλ	1.012
Spring Kron	1.66	1.32	23.67	81.36	43° 07′ 43″ Nφ 20° 11′ 57″ Eλ	1.027
Spring Jaz	1.69	1.77	55.50	46.89	43° 07′ 15″ Nφ 20° 12′ 19″ Eλ	1.018
Gujsko spring	1.25	1.64	37.50	46.27	43° 06′ 15″ Nφ 20° 13′ 48″ Eλ	1.023
Studenac	153	0.58	39.17	7.77	43° 04′ 25″ Nφ 20° 14′ 14″ Eλ	1.157
Crvena voda	1.29	0.64	30.00	8.35	43° 04′ 38″ Nφ 20° 14′ 20″ Eλ	1.133
Stolarc	2.38	1.18	36.6	4.69	43° 06′ 19″ Nφ 20° 22′ 06″ Eλ	945

(continued)

Table 1 (continued)

Biological Parameters for Assessing Ecological Status						
Name of sampling site	Saprobic Index (method by Zelinka & Marvan)	Diversity Index (Shannon–Weaver method)	BMWP	EPT Index	Geographic coordinates	Altitude (m)
Dobri Dub	1.71	0.85	40.00	9.87	43° 04′ 15″ Nφ 20° 18′ 52″ Eλ	1.124
Promuklice	1.79	1.66	38.83	31.35	43° 01′ 52″ Nφ 20° 18′ 33″ Eλ	985
Kovačko spring	1.22	0.97	53.50	62.18	43° 01′ 54″ Nφ 20° 22′ 36″ Eλ	952

At most sites—such as Peskovito Vrelo, Kladanac, Kurtovsko Vrelo, Đurđevića, Vidrenjak II and III, as well as the springs Studenac, Crvena Voda, and Dobri Dub—no statistically significant differences were recorded in the index values in relation to the season or year of sampling, which indicates a certain degree of ecological stability at these sites over time.

In contrast, at sites such as Česmica, Noga, Lokve Karajukića Bunare, and Stolare, statistically significant seasonal or annual variations were observed, particularly for indices such as the Saprobic index, BMWP, and EPT. These differences are mainly attributed to the ephemeral nature of some springs and frequent drying up during drought periods, which impact the structure of the communities, especially sensitive groups like EPT.

Although most of the springs are located relatively close to each other, the ANOVA results show statistically significant differences in the index values between different springs. This suggests that geographic proximity alone is not the determining factor in shaping ecological characteristics; rather, local hydrological and morphological conditions play a more decisive role. The exception is the diversity index, for which no statistically significant differences were found between springs.

4 Discussion

The springs examined on the Pešter Plateau included both ephemeral and permanent sources, reflecting diverse hydroecological conditions. Springs often serve as stable ecosystems due to relatively constant physicochemical parameters; however, substrate variability can lead to high species diversity (Piechocki 2000). Karst springs are particularly sensitive to abiotic fluctuations and respond quickly to wet and dry periods (Zollhöfer et al. 2000). The structure of benthic fauna is strongly shaped by substrate type, especially its granulometric characteristics (Smith et al. 2003; Marković 1998; Fumetti et al. 2006; Sambugar et al. 2006; Dumnicka et al. 2007; Piechocki 2000).

According to Smith et al. (2003), karst springs commonly host Hirudinea, Heteroptera, and large Lymnaeidae snails. However, this is not universal. In the current study, as well as in Dumnicka et al. (2007) on the Kraków-Częstochowa Upland, these taxa were absent.

Dumnicka et al. (2007) analyzed macrozoobenthos composition and substrate influence, identifying 49 families or subfamilies and four higher taxa (Hydrozoa, Turbellaria, Nematoda, Hydracarina). Non-insect invertebrates dominated the samples (86%), particularly Gammaridae and Bythinellinae (52%). Similarly, in our study, non-insect taxa accounted for 64.5%, with Gammaridae being the most abundant (40.17%), followed by Gastropoda (9.93%).

Although all studied rheocrene springs had similar substrates, discharge and flow velocity varied across sites and seasons. In Dumnicka's research, the spring with the highest discharge showed the greatest taxon richness (27 taxa). Springs with coarse substrates had more taxa, though dominant species occurred on both fine and coarse substrates.

Certain taxa showed clear substrate preferences. Orthocladiinae, Cyprididae, and Turbellaria preferred coarse sediments, whereas Chironominae, Limnephilidae, and Bythinellinae were more common in fine ones. In our study, Orthocladiinae and Cyprididae were absent. Turbellaria favored coarse sediments, while Limnephilidae and Bythinellinae preferred finer ones. Chironominae were widespread but more abundant in fine sediments with high organic content.

The benthic fauna was predominantly composed of non-insect taxa, consistent with findings from other karst regions. *Gammarus* (Amphipoda) and *Bythinella* (Gastropoda) were dominant, as also reported by Dumnicka et al. (2007). *Gammarus* showed no clear substrate preference, but body size and mobility varied by habitat. At Đerekarsko Vrelo, a spring with moss-covered stones, *Gammarus* made up 60.3% of individuals.

Bythinella, which thrives in oxygen- and calcium-rich waters (Falniowski 1987), was identified as *B. pesterica* at Pešter sites. It was more common in fine sediments but occurred in both substrate types. Similar amphipod and gastropod communities are reported in karst springs across Europe (Smith et al. 2003) and the USA (Glazier and Gooch 1987).

Zollhöfer (Zollhöfer et al. 2000; Zollhöfer 1999) noted *Bythiospeum* and *Bythinella* as typical karst taxa, which aligns with our results. Fumetti et al. (2006), studying 20 Swiss karst springs, recorded 82 taxa in 155 samples (average: 20 per spring), fewer than in our study. Hirudinea, Heteroptera, and Lymnaeidae were also not dominant in their findings.

Variations in community composition are likely influenced by spring size and discharge. Gammarus was frequent in both Swiss and Pešter springs, and in Polish karst springs. Swiss studies reported 22 Trichoptera species, while we recorded 45. *Crunoecia irrorata* and *Synagapetus dubitans* were common in Switzerland; *Chaetopteryx villosa* was most frequent at Pešter, with abundance affected by flow, depth, and substrate.

Stygobionts like *Bythiospeum* and *Niphargus* were found at two Swiss sites. Nemouridae occurred in 75% of their springs. However, each karst spring is ecologically unique, limiting generalization. *Crunoecia irrorata* and *Synagapetus dubitans* were absent from both Pešter and Kraków-Częstochowa sites. *Niphargus sp.* was present at three Pešter springs during high water but was not observed by Dumnicka et al.

Swiss studies indicate that stable physicochemical conditions during sampling had little effect on macrozoobenthos structure (Smith et al. 2003; Dumnicka et al. 2007), but extreme flow events can disrupt communities. Flow indirectly influences fauna by shaping the substrate, which then determines community composition.

Our findings agree: springs with stable, high discharge had the richest taxa. Pearson correlations and regressions showed that certain parameters—such as carbonate hardness and oxygen saturation—significantly influenced specific macrozoobenthic groups. These variables help explain community patterns and support predictive modeling in karst systems.

Earlier research often included both spring and adjacent habitats. In contrast, studies focused solely on spring macrofauna—especially in the Pešter region—remain rare (Cantonati et al. 2006).

Macrozoobenthos, the main faunal group in these habitats, has been underexplored. Prior studies at Đerekarsko Vrelo by Protić and Živić (2013) recorded *Hesperocorixa parallela* (Fieber 1860) and *Notonecta maculata* (Fabricius, 1794), but these species were not found in our study at any Pešter site.

Marković (1998) conducted a three-year study of 20 karst springs in hilly/mountainous Serbia. Similar to our findings, no dominance of Hirudinea, Heteroptera, or Lymnaeidae was observed—contrary to earlier reports (Smith et al. 2003; Dumnicka et al. 2007).

Our results confirm Marković's observations: amphipods were dominant, with *Gammarus balcanicus* (Schaferna 1922) and *G. fossarum* (Koch 1835) as the most abundant species. They did not show strong substrate preferences, though size and mobility varied by spring type.

Unlike Fumetti et al. (2006) and Dumnicka et al. (2007), Pešter springs included Ephemeroptera such as *Rhithrogena semicolorata* (Curtis 1834), *Baetis sp., Leuctra sp., Nemoura* sp. *Rhithrogena* sp. and *Baetis* sp. rocky substrates, while *Leuctra* sp. and *Nemoura* sp. showed broader tolerance—patterns consistent with Marković's findings.

Elmis aenea (Müller 1806) preferred stony and gravel habitats and was most abundant in such sites, echoing results from Marković (1998).

5 Conclusion

The ecological status of watercourses on the Pešter Plateau cannot yet be formally assessed, as water bodies located above 1000 m in karst regions are not included in the existing national categorization framework. Nevertheless, the studied springs should be considered as distinct ecological units. Despite sharing relatively stable environmental conditions—such as water temperature, substrate composition, and certain chemical parameters—variability in physico-chemical factors significantly affects the distribution and interactions of macrozoobenthic taxa.

The observed correlations between specific taxa and environmental parameters varied across springs, indicating that each spring functions as an isolated and unique ecosystem. Factors such as spatial distribution, subsurface hydrological connectivity, and seasonal variability further enhance the ecological distinctiveness of each locality.

Given the sensitivity and ecological value of these habitats, it is essential to develop an integrated water management plan adopted to the Pešter Plateau. This plan should focus on preserving and improving water quality through coordinated efforts of relevant institutions, local authorities, water users, and the local population. Raising awareness and encouraging shared decision-making are key to ensuring the long-term protection and sustainable use of the region's water resources.

Acknowledgements This work was financed by the Ministry of Science, Technological Development and Innovation of the Republic of Serbia (Grants No. 451-03-137/2025-03/ 200125 & 451-03-136/2025-03/ 200125).

References

Beierkuhnlein, C., Gollan, T.: Ökologie silikatischer Waldquellen in Mitteleuropa. Bayreuther Forum Ökologie **71**, 1–256 (1999)

Cantonati, M., Gerecke, R., Bertuzzi, E.: Springs of the Alps—sensitive ecosystems to environmental change. Hydrobiologia **562**, 59–96 (2006)

Dale, V.H., Beyeler, S.C.: Challenges in the development and use of ecological indicators. Ecol. Ind. **262**, 201–204 (2001)

Dumnicka, E., Galas, J., Koperski, P.: Benthic invertebrates in karst springs: does substratum or location define communities? Internat. Rev. Hydrobiol. **92**, 452–464 (2007)

Dziock, F., Henle, K., Foeckler, F., Follner, K., Scholz, M.: Biological indicator systems in floodplains—a review. Int. Rev. Hydrobiol. **91**, 271–291 (2006)

Falniowski, A.: Hydrobioidea of Poland (Prosobranchia: Gastropoda). Folia Malacologica **1**, 11–122 (1987)

Field, M.S.: In: A Lexicon of Cave and Karst Terminology with Special Reference to Environmental Karst Hydrology. USEPA, pp. 1–221. (2002)

Ford, D.C., Williams, P.: In: Karst Hydrogeology and Geomorphology. Wiley and Sons, England (2007). https://doi.org/10.1002/9781118684986

Glazier, D.S., Gooch, J.L.: Macroinvertebrate assemblages in Pennsylvania (U.S.A.) springs. Hydrobiologia **150**, 33–43 (1987)

Hellawell, J.M.: In: Biological Indicators of Freshwater Pollution and Environmental Management. Springer Science and Business Media (2012)
Knoben, R.A.E., Roos, C., Van Oirschot, M.C.M.: Biological assessment methods for watercourses. RIZA Report 95, Lelystad, Netherlands, pp. 86. (1995
Kurbanov, R., Murray, A., Thompson, W., Svistunov, M., Taratunina, N., Yanina, T.: First reliable chronology for the early Khvalynian Caspian sea transgression in the lower Volga river valley. Boreas **50**(1), 134–146 (2021)
Marković, Z.: Izvori brdsko-planinskih područja Srbije: ekološka studija makrozoobentosa. Doktorska disertacija, Univerzitet u Beogradu, Biološki fakultet, Beograd (1998)
Meyer, A., Meyer, E.I., Meyer, C.: Lotic communities of two small temporary karstic stream systems (East Westphalia, Germany) along a longitudinal gradient of hydrological intermittency. Limnologica **33**, 271–279 (2003)
Niemi, G.J., McDonald, M.E.: Application of ecological indicators. Annu. Rev. Ecol. Evol. Syst. **35**, 89–111 (2004)
Numanović, M., Miljanović, B., Živković, M.: Ekološki status Đerekarske reke. In: Voda, Zbornik radova 50. godišnje konferencije, Zlatibor 216–226 (2021)
Numanović, M., Stojanović, K., Novaković, B., Živković, M., Živić, N., Miljanović, B.: Structure and function of benthic invertebrate assemblages of the Đerekarska River (southwestern Serbia, Pešter plateau). Archiv. Biol. Sci. 437–448 (2023)
Pander, J., Geist, J.: Ecological indicators for stream restoration success. Ecol. Ind. **30**, 106–118 (2013)
Piechocki, A.: The fauna of Niebieskie Źródła Springs. Acta Univ. Lodz. Folia Limnol. **7**, 3–123 (2000)
Protić, Lj., Živić, I.: A contribution to the knowledge of aquatic Heteroptera of Šar planina Mts. and Pešter plateau (Serbia). Acta Entomol. Slov. **21**(1), 65–76 (2013)
Sambugar, B., Dessi, G., Stapelza, A., Stenico, A., Thaler, B., Veneri, A.: Südtirolen Quellfauna. Autonome Provinz Bozen—Südtirol (2006)
Sandin, L., Hering, D.: Comparing macroinvertebrate indices to detect organic pollution across Europe. Hydrobiologia **516**, 55–68 (2004)
Smith, H., Wood, P.J., Gunn, J.: The influence of habitat structure and flow permanence on invertebrate communities in karst spring systems. Hydrobiologia **510**, 53–66 (2003)
Stojanović, K.: Uticaj pastrmskih ribnjaka na zajednice makrozoobentosa tekućica sa posebnim osvrtom na larve roda Baetis (Ephemeroptera, Insecta). Doktorska disertacija, Biološki fakultet, Beograd (2017)
Von Fumetti, S., Nagel, P., Scheifhacken, N., Baltes, B.: Factors governing macrozoobenthic assemblages in perennial springs in north-western Switzerland. Hydrobiologia **568**, 467–475 (2006)
Williams, D.D.: The interface between groundwater and lotic faunas and its use in assessing groundwater quality. Verh. Int. Verein. Limnol. **24**, 1621–1624 (1991)
Zalewski, M., Janauer, G.A., Jolánkai, G.: Ecohydrology: A new paradigm for the sustainable use of aquatic resources. Technical Document in Hydrology No. 7, UNESCO, Paris, pp. 1–58. (1997)
Zollhöfer, J.M.: Spring Biotopes in Northern Switzerland: Habitat Heterogeneity, Zoobenthic Communities and Colonisation Dynamics. PhD Thesis, ETH Zürich (1999)
Zollhöfer, J.M., Brunke, A., Gonser, T.: A typology of springs in Switzerland by integrating habitat variables and fauna. Arch. Hydrobiol. Suppl. **121**, 349–376 (2000)

Photocatalytic Prospects of Mushroom-Mediated ZnO Nanostructures for Wastewater Recycling in Recirculating Fish Farming

Tamara Ivetić, Milena Rašeta, Melisa Curić, Nemanja Pankov, and Branko Miljanović

Abstract In this work, we investigate the potential application of a novel biogenic ZnO nano-photocatalyst (NPc) for the sustainable photocatalytic recycling and reuse of aquaculture wastewater. The ZnO nanostructures were synthesized via a green chemical precipitation method using an aqueous solution of non-toxic precursors (dried extract of the edible mushroom *Cyclocybe aegerita* (V. Brig.) Vizzini 2014 and zinc acetate) under mild conditions by stirring at 80 °C for two hours with pH adjusted to 12 using 2 M NaOH. The resulting flower-like ZnO NPc was characterized using High-Resolution Scanning Electron Microscopy, X-ray Diffraction, UV–Vis Diffuse Reflectance and Fourier Transform Infrared Spectroscopy confirming its nanostructured morphology, crystalline phase, and functional group interactions. The unique properties, including the appropriate valence and conduction band redox potentials make this biogenic ZnO NPc a promising candidate for photocatalytic control of (de)nitrification in recirculating fish farming systems. The conditions for successful simultaneous (de)nitrification of nitrite and ammonia in a simple two-stage photocatalytic experimental setup were investigated. This mushroom-mediated synthesis not only enhances the environmental compatibility of the NPc but also supports circular bioeconomy principles. Our findings propose a cost-effective, environmentally friendly, and efficient solution for managing nitrogen species in fish farming effluents, contributing to cleaner aquaculture practices and sustainable water resource management.

T. Ivetić (✉)
Department of Physics, Faculty of Sciences, University of Novi Sad, Novi Sad, Serbia
e-mail: tamara.ivetic@df.uns.ac.rs

M. Rašeta
Department of Chemistry, Biochemistry and Environmental Protection, Faculty of Sciences, University of Novi Sad, Novi Sad, Serbia

M. Curić · N. Pankov · B. Miljanović
Department of Biology and Ecology, Faculty of Sciences, University of Novi Sad, Novi Sad, Serbia

H. Xu (ed.), *Proceedings of the 8th International Symposium on Water Resource and Environmental Management*, Environmental Science and Engineering,
https://doi.org/10.1007/978-3-032-19088-8_3

Keywords ZnO nanoflowers · Clean water purification technology · Photocatalytic denitrification/nitrification · Recirculating fish farming

1 Introduction

Zinc oxide (ZnO) is a common metal oxide that has been used for centuries in diverse applications, from medical ointments in ancient civilizations to pigment production in the eighteenth and nineteenth century (Zhu and Wang 2025; Nehru et al. 2023). Since the early twentieth century, scientific interest in ZnO has grown substantially driven by its semiconductor, piezoelectric, optical, and catalytic properties. More recently, ZnO has attracted particular attention in nanotechnology, renewable energy, and biomedical research, where its multifunctional properties are being exploited for advanced applications (Sharma et al. 2022). Owing to its high chemical stability, broad absorption spectrum, photostability, large exciton binding energy, and wide band gap, ZnO is now considered a highly versatile functional material with applications spanning electronics, optoelectronics, and environmental remediation. Recent research highlights the promise of ZnO nanostructures in photocatalytic wastewater treatment (Sharma et al. 2022). Conventional methods for nanoparticles (NPs) synthesis, however, often rely on toxic reagents and energy-intensive processes, raising significant environmental concerns. Green nanotechnology has emerged as a sustainable alternative, utilizing biological systems including plants, fungi, bacteria, and algae as reducing and stabilizing agents. In this context, mushroom-mediated synthesis of ZnO NPs is particularly innovative, as fungal metabolites such as polysacharides, proteins, and phenolic compounds act not only as eco-friendly reducing agents but also as natural capping and functionalizing molecules, potentially enhancing the photocatalytic performance of the resulting nanomaterials (Nehru et al. 2023; Ramesh et al. 2021). This dual role of mushrooms, as both biofactories and sources of bioactive molecules, represents a significant step forward in developing multifunctional NPs tailored for environmental remediation. At the same time, the increasing global demand for aquaculture products underscores the importance of developing efficient wastewater treatment strategies in recirculating aquaculture systems (RAS) for fish farming (Almeida et al. 2020; Bregnballe 2015). Such systems significantly reduce freshwater consumption but suffer from the accumulation of nitrogenous wastes, particularly ammonia and nitrate, which are toxic to aquatic organisms at elevated concentrations (Lee et al. 2000). In the European Union, aquaculture is shaped by strict environmental regulations, such as the European Green Deal's zero-pollution ambition, which promotes the adoption of RAS to minimize effluent discharge and safeguard freshwater resources (Commission 2021). Countries like Denmark and Norway have pioneered commercial-scale RAS for salmon production, while Balkan countries such as Serbia, Croatia, and Greece rely on carp and trout farming, where the uptake of RAS technologies is gradually increasing under EU policy incentives (Commission 2023). In contrast, China, the world's largest aquaculture producer, contributing over 60% of global output, has traditionally relied on pond-based and open-water systems (FAO 2022; Chen et al. 2023). However, rising environmental pressures and new development policies have driven China to increasingly invest in RAS technologies, reflecting a transition toward sustainable aquaculture (Ministry of Agriculture and Rural Affairs of China 2021). These regional contrasts emphasize the shared need for innovative, environmentally benign solutions to nitrogen management in aquaculture. Photocatalysis using nanomaterials represents a promising alternative to conventional biological treatment, enabling selective and eco-friendly conversion of nitrogenous species. Here, we propose mushroom-mediated ZnO NPs as a novel class of green photocatalysts for (de)nitrification in recirculating fish farming, with the advanced functional performance of ZnO, opening new perspectives for climate-smart aquaculture and sustainable nutrient recycling (Tugaoen et al. 2017).

2 Materials and Methods

2.1 Synthesis and Characterization of Biogenic ZnO NPc

Three-dimensional (3D) ZnO NPc were synthesized via a green chemical precipitation method, in which an extract of *Cyclocybe aegerita* (V. Brig.) Vizzini 2014, an edible medicinal mushroom, served as a biogenic reducing and stabilizing agent. Zinc acetate dihydrate ($C_4H_6O_4Zn \times 2H_2O$, Sigma-Aldrich, ACS reagent, $\geq$ 98%) was used as the Zn precursor. The reaction mixture was continuously stirred at 80 °C for 2 h, and the pH was adjusted to 12 with 2 M solution of sodium hydroxide (NaOH, reagent grade, $\geq$ 98%, anhydrous pellets) (Fig. 1).

X-Ray Diffraction (XRD) pattern was recorded using a Rigaku Miniflex 600 X-ray diffractometer and Cu-Kα radiation in the 2θ range from 20° to 80°, with a step of 0.02° and a speed of 1°/min. Morphological features were examined by

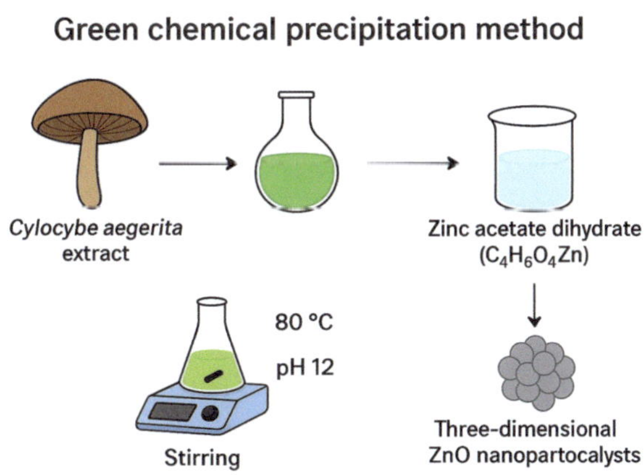

Fig. 1 Novel biogenic synthesis of ZnO NPc using *C. aegerita* extract

high-resolution scanning electron microscopy (HR-SEM) Thermo Fischer Apreo 2C instrument. Fourier-transform infrared (FTIR) spectroscopy was conducted at room temperature to investigate the functional surface groups of nano-sized ZnO. Spectrum was recorded in the range of 400–4000 $cm^{-}1$ using a Thermo NicoletTM iS20 FTIR spectrometer with a transmission technique using KBr and a DTGS (Deuterated Triglycine Sulfate) detector. Ultraviolet—Visible (UV–Vis) diffuse reflectance spectroscopy (DRS) was recorded by a double-beam UV/VIS/NIR spectrophotometer (Perkin Elmer Lambda-950), in the spectral range from 300 to 900 nm using an integrating sphere. Spectral parameters were employed to estimate the optical bandgap energy. The Kubelka–Munk (K–M) function was applied to transform reflectance spectra, and bandgap value was determined using the Tauc relation for direct bandgap transitions (n = 1/2).

2.2 Design of the Photocatalytic (De)Nitrification Reactor

A photocatalytic reactor was designed to simulate nitrogen removal processes relevant to RAS. The setup consists of a commercial borosilicate glass photoreactor (Shanghai Binlong Instrument Co. Ltd.) contained of a vessel (around 250 mL) for the reaction mixture and equipped with a 300 W high-pressure mercury lamp surrounded by a circulating water jacket to maintain the reaction temperature at 25 °C. Continuous mixing of the suspension is achieved using a magnetic stirrer, while the solution pH was maintained at 9.5 by automatic titration with 1 M NaOH (Fig. 2).

In the first stage of the experiment, a suspension containing the NPc and aqueous ammonia is purged with nitrogen gas (N_2 at 0.3 L/min) for 30 min to establish anaerobic conditions and eliminate dissolved oxygen. The reactor is subsequently irradiated with UV light, and the concentrations of ammonia, nitrite, and nitrate are

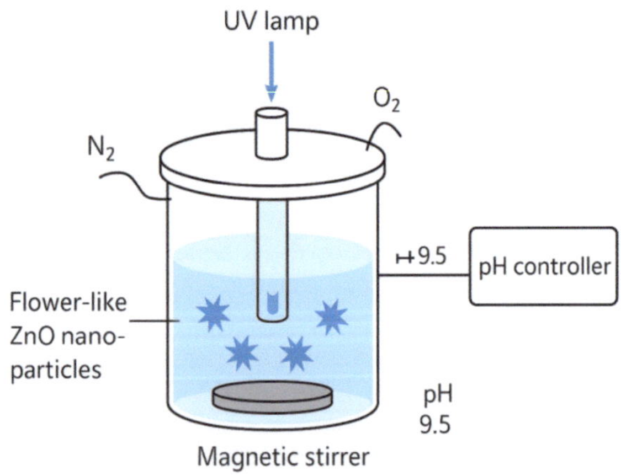

Fig. 2 Design of the photocatalytic reactor with ZnO NPc

periodically monitored. In the second stage, oxygen gas at a flow rate of 0.2–0.3 L/min is introduced into the system under identical pH and irradiation conditions to promote aerobic photocatalytic conversion of the remaining nitrogen species. This two-step configuration is designed to mimic sequential denitrification and nitrification processes, thereby enabling the simultaneous photocatalytic removal of ammonia and nitrite (Ye et al. 2019a). The reactor experiments remain under laboratory investigation, and further studies are required to evaluate long-term performance, NPc stability, and scalability. The mushroom-mediated ZnO NPc with favorable bandgap and redox potential alignment upon UV irradiation generate, in principle, the conduction band electrons which can reduce nitrate and nitrite, while their valence band holes (via ·OH radicals) can oxidize ammonia, with N_2 as the expected final product. However, experimental confirmation of gaseous N_2 formation requires gas-phase analysis; in this study, 'removal' refers to decreases in the aqueous concentrations of NH_4^+, NO_2^-, and NO_3^-. Partial degradation of residual organics furthermore contributes to reductions in BOD and COD, consistent with previous studies on hybrid ZnO systems (Ye et al. 2019a, b; Panahi et al. 2025; Silver et al. 2025).

3 Results and Discussion

XRD analysis (Fig. 3) confirmed the successful formation of crystalline ZnO NPs with an average crystallite size of approximately 13 nm, consistent with the Debye–Scherrer calculation.

The diffraction peaks corresponded well with the hexagonal wurtzite ZnO structure (Joint Committee on Powder Diffraction Standards–JCPDS card no. 36–1451), confirming high crystallinity. HR-SEM imaging (Fig. 4) revealed flower-like 3D morphologies composed of aggregated nanocrystallites, which are advantageous for photocatalysis due to their high surface-to-volume ratio.

FTIR analysis (Fig. 5) of nano-sized ZnO showed absorption bands at 3424, 2925, 2854, 1635, 1456, 1408, 1042, 923, 498, and 426 cm-1.

The broad band at 3424 cm^{-1} corresponds to O–H stretching of hydroxyl groups and adsorbed water, while the peaks at 2925 and 2854 cm^{-1} are attributed to C–H stretching vibrations of residual organics. The signal at 1635 cm^{-1} arises from H–O–H bending of molecular water, and those at 1456 and 1408 cm^{-1} indicate C–H bending or carbonate species. The 1042 and 923 cm^{-1} bands are assigned to C–O vibrations, whereas the strong absorptions at 498 and 426 cm^{-1} correspond to Zn–O stretching modes, confirming the ZnO lattice structure (Azam et al. 2009). Therefore, FTIR spectroscopy validated the successful capping and stabilization by biomolecules.

The UV–Vis DRS showed strong absorption in the UV region (Fig. 6, left), with the calculated bandgap energy of the biogenic ZnO NPc being 3.24 eV (Fig. 6, right), closely aligned with reported values for pristine ZnO.

Such a bandgap is suitable for UV-driven photocatalysis and suggests potential for enhanced electron–hole pair generation and redox activity. The band edge positions of

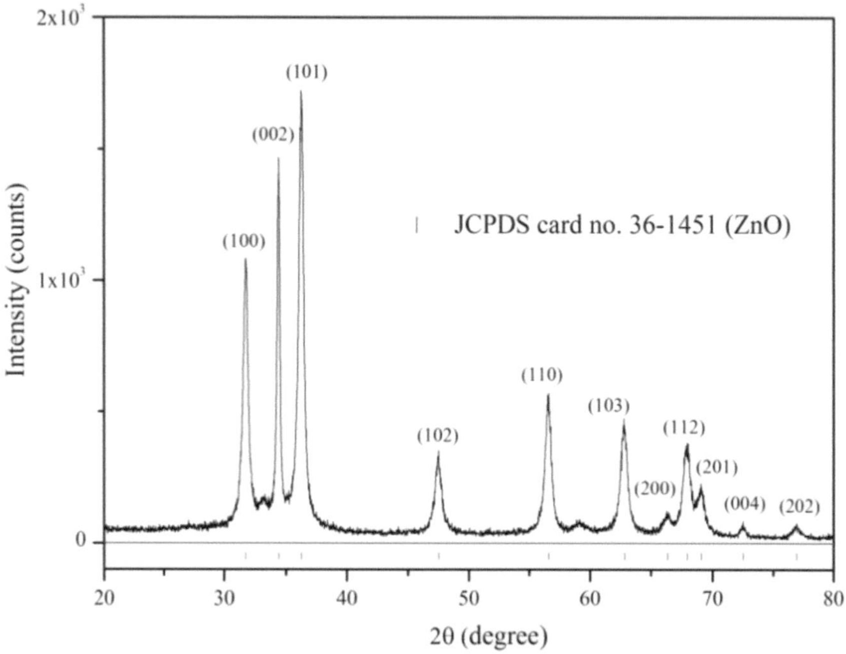

Fig. 3 XRD pattern of ZnO NPc

Fig. 4 HR-SEM image of ZnO NPc

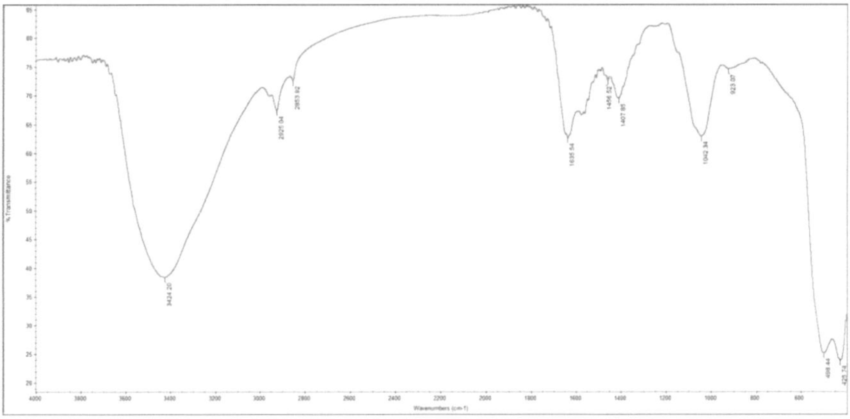

Fig. 5 FTIR spectrum of ZnO NPc

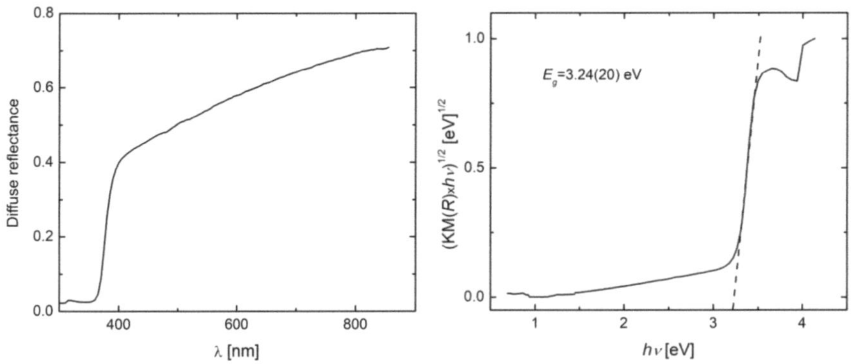

Fig. 6 UV–Vis DRS (left) and K-M transformation (right) of ZnO NPc

the ZnO nanostructures were further evaluated using the Mulliken electronegativity approach, which relates the conduction and valence band potentials to the absolute electronegativity of the semiconductor (Xu and Schoonen 2000; Butler and Ginley 1978). Based on the reported electronegativity of ZnO ($\chi = 5.79$ eV) and the bandgap energy determined from DRS ($E_g = 3.24$ eV), the conduction band (ECB) and valence band (EVB) edge potentials were calculated using the relations (Panahi et al. 2025): $\mathrm{ECB} = \chi - E_e - \frac{1}{2}E_g$, $\mathrm{EVB} = \mathrm{ECB} + E_g$, where E_e is the energy of free electrons on the hydrogen scale (4.50 eV). The calculated values were − 0.33 V (ECB) and + 2.91 V (EVB) versus NHE at pH 7. After correcting for the Nernstian shift of − 59 mV per pH unit, the band edge positions at pH 9.5 were estimated to be − 0.48 V (ECB) and + 2.76 V (EVB). These potentials are thermodynamically favorable for photocatalytic denitrification and nitrification, as the conduction band electrons can reduce nitrate and nitrite intermediates to N_2, while the valence band holes (and derived $^{\bullet}OH$ radicals) can oxidize ammonia to nitrogen gas. This alignment is schematically illustrated in Fig. 7, which shows the ZnO band structure relative to key redox couples for nitrogen transformations.

These results demonstrate that biogenic ZnO NPc can support both denitrification and nitrification processes, addressing the limitations of conventional photocatalysts such as TiO_2, which often suffer from incomplete selectivity and accumulation of toxic intermediates. The mushroom-mediated ZnO NPc developed in this study was

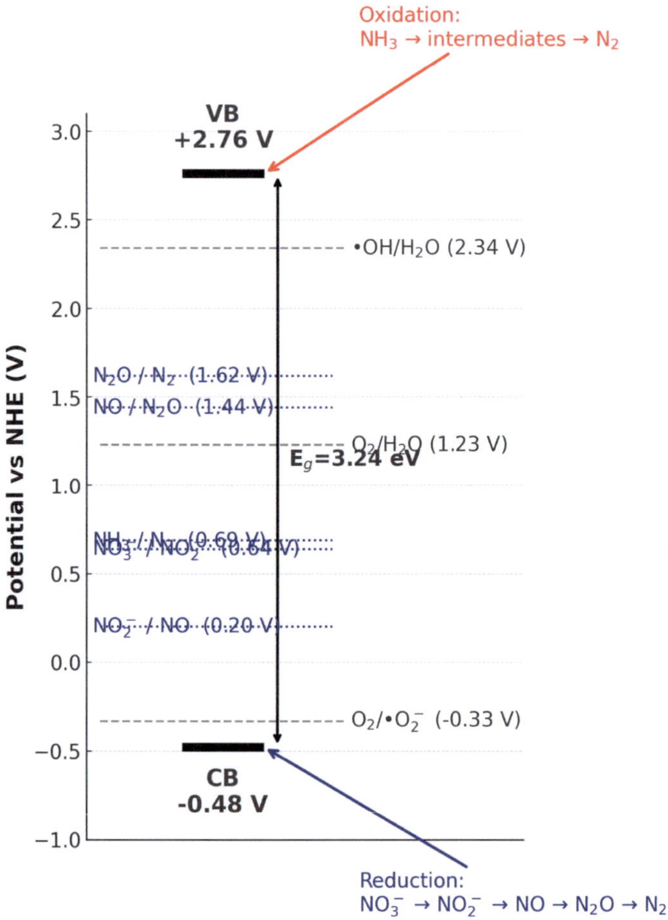

Fig. 7 Schematic illustration of the band alignment of biogenic ZnO NPc at pH 9.5, determined using the Mulliken electronegativity method. The conduction band edge (ECB = − 0.48 V vs. NHE) and valence band edge (EVB = + 2.76 V vs. NHE) were calculated based on the measured bandgap energy (Eg = 3.24 eV). The positions of relevant redox couples are indicated, including $O_2/^{\cdot}O_2^{-}$, O_2/H_2O, and $^{\cdot}OH/H_2O$, as well as key nitrogen transformations (NO_3^{-}/NO_2^{-}, NO_2^{-}/NO, NO/N_2O, N_2O/N_2, NH_3/N_2). The band structure demonstrates the thermodynamic feasibility of photocatalytic denitrification via conduction band electron-driven reduction of nitrate and nitrite to N_2, and simultaneous photocatalytic nitrification/oxidation of ammonia through valence band holes and hydroxyl radicals

further compared to results found in recent literature on ZnO-based photocatalysts for nitrogen removal in wastewater and aquaculture applications. As summarized in Table 1, previous studies have explored chemical (Zhu and Wang 2025), plant-mediated (Ramesh et al. 2021), and fungal-assisted (Nehru et al. 2023) synthesis routes, producing NPs with varied morphologies, bandgap energies, and efficiencies toward nitrate, ammonia, nitrite, and COD removal. Photocatalytic hybrid and immobilized systems have also shown high nitrogen removal efficiencies, including suspended and immobilized ZnO NPs (Panahi et al. 2025), Ag-modified ZnO NPs (Silver et al. 2025), and comprehensive review strategies (Abou Zeid and Leprince-Wang 2024). This comparative analysis demonstrates that the proposed mushroom-mediated approach offers a competitive, environmentally friendly, and scalable alternative to conventional ZnO photocatalysts, supporting its potential application in recirculating aquaculture systems (RAS) and broader wastewater treatment contexts.

4 Conclusion

In this study, 3D ZnO NPc were synthesized via a green precipitation method using *C. aegerita* extract, yielding flower-like morphologies (~13 nm crystallite size) with excellent biogenic stabilization. Spectroscopic and optical analyses indicated a direct bandgap of 3.24 eV and well-aligned conduction and valence band edges (−0.48 V and + 2.76 V vs. NHE at pH 9.5), conducive to efficient charge separation and photocatalytic activity. A custom-designed reactor was considered to simulate sequential denitrification and nitrification under UV illumination, wherein conduction band electrons reduced nitrate/nitrite to N_2 and valence band holes (via ·OH radicals) oxidized ammonia to N_2 gas. Mushroom-mediated ZnO NPs offer a promising hybrid photocatalytic solution for nitrogen removal tailored to diverse aquaculture contexts: from EU-regulated RAS systems, through traditional and budget-limited farms in the Balkans, to large-scale Chinese operations adopting modernization. This technology presents a globally scalable pathway toward sustainable aquaculture, aligning with the United Nations Sustainable Development Goals (SDGs) on clean water, responsible consumption, and climate resilience. Future studies should focus on several aspects to enhance the applicability and sustainability of mushroom-mediated ZnO NPc in aquaculture wastewater treatment. These include: (i) exploring visible-light activation of the ZnO NPs to improve solar utilization and reduce energy requirements; (ii) conducting long-term pilot-scale experiments in RAS to assess the stability, reusability, and regeneration of the NPs under realistic conditions; (iii) integrating photocatalytic treatment with biological units to achieve synergistic removal of nitrogen species and organic matter; and (iv) evaluating potential ecotoxicological impacts of residual NPs on aquatic organisms to ensure environmental safety and regulatory compliance.

Table 1 Literature review of ZnO-based photocatalysts for nitrogen removal and wastewater treatment

Ref. (year)	Photocatalyst	Synthesis	Morphology	Bandgap (eV)	Target pollutants/ Process	Reported efficiency
This study (2025)	ZnO nanoflowers (*C. aegerita*-mediated)	Biogenic precipitation	3D flower-like (~13 nm crystallite)	3.24/UV	Photocatalytic (de)nitrification	Favorable band alignment
Zhu and Wang (2025)	ZnO NPs	Chemical precipitation	Spherical NPs	3.21/UV	Nitrate reduction/ dye degradation	72% nitrate removal
Panahi et al. (2025)	ZnO NPs	Traditional precipitation	Nanoscale particles	UV	Nitrate removal	Suspended: ~ 81%; Immobilized: ~ 90% (150 min)
Silver et al. (2025)	ZnO/Ag NPs	Standard NP mixing	Mixed NPs in solution	UV/Vis	NH_4^+-N, NO_2^--N, COD	NH_4^+-N: ~ 98%; NO_2^--N: ~ 94.5%; COD: ~ 69%
Abou Zeid and Leprince-Wang (2024)	ZnO-based photocatalysts	Multiple	Multiple morphologies	UV/Vis	Water treatment applications	Review of strategies
Nehru et al. (2023)	ZnO–fungal extract	Green synthesis (*Xylaria arbuscula*)	Irregular nanostructures	3.18/UV	Antibacterial activity, wastewater	High antimicrobial/ photocatalytic activity
Ramesh et al. (2021)	ZnO–plant mediated	Phytosynthesis	Rod-like	3.25/UV	Organic pollutants, COD reduction	~ 65% COD reduction in 4 h
Ye et al. 2019a) (2019)	$ZnFe_2O_4$/AC hybrid	Hydrothermal impregnation	Mixed porous	2.85/UV–Vis	Simultaneous NH_3 and NO_3^- removal	80% ammonia, 70% nitrate removal
Ye et al. (2019b)	Nano ZnO	Sol–gel	Nanoplates	3.22/UV	Denitrification/ nitrification	Effective conversion of NO_3^-/NH_3 to N_2

Acknowledgements This work was supported by the Provincial Secretariat for Higher Education and Scientific Research of Vojvodina (Project: *Development of new highly-sensitive sensors for monitoring of gas pollution and humidity in Vojvodina*, Contract No. 003075795 2024 09418 003 000 000 001/2). The authors also gratefully acknowledge financial support from the Ministry of Science, Technological Development and Innovation of the Republic of Serbia (Grants No. 451-03-137/2025-03/200125 and 451-03-136/2025-03/200125).

References

Abou Zeid, S., Leprince-Wang, Y.: Advancements in ZnO-based photocatalysts for water treatment: a comprehensive review. Crystals **14**, 611 (2024)

Almeida, P., Dewasme, L., Wouwer, A.V.: Denitrification control in a recirculating aquaculture system-a simulation study. Processes **8**(1–21), 1306 (2020)

Azam, A., Ahmed, F., Arshi, N., Chaman, M., Naqvi, A.H.: Low temperature synthesis of ZnO nanoparticles using mechanochemical route: a green chemistry approach. Int. J. Theor. Appl. Sci. **1**(2), 12–14 (2009)

Bregnballe, J.: A Guide to Recirculation Aquaculture, Published by the Food and Agriculture Organization of the United Nations (FAO), Budapest, Hungary, and Eurofish International Organisation. Denmark, Copenhagen (2015)

Butler, M.A., Ginley, D.S.: Prediction of flatband potentials at semiconductor–electrolyte interfaces from atomic electronegativities. J. Electrochem. Soc. **125**, 228–232 (1978)

Chen, T., Zhang, J., Huang, X.: Evaluation of the green development efficiency of marine fish culture in China. Front. Sustain. Food Syst. **7**, 112345 (2023)

Commission, E.: A Zero Pollution Ambition for a Toxic-Free Environment—The European Green Deal. Publications Office of the European Union, Luxembourg (2021)

Commission, E.: EU Aquaculture Assistance Mechanism: Aquaculture in the Western Balkans: Current Status and Future Development. Belgium, Brussels (2023)

FAO: The State of World Fisheries and Aquaculture 2022—Towards Blue Transformation. Food and Agriculture Organization of the United Nations, Rome (2022)

Lee, P.G., Lea, R.N., Dohmann, E., Prebilsky, W., Turk, P.E., Ying, H., Whitson, J.L.: Denitrification in aquaculture systems: an example of a fuzzy logic control problem. Aquacult. Eng. **23**, 37–59 (2000)

Ministry of Agriculture and Rural Affairs of China: China's 14th Five-Year Plan for Fisheries Development. Beijing, China (2021)

Nehru, L., Kandasamy, G.D., Sekar, V., Alshehri, M.A., Panneerselvam, C., Alasmari, A., Kathirvel, P.: Green synthesis of ZnO-NPs using endophytic fungal extract of *Xylaria arbuscula* from *Blumea axillaris* and its biological applications. Artif. Cells, Nanomed., Biotechnol. **51**(1), 318–333 (2023)

Panahi, H., Karami, S., Fattahi, A., Mohammadi, R.: Photocatalytic removal of nitrate from aqueous solutions using suspended and immobilized Zinc Oxide nanoparticles under UV irradiation. Int. J. Environ. Health Res. **15**, 120–131 (2025)

Ramesh, P., Saravanan, K., Manogar, P., Johnson, J., Vinoth, E., Mayakannan, M.: Green synthesis and characterization of biocompatible zinc oxide nanoparticles and evaluation of its antibacterial potential. Sens. Bio-Sens. Res. **31**, 100399 (2021)

Sharma, P., Hasan, M.R., Mehto, N.K., Bishoyi, A., Narang, J.: 92 Years of zinc oxide: has been studied by the scientific community since the 1930s-an overview. Sens. Int. **3**, 100182 (2022)

Silver, R., Zhang, L., Kumar, P.: Enhanced photocatalytic nitrogen removal from aquaculture wastewater using ZnO/Ag nanoparticles. Nanomaterials **15**, 559 (2025)

Tugaoen, H.O., Garcia-Segura, S., Hristovski, K., Westerhoff, P.: Challenges in photocatalytic reduction of nitrate as a water treatment technology. Sci. Total. Environ. **599–600**, 1524–1551 (2017)

Xu, Y., Schoonen, M.A.A.: The absolute energy positions of conduction and valence bands of selected semiconducting minerals. Am. Miner. **85**, 543–556 (2000)

Ye, J., Liu, S.-Q., Liu, W.-X., Meng, Z.-D., Luo, L., Chen, F., Zhou, J.: Photocatalytic simultaneous removal of nitrate and ammonia via a zinc ferrite/activated carbon hybrid catalyst under UV-visible irradiation. ACS Omega **4**, 6411–6420 (2019a)

Ye, Y., Brillas, E., Centellas, F., Cabot, P.L., Sirés, I.: Photoelectrocatalytic denitrification and nitrification of wastewaters by nanostructured ZnO: reaction pathways and mechanisms. ACS Omega **4**, 6411–6420 (2019b)

Zhu, C., Wang, X.: Nanomaterial ZnO synthesis and its photocatalytic applications: a review. Nanomaterials **15**(9), 682 (2025)

Accumulation of Toxic Heavy Metals in Groundwater Around Botshabelo Non-engineered Landfill Site, South Africa

Innocent Mugudamani, Saheed Adeyinka Oke, Thandi Patricia Gumede, Moeketsi Sesing, and Silent Ruzvidzo

Abstract Non-engineered municipal solid waste landfills pose significant threats to the surrounding environment. Therefore, the purpose of this study was to assess the concentrations of toxic heavy metals in groundwater surrounding the Botshabelo non-engineered landfill site in South Africa. Four groundwater samples were collected randomly during early summer season. The levels of toxic heavy metals in groundwater sources were determined using inductively coupled plasma mass spectrometry and ion chromatography. The concentrations of heavy metals were as Al (0.08), As (<LOD), Sb (<LOD), Ba (0.06), B (0.03), Cd (<LOD), Co (<LOD), Cr (0.02), Cu (0.02), Fe (0.03), Mn (0.24), Ni (<LOD), Mo (0.04), Pb (<LOQ), Se (<LOD), U (0.06), V (0.02), Zn (0.01) and F (0.29) mg/l. They were descending as F > Mn > Al = U > Ba > Mo > B = Fe > Cr = Cu = V > Zn while As, Sb, Cd, Co, Ni, Pb, and Se were below limit of detection. When compared to the national standards (SANS and DWAF) and international standards (WHO), the mean values of toxic heavy metals such as Mn and U were above all drinking water standards (SANS and WHO) and irrigation standards (DWAF) while Mo was above DWAF acceptable standard for irrigation. To lessen water contamination in the surrounding area, better waste management procedures and well-designed landfill sites must be built. It is also recommended that proper segregation of biodegradable, non-biodegradable and recyclable materials such as plastics and metals be implemented. The outcomes of the study may assist local government bodies in determining appropriate treatment techniques and developing waste management policies and strategies.

I. Mugudamani · S. A. Oke (✉)
Department of Civil Engineering, Faculty of Engineering Built Environment and Information Technology, Centre for Sustainable Smart Cities, Central University of Technology, Free State, Bloemfontein, South Africa
e-mail: soke@cut.ac.za

T. P. Gumede · M. Sesing · S. Ruzvidzo
Department of Life Science, Central University of Technology, Free State, Bloemfontein, South Africa
e-mail: tgumede@cut.ac.za

H. Xu (ed.), *Proceedings of the 8th International Symposium on Water Resource and Environmental Management*, Environmental Science and Engineering,
https://doi.org/10.1007/978-3-032-19088-8_4

Keywords Landfill · water quality · Toxic heavy metals · water pollution · Health risks · Water sources

1 Introduction

In both developed and developing countries, landfilling remains the most straightforward, cost-effective, and economical method of waste disposal (Ehsan et al. 2025). In many developing countries, including South Africa, the implementation of landfill systems for waste disposal often falls short of established standards and guidelines (Wachirawongsakorn and Sangyoka 2013). Most developing countries dispose of solid waste in either engineered or non-engineered landfills (Iqbal et al. 2020). Non-engineered municipal solid waste landfills pose a significant threat to the surrounding environment (Mishra et al. 2018). Materials such as electroplating waste, used batteries, electronic waste, pesticide containers, and paint residues are frequently mishandled and mixed with municipal solid waste. This improper disposal contributes to toxic heavy metal contamination at landfill sites, posing significant environmental risks (Okafor and Obaze 2025). According to Ololade et al. (Ololade et al. 2019), non-engineered municipal solid waste landfills pose a significant threat to various aspects of the environment as waste dumped in landfills is subject to groundwater infiltration from precipitation or erosion to surface water sources. The resulting contaminated water is known as "leachate" (Amano et al. 2020). Leachate generated from landfill sites often contains elevated concentrations of pollutants, including heavy metals (Xiang et al. 2019). Leachate from non-engineered municipal solid waste landfills can pollute water sources for decades or centuries, posing a serious threat to their security (Chen et al. 2019). Exposure to high concentrations of toxic heavy metals can lead to health problems such as cardiovascular diseases, thyroid dysfunction, kidney failure, dental problems, neurological disorders, and renal impairment (Ehsan et al. 2025).

Groundwater sources are valuable natural resources that are primarily used for domestic, agricultural, and industrial purposes. Hence, they should be accessible and free of pollution (Palomeras et al. 2020). However, because the majority of non-engineered municipal solid waste landfills in many developing countries lack leachate collection and treatment systems, groundwater contamination is more severe (Arunbabu et al. 2017). In these cases, the lack of an underpinning barrier and leachate collection systems at the chosen landfill site increases the likelihood of adversely affecting the quality of water sources. Many researchers concur that these conditions in non-engineering landfills can be a significant threat to the environment and the ecosystem as dissipation of leachate may contaminate groundwater due to the dissipation of leachate through the soil (Ololade et al. 2019; Makhadi et al. 2020; Fattahzadeh et al. 2021; Osman et al. 2024). Since groundwater is typically thought to be of good natural quality due to its geological environment, evaluating the level of heavy metals in time and space is especially important to reduce the risk of groundwater pollution by leachates from landfills (Abiriga et al. 2020). According to Przydatek and Kanownik (2021) the monitoring of landfill leachates offers early risk identification, which may help limit harm to humans and the environment. In regions like Botshabelo, where solid waste is disposed of in non-engineered landfill sites, there has been no assessment of toxic heavy metal levels in nearby water sources. Thus, it is critical to comprehend how the chemistry of groundwater is affected by the leachate that emerges from the non-engineered municipal solid waste landfill. To achieve this, various toxic heavy metals in groundwater around Botshabelo non-engineered landfill site, South Africa were monitored. The outcomes of the study may be able to assist local government bodies in determining appropriate treatment techniques and developing waste management policies.

2 Material and Methods

2.1 Description of the Study Area

The landfill site is located in Botshabelo Township at S 29°14′44.72″ and E 26°44′57.46″ (Fig. 1). The Township is the largest and fastest-growing part of the Mangaung Metropolitan Municipality in South Africa's Free State province. It is located approximately 55 km east of Bloemfontein. The area is regarded as semi-arid region with a Highveld climate. The region is thought to have a Highveld climate and be semi-arid. Winter lasts from May to August, and summer lasts from October to February. Its average summer temperature is 26 °C, and its average winter temperature is 16 °C. The region occasionally experiences severe frost. The study area is located in a region that receives summer rainfall, with 600–750 mm of rain falling on average annually.

The dump site is a non-engineered landfill that covers 15 hectares and was designed to take only common non-hazardous garbage. It has been designated as a general, medium-sized, class B landfill. General waste includes tyres disposed of at randomly, construction debris, garden greens, and home waste. Approximately 747 tonnes of waste are delivered to the Botshabelo landfill site each month on average. Most of the wastes disposed of here are from commercial and residential sources. General non-hazardous waste, including food, garden, paper and cardboard, plastic film, plastic dense, metals, glass, textiles, treated wood, and other items, make up the consolidated generated municipal area composition (Mangaung Metropolitan Municipality 2016).

The sampled area is located in the Modder River catchment, which is distinguished by the Ecca and Beaufort groups, two Karoo Supergroup geological groups. There are two common aquifer types in the region: intergranular and fractured aquifers. Dolerite intrusions have been found throughout the sedimentary rocks that make up the intergranular aquifers. The dolerite dykes have an average borehole yield of

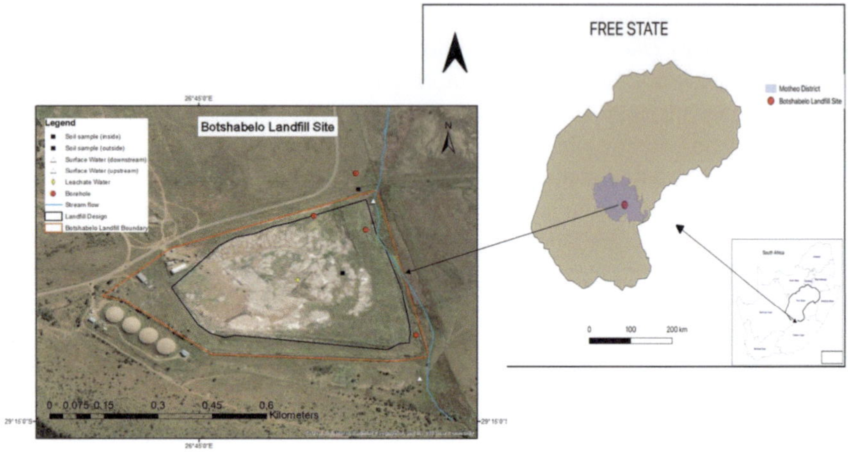

Fig. 1 Botshabelo non-engineered landfill site

between 0.5 L/s and 5 L/s (Makhadi et al. 2020). The dolerite dykes typically produce borehole yields ranging from 0.5 L/s to 5 L/s (Makhadi et al. 2020). The soils are predominantly black and red, highly structured clays with a high base status (Ololade et al. 2019).

2.2 Sample Collection and Analysis

To assess the impact of a non-engineered landfill on water sources, groundwater samples were gathered in the early summer season. Using thoroughly cleaned 750 ml plastic bottles, samples were randomly collected (Fig. 1) from boreholes located near the Botshabelo landfill site. Four groundwater samples (BH1, BH2, BH3 and BH4) were collected in this study. The water level in the study area ranges from 3 to 10 m for all boreholes. For borehole BH1, BH2, BH3, and BH4 the water table is at 3.0 m, 4.1 m, 3.1 m, and 10.37 m deep respectively. All the boreholes were drilled up to $\pm$ 30 m deep. The collection, preservation, and measurement of samples followed the Standard Procedures for the Examination of Water and Wastewater (2005). At the Institute for Groundwater Studies (University of the Free State), analyses for toxic heavy metals were conducted using inductively coupled plasma mass spectrometry (ICP-MS), ion chromatography, and the IDEXX (Colilert18) Quanti-Tray™ (Westbrook, Maine, USA) method. Similar methodologies were reported by Makhadi et al. (2020) and Ololade et al. (2019). Non-target analysis was used to identify and characterize a wider range of toxic heavy metals such as aluminium (Al), arsenic (As), antimony (Sb), barium (Ba), boron (B), cadmium (Cd), cobalt (Co), chromium (Cr), copper (Cu), iron (Fe), manganese (Mn), nickel (Ni), molybdenum (Mo), lead (Pb), selenium (Se), uranium (U), vanadium (V), zinc (Zn), and fluorine (F). Throughout the analysis, double-distilled water was used alongside blank and duplicate samples to maintain data quality assurance. The analytical error was calculated to be under 10%, with data reproducibility within 5%.

3 Results and Discussion

3.1 Concentration of Toxic Heavy Metals in Groundwater Around Landfill Site

Heavy metals concentrations in water sources are very important as they also determine how polluted and toxic water sources are (Amano et al. 2020). Therefore, availability and levels of toxic heavy metals in non-engineered landfills sites should be monitored to prevent contamination of surrounding water resources. The concentration of groundwater around Botshabelo landfill site are presented in Table 1. The level of toxic heavy metals in groundwater were descending as F > Mn > Al = U > Ba > Mo > B = Fe > Cr = Cu = V > Zn while As, Sb, Cd, Co, Ni, Pb, and Se were below limit of detection. The occurrence of toxic metals in groundwater sources around landfill sites were also reported in other studies (Makhadi et al. 2020; Fattahzadeh et al. 2021; Osman et al. 2024; Saha et al. 2019; Podlasek et al. 2021; Edokpayi et al. 2018).

When compared to the national (SANS and DWAF) and international standards (WHO), the mean value of toxic heavy metals such as Mn was above WHO drinking water standard and DWAF irrigation standard. Uranium mean value was above all drinking water standards (SANS and WHO) and irrigation standards (DWAF) while Mo was above DWAF acceptable standard for irrigation. In other studies, such as in Revdalen landfill, Norway (Abiriga et al. 2020), in New Delhi, India (Ahamad et al. 2018) and in Subotica, Serbia (Krčmar et al. 2017), Mn in groundwater was also above the acceptable standard. The present of Mn in groundwater may be attributed to industrial wastes and weathering of Mn-bearing minerals and rocks (El-Salam and Abu-Zuid 2014). The mean concentration of Mo above the acceptable standards may be linked to landfill waste from ceramics, glass, contact lens solutions, metallurgy, lubricants, pigments, catalysts, electronics, and cosmetic products (Tambat et al. 2023). The occurrence of U at high concentration in this study can be attributed to local geology i.e., dominated by sedimentary rocks which may contain traces of uranium. As a result of the breakdown (weathering) of rocks, it may be introduced into surrounding water sources. Furthermore, agricultural wastes such as phosphate fertilisers dumped in landfill sites could also be a source of U in groundwater (Balaram et al. 2022).

The concentrations of other toxic heavy metals were found to be within permissible limits. Similar findings of low toxic heavy metal levels have been reported in several studies (Makhadi et al. 2020; Abiriga et al. 2020; Przydatek and Kanownik 2021). Such low concentrations may be attributed to strong attenuation processes

Table 1 Comparation of heavy metals concentration in groundwater around Botshabelo landfill site with WHO (2015), SANS (2015) and DWAF (1996) standards

Parameters	Unit	Groundwater		Water Standards		
		Min–Max	Mean ± SD	WHO	SANS	DWAF
Al	mg/l	0.06–0.10	0.08 ± 0.02	< 0.1	≤ 0.3	≤ 5
As	mg/l	–	–	0.01	≤ 0.01	≤ 0.1
Sb	mg/l	–	–	0.02	≤ 0.02	–
Ba	mg/l	0.04–0.09	0.06 ± 0.02	1.3	≤ 0.7	–
B	mg/l	0.02–0.04	0.03 ± 0.01	2.4	≤ 2.4	≤ 0.5
Cd	mg/l	–	–	0.003	≤ 0.003	≤ 0.01
Co	mg/l	–	–	–	–	≤ 0.05
Cr	mg/l	0.01–0.02	0.02 ± 0.01	0.05	≤ 0.05	≤ 0.1
Cu	mg/l	0.00–0.07	0.02 ± 0.03	2	≤ 2	≤ 0.2
Fe	mg/l	0.00–0.07	0.03 ± 0.03	–	≤ 2	≤ 5
Mn	mg/l	0.00–0.86	0.24 ± 0.41	0.08	≤ 0.4	≤ 0.02
Ni	mg/l	–	–	0.07	≤ 0.07	≤ 0.20
Mo	mg/l	0.01–0.12	0.04 ± 0.05	–	–	≤ 0.01
Pb	mg/l	–	–	0.1	≤ 0.01	≤ 0.2
Se	mg/l	–	–	0.04	≤ 0.04	≤ 0.02
U	mg/l	0.01–0.11	0.06 ± 0.05	0.03	≤ 0.03	≤ 0.01
V	mg/l	0.00–0.07	0.02 ± 0.04	–	–	≤ 0.10
Zn	mg/l	0.00–0.01	0.01 ± 0.00	3	≤ 0.01	≤ 1
F	mg/l	0.25–0.38	0.29 ± 0.06	1.5	≤ 1.5	≤ 2

Notation: mg/l: milligram per litre; –: below limit of detection; ±: standard deviation; WHO: world health organisation; SANS: South African National Standards; DWAF: Department of water affairs

occurring both inside the landfill and in nearby water sources. These processes include direct reduction to elemental form, sorption, dilution, precipitation as sulphides or carbonates, and the limited complexing capacity of organic matter. It is important to note that the actual concentrations of toxic heavy metals can vary significantly depending on the composition of the waste. Therefore, the observed low levels of toxic heavy metals may also indicate a low presence of hazardous waste in the landfill, as no records of waste composition were available (Abiriga et al. 2020). Erosion and leachate infiltration or permeation through water can exacerbate groundwater quality, particularly during the rainy season (Amano et al. 2020). Although the majority of toxic heavy metals in this study have concentrations that are within the permissible limits for drinking water quality and irrigation, they are also regarded as a significant threat to surrounding water bodies. It is also expected that during heavy rains, there will be very high levels of groundwater and runoff, with the greatest possibility of leachate penetrating groundwater, thus affecting water quality (Ancic et al. 2020).

3.2 Comparation of Toxic Heavy Metals in Groundwater Around Botshabelo Landfill Site with Other Global Studies

The mean concentrations of toxic heavy metals in this study were compared to the others studies around the world (Table 2). The majority of toxic heavy metals identified in this study were not detected in other studies. However, the mean concentration of Cr, Cu, Fe and Zn in this study were lower than those identified in other studies while Mn concentration was above all other studies (Ehsan et al. 2025; Fattahzadeh et al. 2021; Boateng et al. 2019; Abdelwaheb et al. 2012; Olagunju et al. 2020; Alghamdi et al. 2022). In general, Zn, Ni and Fe recorded the highest concentrations in Iran (Fattahzadeh et al. 2021). The difference in mean concentrations of toxic heavy metals in groundwater around landfill sites may be attributed to many factors such as location of landfill, industrial activities, population size, sampling methods, type of waste, and geology of the area. Moreover, the distance between the boreholes and landfill, type of landfill site i.e., engineered or non-engineered may also play a crucial role in pollution level of toxic heavy metals in groundwater (Nevondo et al. 2019).

4 Conclusion

The study's findings revealed that the Botshabelo non-engineered landfill site has the potential to cause impact on the quality of groundwater. The presence of elevated levels of heavy metals (Mn, Mo, and U) in groundwater above acceptable standards indicates that a lack of leachate collection system, as well as poor management of a non-engineered landfill site, will have serious consequences on the quality of water sources in the area over time as a result of uncontrolled leachate accumulation at the landfill base. These findings showed that water sources should not be relied on for drinking water supply purposes, and disinfection is recommended before use. There is also a need for extensive and frequent monitoring of water sources near the site to track leachate migration and potential contamination of groundwater. Moreover, enhancing waste management practices and establishing a properly engineered landfill site are essential to minimizing water contamination. It is also recommended that proper segregation of biodegradable, non-biodegradable and recyclable materials such as plastics and metals be implemented. Future studies should assess human health risks of heavy metals in soil, ground and surface water around the landfill site.

Table 2 Comparation of heavy metals in groundwater around Botshabelo landfill site with other studies

Toxic heavy metals	Current study	Pakistan	Ghana	Iran	Tunisia	Nigeria	Saudi Arabia
Al	0.08	–	–	–	–	–	–
As	–	–	–	3.02	–	–	–
Sb	–	–	–	–	–	–	–
Ba	0.06	–	–	–	–	–	–
B	0.03	–	–	–	–	–	–
Cd	–	–	0.033	0. 48	0.04	0.04	0.004
Co	–	1.22	–	1. 41	–	–	–
Cr	0.02	0.20	0.039	2. 15	1.17	0.02	0.064
Cu	0.02	–	0.247	9. 39	0.80	0.05	–
Fe	0.03	–	2.292	12. 13	–	–	–
Mn	0.24	–	–	–	0.05	–	0.007
Ni	–	0.06	–	14.18	0.23	0.03	0.006
Mo	0.04		–	–	–	–	–
Pb	–	0.60	0.011	3.33	0.35	0.05	0.027
Se	–	–	–	–	–	–	–
U	0.06	–	–	–	–	–	–
V	0.02	–	–	–	–	–	–
Zn	0.01	4.39	0.237	75.24	0.75	0.08	0.061
F	0.29	–	–	–	–	–	–
References	Current Study	Ehsan et al. (2025)	Boateng et al. (2019)	Fattahzadeh et al. (2021)	Abdelwaheb et al. (2012)	Olagunju et al. (2020)	Alghamdi et al. (2022)

References

Abdelwaheb, A., Moncef, Z., Hamed, B.D.: Landfill leachate generation and its impact on water at an urban landfill (Jebel Chakir, Tunisia). Hydrol.: Curr. Res. **3**, 1–6 (2012)

Abiriga, D., Vestgarden, L.S., Klempe, H.: Groundwater contamination from a municipal landfill: effect of age, landfill closure, and season on groundwater chemistry. Sci. Total. Environ. **737**, 140307 (2020)

Ahamad, N.A., Raju, J., Madhav, S., Wycisk, W.G.P.: Impact of non-engineered Bhalswa landfill on groundwater from quaternary alluvium in Yamuna flood plain and potential human health risk, New Delhi, India. Quat. Int. **507**, 352–369 (2018)

Alghamdi, A.G., EL-Saeid, M.H., Alzahrani, A.J., Ibrahim, H.M., Bashir, S.: Heavy metal pollution and associated health risk assessment of urban dust in Riyadh. Saudi Arabia. Plos one **17**(1), e0261957 (2022)

Amano, K.O.A., Danso-Boateng, E., Adom, E., Nkansah, D.K., Amoamah, E.S., Appiah-Danquah, E.: Effect of waste landfill site on surface and ground water drinking quality. Water Environ. J. **35**(2021), 715–729 (2020)

Ancic, M., Hudek, A., Rihtaric, I., Cazar, M., Bacun-Druzina, V., Kopjar, N., Durgo, K.: Physico-chemical properties and toxicological effect of landfill groundwaters and leachates. Chemosphere **238**, 124574 (2020)

Apha: Standard Methods for the Examination of Water and Wastewater, 21st ed.; American Public Health Association; American Water Works Association; Water Environment Federation; Government Printing Office: Washington, DC, USA (2005)

Arunbabu, V., Indu, K.S., Ramasamy, E.V.: Leachate pollution index as an effective tool in determining the phytotoxicity of municipal solid waste leachate. Waste Manage. **68**, 329–336 (2017)

Balaram, V., Rani, A., Rathore, D.P.S.: Uranium in groundwater in parts of India and world: a comprehensive review of sources, impact to the environment and human health, analytical techniques, and mitigation technologies. Geosyst. Geoenviron. **1**(2), 100043 (2022)

Boateng, T.K., Opoku, F., Akoto, O.: Heavy metal contamination assessment of groundwater quality: a case study of Oti landfill site, Kumasi. Appl Water Sci **9**, 33 (2019)

Chen, R., Teng, Y., Chen, H., Hu, B., Yue, W.: Groundwater pollution and risk assessment based on source apportionment in a typical cold agricultural region in Northeastern China. Sci. Total. Environ. **696**, 133972 (2019)

Department of water affairs (DWAF): South African Water Quality Guidelines, 2nd ed.; Volume 4: Agricultural use: Irrigation; Department of Water Affairs and Forestry: Pretoria, South Africa, p. 196 (1996). www.dws.gov.za. Accessed 23 July 2025

Edokpayi, J.N., Durowoju, O., Odiyo, J.O.: Assessment of heavy metals in landfill leachate: a case study of Thohoyandou landfill, Limpopo Province, South Africa. In: Saleh, H.E.-D.M., Aglan, R.F. (eds.) Chapter 13, Heavy Metals, pp. 219–231 (2018)

Ehsan, N., Dawood, A., Waheed, F., Nasir, R., Rebouh, N.Y., Rizvi, S.I., Muzamil, A., Zaman, Q.: Human and ecological health risks from heavy metal contamination in groundwater aquifers. Sci. Rep. **15**, 22826 (2025)

El-Salam, M.M.A., Abu-Zuid, G.I.: Impact of landfill leachate on the groundwater quality: a case study in Egypt. J. Adv. Res. **6**, 579–586 (2014)

Fattahzadeh, M., Hoshyari, E., Parang, S., Fereidoni, H., Khoshbakht, R., Rajmjooe, J.H., Charkhestani, A.: Assessment of heavy metal concentration and their source in the groundwater near the landfill site: case study (Shiraz landfill). J. Matter. Environ. Sci. **12**(11), 1430–1443 (2021)

Iqbal, A., Liu, X., Chen, G.H.: Municipal solid waste: review of best practices in application of life cycle assessment and sustainable management techniques. Sci. Total. Environ. **729**, 138622 (2020)

Krčmar, D., Tenodi, S., Grba, N., Kerkez, D., Watson, M., Rončević, S., Božo Dalmacija, B.: Preremedial assessment of the municipal landfill pollution impact on soil and shallow groundwater in Subotica, Serbia. Sci. Total. Environ. **615**, 1341–1354 (2017)

Makhadi, R., Oke, S.A., Ololade, O.O.: The influence of non-engineered municipal landfills on groundwater chemistry and quality in Bloemfontein, South Africa. Molecules **25**, 5599 (2020)

Mangaung Metropolitan Municipality.: Integrated waste management plan: review of integrated waste management plan; Mangaung metropolitan municipality: Bloemfontein, South Africa, (2016)

Mishra, S., Tiwary, D., Ohri, A., Agnihotri, A.K.: Impact of municipal solid waste landfill leachate on groundwater quality in Varanasi, India. Groundwater Sustain. Dev. **9**, 100230 (2018)

Nevondo, V., Malehase, T., Daso, A.P., Okonkwo, O.J.: Leachate seepage from landfill: a source of groundwater mercury contamination in South Africa. Water SA **45**(2) (2019)

Okafor, O.C., Obaze, W.O.: Assessment of heavy metal contents of surface water around waste dumpsites in Ebonyi, Enugu and Anambra States, Southeastern Nigeria. Environ. Forensics **26**(4), 545–562 (2025)

Olagunju, T., Olagunju, A., Akawu, I., Ugokwe, C.: Quantification and risk assessment of heavy metals in groundwater and soil of residential areas around Awotan landfill, Ibadan, Southwest Nigeria. J. Toxicol. Risk Assess **6**, 33 (2020)

Ololade, O.O., Mavimbela, S., Oke, S.A., Makhadi, R.: Impact of leachate from northern landfill site in Bloemfontein on water and soil quality: implications for water and food security. Sustainability **11**, 4238 (2019)

Osman, O.A., Ochieng, G.M., Rwanga, S.: Predictive simulation of groundwater contamination due to landfill leachate: a case study on the Robinson deep landfill, Johannesburg, South Africa. JDFEWS **5**(2), 65–82 (2024)

Palomeras, R.J.B., Sanches, R.L., Fermilan, C.P., Ballon, J.E., Lao, C.B., Malabarbas, G.T.: Environmental impact of sanitary landfill to the groundwater and surface water quality. Eco. Env. Cons. **27**, S73–S79 (2020)

Podlasek, A., Jakimiuk, A., Vaverková, M.D., Koda, E.: Monitoring and assessment of groundwater quality at landfill sites: selected case studies of Poland and the Czech Republic. Sustainability **13**, 7769 (2021)

Przydatek, G., Kanownik, W.: Physicochemical indicators of the influence of a lined municipal landfill on groundwater quality: a case study from Poland. Environ. Earth Sci. **80**, 456 (2021)

Saha, S., Reza, A.H.M.S., Roy, M.K.: Hydrochemical evaluation of groundwater quality of the Tista floodplain, Rangpur, Bangladesh. Appl Water Sci **9**, 198 (2019)

South African National Standard 241–1: Drinking Water, Part 1: Microbiological, Physical, Aesthetic and Chemical Determinants; 241–2:2015 drinking water; SABS: Pretoria, South Africa (2015)

Tambat, V.S., Tseng, Y., Kumar, P., Chen, C., Singhania, R.R., Chang, J., Dong, C., Patel, A.K.: Effective and sustainable bioremediation of molybdenum pollutants from wastewaters by potential microalgae. Environ. Technol. Innov. **20**, 103091 (2023)

Wachirawongsakorn, P., Sangyoka, S.: Assessment of heavy metal distribution in soil and groundwater surrounding municipal solid waste dumpsite in Nai Muang sub-district administrative organization, Amphur Phichai, Uttaradit. NU Sci. J. **10**(1), 18–29 (2013)

World Health Organization (WHO): Guidelines for Drinking Water Quality; Final Task Group Meeting; WHO: Geneva, Switzerland, (2015)

Xiang, R., Xu, Y., Liu, Y.Q., Lei, G.Y., Liu, J.C., Huang, Q.F.: Isolation distance between municipal solid waste landfills and drinking water wells for bacteria attenuation and safe drinking. Sci. Rep. **9**(1), 1–11 (2019)

Performance Evaluation of a High-Efficiency Water Disinfection Module Based on UV-C LED Technology

Eason Liao, Xiaoxiao Wang, Qing Zhang, Jiancheng Wang, and Shuzhong Li

Abstract A study evaluated a UV-C LED water disinfection module using 270–280 nm GaN high-power chips, testing various configurations, structural parameters, and pressure conditions. At 9 L/min flow rate, dual-chip configurations (5-series-2-parallel, 4-series-2-parallel, 3-series-2-parallel) achieved irradiation intensities of 377, 320, and 273 $\mu W/cm^2$, respectively, with a 99.999% Escherichia coli inactivation rate despite a 40% reduction in beads and 27.5% intensity drop. Structural optimizations included switching to a copper substrate (reducing optical power attenuation from 54 to 42%), increasing quartz tube diameter from 6 to 16 mm for longer UV exposure, and improving light utilization with high-reflectivity materials. Single-chip configurations (10, 8, 6 beads) yielded 305, 250, and 210 $\mu W/cm^2$, respectively, 19%–23% lower than dual-chip setups, yet maintained 99.999% sterilization. The module, sealed with food-grade silicone, operated at 9 L/min under 0.12 MPa inlet pressure, with a maximum tolerance of 0.45 MPa, showing no leakage and high stability. These findings support the development of efficient, cost-effective UV-C LED water disinfection systems for practical applications.

Keywords Water disinfection · UV-C LED · Escherichia coli · Sterilization rate · Mercury-free

1 Introduction

Drinking water safety is a critical criterion for assessing human health in daily life. However, drinking water used in everyday settings may become contaminated with pathogenic bacteria, protozoa, and viruses that cause waterborne diseases in humans and animals due to factors such as insufficient disinfection time, limited disinfection methods, or prolonged storage (Griffiths 2017; Lanrewaju et al. 2022a, 2022b; Tamele and Vasconcelos 2020). Secondary sterilization of domestic water before use

E. Liao (✉) · X. Wang · Q. Zhang · J. Wang · S. Li
MASSPHOTON Limited Hong Kong, Hong Kong, China
e-mail: eason@massphoton.com

H. Xu (ed.), *Proceedings of the 8th International Symposium on Water Resource and Environmental Management*, Environmental Science and Engineering,
https://doi.org/10.1007/978-3-032-19088-8_5

can effectively reduce the transmission of waterborne diseases. Currently, common disinfection methods for inactivating viruses in water include traditional chemical chlorination, as well as physical disinfection techniques such as membrane filtration, ozonation, and ultraviolet (UV) irradiation (Collivignarelli et al. 2018; Ngwenya et al. 2013; Li et al. 2023). Among these, ozone is the most effective disinfectant for inactivating bacteria, viruses, and protozoa, while chloramines exhibit the lowest efficacy but are more suitable for secondary water disinfection due to their slower reaction rates compared to chlorine and longer persistence in distribution systems. Additionally, chloramines produce lower levels of disinfection byproducts (DBPs) than chlorine, although microbial activity in distribution systems may generate nitrate ions when chloramines are used as residual disinfectants (Ngwenya et al. 2013).

UV-C, as a chemical-free physical disinfection technology that does not produce disinfection byproducts, has garnered significant attention in microbial inactivation (Oguma and Rattanakul 2021). Although widely applied in air, water, and surface disinfection (Montazeri et al. 2025; Muhammad et al. 2024), its limitations include low efficiency, frequent lamp replacement, and the environmental pollution caused by mercury-containing materials, which increase economic costs for users (Čizmić et al. 2019). UV-C light-emitting diode (LED) technology based on gallium nitride (GaN) materials has emerged as a solution to address these shortcomings (Würtele et al. 2011). In addition to overcoming the drawbacks of mercury-based lamps, UV-C LEDs offer advantages such as long lifespan, high stability, compact size, and short warm-up time (Pelayo et al. 2024, 2023). These characteristics make UV-C LEDs an ideal choice for next-generation water disinfection technologies.

2 Material and Method

2.1 UV-C LED Beads

This study utilized high-power 270–280 nm gallium nitride (GaN)-based UV-C LED chips independently developed and produced by MASSPHOTON. Through an optimized eutectic bonding process, high-quality bonding between the chip and the ceramic substrate was achieved, reducing the interfacial void fraction and enhancing heat dissipation efficiency. Figure 1a illustrates a single-chip packaged LED bead with an irradiation intensity of up to 30 mW. When two chips were packaged, the irradiation intensity of the UV-C LED bead reached 50 mW, representing an overall efficiency improvement of 66.6%. This optimized packaging approach provides reliable technical support for the application of deep-ultraviolet LEDs in sterilization and disinfection fields.

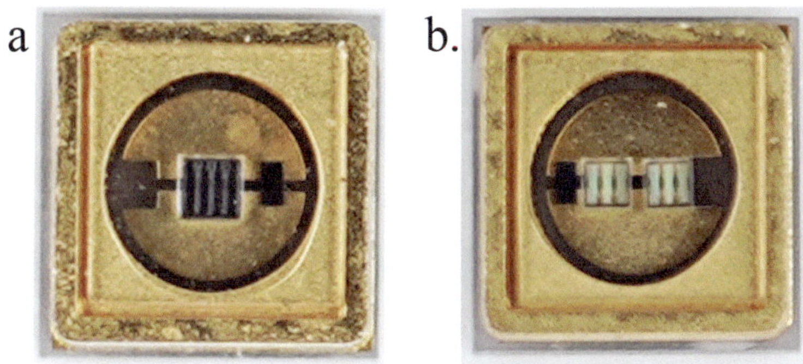

Fig. 1 MASSPHOTON LED beads: **a** Single-chip LED bead; **b** Dual-chip LED bead

2.2 *Test Strain*

The second generation of Escherichia coli (8099) was streaked onto a culture plate for activation and set aside for use. Fresh E. coli slant cultures from the third to seventh generations, incubated at 36 ± 1 °C for 18–24 h, were inoculated into LB liquid medium and cultured in a constant-temperature incubator at 36 ± 0.5 °C for 16–20 h. After significant bacterial colony expansion, the cultures were stored at room temperature for subsequent use.

2.3 *Detection Method*

The detection method and calculation of log inactivation rate were performed in accordance with Appendix E of GB28235-2020, Hygienic Requirements for Ultraviolet Appliance of Disinfection, which outlines the laboratory method for microbial inactivation testing in water disinfection.

3 Results and Discussions

3.1 *Introduction to Water Module Structure*

Aluminum rapidly forms a dense and robust aluminum oxide (Al_2O_3) protective film in air, which is compact and exhibits high chemical stability, preventing further oxidation of the underlying aluminum. Consequently, pure aluminum is highly resistant to rusting in dry air. Even in humid environments, as long as the oxide film remains intact (e.g., not compromised by strong acids or alkalis), its corrosion resistance significantly surpasses that of steel. Additionally, aluminum has a density of

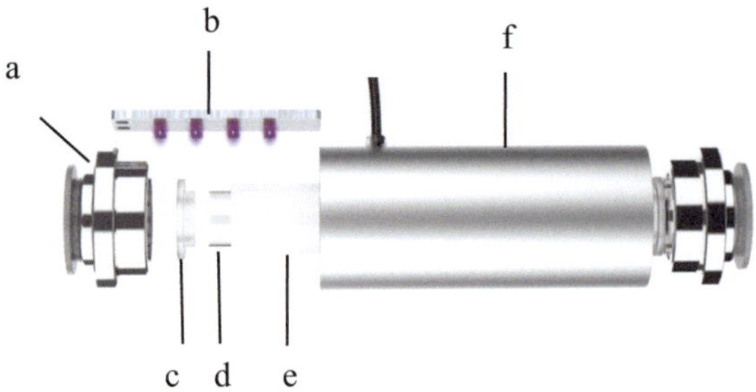

Fig. 2 Product diagram of MASSPHOTON 9 L/min water disinfection module: **a** Metal head **b** UV-C LED board **c** Waterproof silicone ring **d** Quartz tube **e** High-transmittance reflective material **f** Metal sleeve

approximately 2.7 g/cm^3, about one-third that of iron or copper, effectively reducing weight, making it an ideal material for the casing of the MASSPHOTON 9 L/min water disinfection module (Fig. 2). To enhance the oxidation and corrosion resistance of the aluminum-based casing, researchers applied anodization treatment to form a thicker and denser oxide film on the aluminum surface, thereby improving its durability.

3.2 *Study on Sterilization Efficacy of UV-C LED Under Different Bead Configurations*

To investigate the relationship between the number of LED beads and the sterilization rate of flowing water, three configurations of dual-chip UV-C LED beads were tested: 5-series-2-parallel (10 beads), 4-series-2-parallel (8 beads), and 3-series-2-parallel (6 beads). The sterilization efficacy against Escherichia coli (ATCC 8099) was evaluated under identical water module specifications, with each bead supplied with a current of 200 mA.

The experimental results (Table 1) demonstrate that, at a flow rate of 9 L/min and with an unmodified water module structure, the irradiation intensities for configurations using 5-series-2-parallel (10 beads), 4-series-2-parallel (8 beads), and 3-series-2-parallel (6 beads) dual-chip UV-C LED beads were 377 μW/cm^2, 320 μW/cm^2, and 273 μW/cm^2, respectively. Despite reducing the number of beads from 10 to 6, a 40% decrease, and the irradiation intensity from 377 μW/cm^2 to 273 μW/cm^2, a 27.5% reduction, the sterilization rate still achieved 99.999% (Fig. 3), fully meeting the stringent standards for drinking water disinfection.

Table 1 Relationship between the number of dual-chip UV-C LED beads, sterilization performance, and irradiation intensity

Number of lamp beads in the water module	Sterilizing rate (%)	Irradiation intensity (μW/cm^2)
Ten UV-C LED lamp beads with two chips each	99.999	377
Eight UV-C LED lamp beads with two chips each	99.999	320
Six UV-C LED lamp beads with two chips each	99.999	273

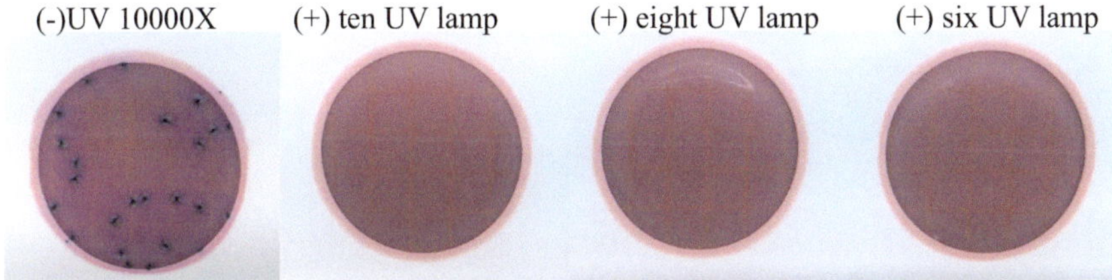

Fig. 3 Relationship between the number of dual-chip UV-C LED beads and sterilization test results

3.3 *Study on the Impact of Quartz Tube Diameter and Chip Quantity on Sterilization Efficiency*

To achieve cost reduction while maintaining sterilization efficiency, the research team, based on prior test results showing that copper substrates significantly reduce the optical power attenuation rate of UV-C LEDs compared to aluminum substrates (42% vs. 54%), replaced the original aluminum substrate in the water module with a copper substrate. To extend the residence time of flowing water within the module, the quartz tube diameter was increased from 6 to 16 mm, allowing for prolonged water-UV contact time while maintaining the same flow rate. Additionally, the aperture design of the high-transmittance reflective material at the LED bead positions was modified from elongated slots to individual openings aligned with each bead, thereby enhancing disinfection efficiency. Building on these structural optimizations, subsequent experiments were conducted using single-chip UV-C LED bead configurations of 5-series-2-parallel (10 beads), 4-series-2-parallel (8 beads), and 3-series-2-parallel (6 beads).

The experimental results (Table 2) indicate that, with configurations of 5-series-2-parallel (10 beads), 4-series-2-parallel (8 beads), and 3-series-2-parallel (6 beads) single-chip UV-C LED beads, the corresponding irradiation intensities were 305 μW/cm^2, 250 μW/cm^2, and 210 μW/cm^2, respectively. Compared to dual-chip UV-C LED configurations with the same number of beads, although the total number of beads remained unchanged, the irradiation intensity decreased by 19%, 22%, and

Table 2 Relationship between the number of single-chip UV-C LED beads, sterilization performance, and irradiation intensity

Number of lamp beads in the water module	Sterilizing rate (%)	Irradiation intensity (μW/cm^2)
Ten UV-C LED light beads, each consisting of one chip	99.999	305
Eight UV-C LED light beads, each consisting of one chip	99.999	250
Six UV-C LED light beads, each consisting of one chip	99.999	210

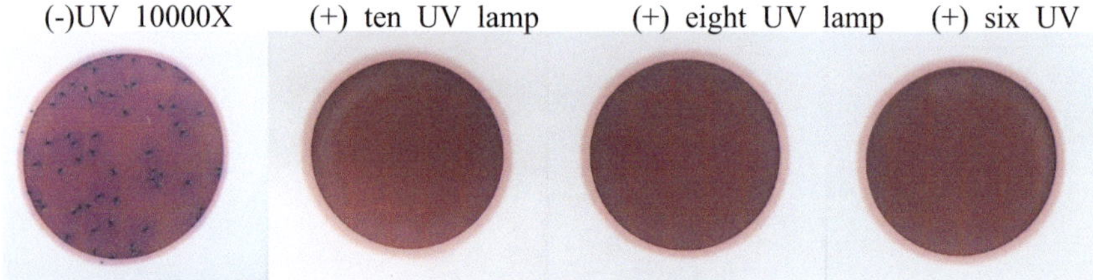

Fig. 4 Relationship between the number of single-chip UV-C LED beads and sterilization test results

23%, respectively, due to the halving of chips per bead. With further optimizations, including an increased quartz tube diameter, replacement with a copper substrate, and the use of additional high-transmittance reflective materials, configurations with 10, 8, and 6 single-chip UV-C LEDs still achieved a sterilization rate of 99.999% (Fig. 4), despite a 40% reduction in bead count and a 27.5% decrease in irradiation intensity, fully meeting the stringent requirements for drinking water disinfection.

3.4 *Water Flow Rate Variations Under Different Pressures*

As a core component in direct contact with water, the watertight sealing and pressure resistance stability of the MASSPHOTON 9 L/min water disinfection module are critical prerequisites for ensuring safe operation. To meet these essential requirements, food-grade silicone seals were used for the inlet and outlet sealing components. This material not only exhibits excellent elasticity and aging resistance, allowing a tight fit with the interface structure to prevent water leakage, but also avoids secondary contamination of water quality due to material leaching during disinfection, balancing both sealing performance and safety. To further validate the reliability of the sealing design, researchers simulated real-world operating conditions, continuously monitoring changes in water flow rate through the module under varying pressure conditions to determine the module's pressure tolerance range.

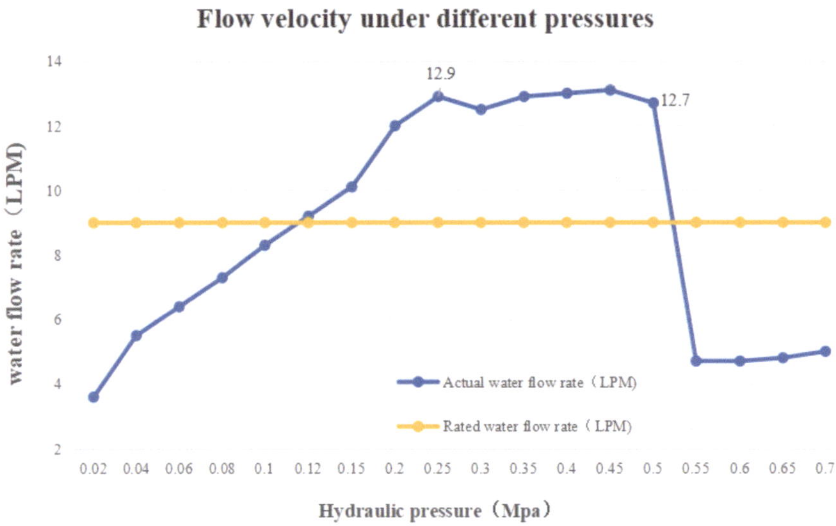

Fig. 5 Water flow rate through the water module under different pressures

As shown in Fig. 5, the current structure of the MASSPHOTON 9 L/min water disinfection module achieves a flow rate of 9 L/min at 0.12 MPa and a maximum flow rate of 12.9 L/min at 0.2 MPa, with no leakage observed during long-term operation tests. When the pressure increases to 0.42 MPa, the flow rate begins to decline, potentially due to deformation of the sealing ring caused by excessive pressure. These results indicate that, with adequate sealing performance, the flow rate remains stable without abnormal attenuation. However, if sealing defects (e.g., interface leakage) are present, the flow rate exhibits significant fluctuations due to pressure loss. The MASSPHOTON 9 L/min water disinfection module successfully passed pressure tests ranging from 0.12 to 0.45 MPa, demonstrating its excellent mechanical sealing performance and pressure resistance stability, ensuring no issues such as water leakage or pressure loss during long-term use.

4 Conclusions

This study systematically evaluates the key performance parameters of a UV-C LED water disinfection module, including the number of packaged chips, LED configuration, irradiation intensity, and pressure resistance, providing critical insights for developing efficient, low-cost, and low-risk UV-C LED sterilization systems. Experimental results demonstrate that replacing the original aluminum substrate with a copper substrate and increasing the quartz tube diameter enables the optimization of the UV-C LED configuration from a 5-series-2-parallel (10 LEDs) dual-core setup to a 3-series-2-parallel (6 LEDs) single-core setup, reducing the total number of chips from 20 to 6, significantly lowering chip costs. Additionally, pressure retention tests on the MASSPHOTON 9 L/min water disinfection module indicate a maximum pressure resistance of 12.9 MPa, with no leakage observed during long-term operation, demonstrating excellent sealing and reliability.

In conclusion, through optimizations in materials and structure, this study achieves significant improvements in cost control and reliability of the UV-C LED water disinfection module while maintaining disinfection performance. Future research may focus on the long-term operational stability of higher-power UV-C LEDs, as well as the impact of industrial high-flow water conditions and varying water qualities on disinfection efficiency, to enhance the adaptability and widespread adoption of this technology in practical applications.

References

Čizmić, M., Ljubas, D., Rožman, M., Ašperger, D., Ćurković, L., Babić, S.: Photocatalytic degradation of azithromycin by nanostructured TiO_2 film: kinetics, degradation products, and toxicity. Materials **6**(12) (2019)

Collivignarelli, M.C., Abbà, A., Benigna, I., Sorlini, S., Torretta, V.: Overview of the main disinfection processes for wastewater and drinking water treatment plants. Sustainability **1**(10), 86 (2018)

Griffiths, J.K.: Waterborne Diseases. International Encyclopedia of Public Health Second Edition, pp. 388–401 (2017)

Lanrewaju, A.A., et al.: A review on disinfection methods for inactivation of waterborne viruses. Front. Microbiol. **13** (2022a)

Lanrewaju, A.A., Enitan-Folami, A.M., Sabiu, S., Lanrewaju, A.A., Enitan-Folami, A.M., Sabiu, S., Edokpayi, J.N., Swalaha, F.M.: Global public health implications of human exposure to viral contaminated water. Front. Microbiol. **13** (2022b)

Li, H., Feng, M., Yu, X.: Qualitative and quantitative analysis of the effects of drinking water disinfection processes on eukaryotic microorganisms: a meta-analysis. Chemosphere 332 (2023)

Montazeri, M.M., Hejazi, S.A., Taghipour, F.: Ultraviolet light-emitting diode (UV-LED) water disinfection photoreactors: a review. J. Environ. Manage. 386 (2025)

Muhammad, S., Muhammad, F., Liao E.: Study of UV-C LED technology on in-door air disinfection: an effective mercury-free and ozone-free method to irradicate airborne microorganisms. In: Proceedings of the 2024 5th International Symposium on Artificial Intelligence for Medicine Science, pp. 859–863 (2024)

Ngwenya, N., Ncube, E.J., Parsons, J.: Recent advances in drinking water disinfection: successes and challenges. Rev. Environ. Contam. Toxicol. 111–70 (2013)

Oguma, K., Rattanakul, S.: UV inactivation of viruses in water: its potential to mitigate current and future threats of viral infectious diseases. Jpn. J. Appl. Phys. **11**(60) (2021)

Pelayo, D., Rivero, M.J., Santos, G., Gómez, P., Ortiz, I.: Techno-economic evaluation of UV light technologies in water remediation. Sci. Total Environ. 868 (2023)

Pelayo, D., Hernández-Pellón, A., Santos, G., Rumayor, M., Ortiz, I., Rivero, M.J.: Performance of high-efficiency UV-C LEDs in water disinfection: experimental, life cycle assessment, and economic analysis of different operational scenarios. J. Environ. Manage. 364 (2024)

Tamele, I.J., Vasconcelos, V.: Microcystin incidence in the drinking water of mozambique: challenges for public health protection. Toxins **12**(6) (2020)

Würtele, M.A., Kolbe, T., Lipsz, M., Külberg, A., Weyers, M., Kneissl, M., Jekel, M.J.W.R.: Application of GaN-based ultraviolet-C light emitting diodes–UV LEDs–for water disinfection. Water Res. **3**(45), 1481–1489 (2011)

Sustainable Lithium Recovery from Wastewater: A Comprehensive Literature Review

Nabiazam Sayed, Dibin Bhasi, Abdul Raheem, Shanid Thayyil, Parth Gupta, Grzegorz Brudecki, Wael Elamin, and Samara Bin Salem

Abstract The global demand for lithium (Li) is rapidly increasing due to its essential role in lithium-ion batteries (LIBs), which power electric vehicles, electronics, and large-scale renewable energy storage systems. Conventional extraction from ores and brines involves multiple steps, including clarification, filtration, concentration, and purification, which consume high energy, utilize chemicals, and have significant environmental impacts. Alternative sources, including municipal and industrial wastewater, geothermal brines, and spent battery leachates, are gaining attention as potential sources of lithium. Although these sources contain lower lithium concentrations, they provide important benefits by supplementing the supply and reducing environmental discharge. Recent technological advances have improved the feasibility of recovering lithium from these unconventional sources. This review focuses on membrane-based recovery technologies, including nanofiltration membranes, supported liquid membranes, ion-imprinted membranes, ion-sieve membranes, membrane distillation crystallization, selective electrodialysis, and capacitive deionization. Each method is analyzed for its separation mechanisms, performance, advantages, and limitations. Overall, while considerable progress has been made in selectivity, efficiency, and environmental compatibility, challenges remain in terms of scalability, cost, and integration into existing systems. Future developments will require hybrid approaches, digital optimization, and collaborative efforts to establish sustainable and commercially viable pathways for lithium recovery.

Keywords Lithium recovery · Wastewater treatment · Membrane-based technologies · Nanofiltration · Ion-imprinted polymers · Membrane distillation · Electrodialysis

---

N. Sayed (✉) · D. Bhasi · A. Raheem · S. Thayyil · P. Gupta · G. Brudecki · W. Elamin
RASID Laboratory, M42 Health, Abu Dhabi, UAE
e-mail: nsayed@m42.ae

S. B. Salem
Abu Dhabi Quality and Conformity Council (ADQCC), Abu Dhabi, UAE

H. Xu (ed.), *Proceedings of the 8th International Symposium on Water Resource and Environmental Management*, Environmental Science and Engineering,
https://doi.org/10.1007/978-3-032-19088-8_6

1 Introduction

Lithium (Li), with an atomic number of 3 and atomic mass of 6.94 g/mol, is an alkali metal widely used as an electrochemical power source due to its high cell voltage, long storage life, low self-discharge rate, and wide temperature range of use (Baudino et al. 2023; Granata et al. 2012). These properties make it ideal for producing lithium-ion batteries (LIBs) that are extensively used in applications such as mobile electronic devices, electric vehicles, and smart grids for energy storage (Narins 2017; International Energy Agency (IEA) 2019).

The increasing global transition to electrified transportation and renewable energy storage has significantly raised the demand for lithium. As illustrated in Fig. 1, according to the U.S. Geological Survey (USGS) data, global lithium consumption has grown rapidly over the years (U.S. Geological Survey 2025). As of 2024, more than 87% of the world's lithium consumption is used for battery production, as shown in Fig. 2 (U.S. Geological Survey 2025). This growth is expected to continue as the adoption of electric vehicles (EVs) expands and grid-scale storage systems become more common.

Lithium does not occur in elemental form naturally. It is typically extracted from mineral ores and brines using multi-step processes involving clarification, filtration, concentration, and purification, producing lithium carbonate or lithium hydroxide (Meng et al. 2019). These conventional methods are energy-intensive, chemically demanding, and environmentally burdensome.

To address these challenges, there is growing interest in recovering lithium from unconventional sources such as wastewater, industrial effluents, geothermal brines, and battery recycling leachates. Lithium recovery from such sources provides a supplementary supply while supporting environmental sustainability by reducing

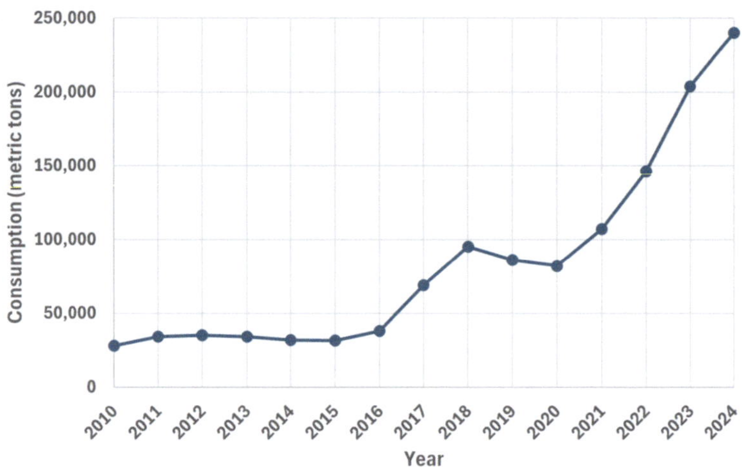

Fig. 1 Global lithium consumption trends from 2010 to 2024 (metric tons) (U.S. Geological Survey 2025)

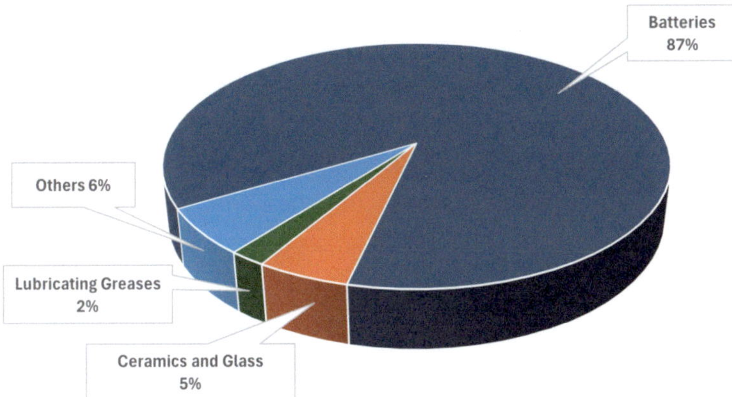

Fig. 2 Global end-use market of lithium, according to USGS data (U.S. Geological Survey 2025)

lithium discharge into ecosystems (Huang et al. 2021; Manjong et al. 2021; Bello et al. 2021; Kim et al. 2018).

2 Sources of Lithium in Wastewater

Lithium contamination in wastewater arises from various industrial, mining, and recycling activities, with concentrations determined by the source and process characteristics. The increasing global demand for lithium, particularly for energy storage applications, has amplified its occurrence in waste streams, emphasizing the need for efficient recovery technologies (Kim et al. 2018). The variability of lithium concentrations across different wastewater sources is graphically represented in Fig. 3.

Mining and brine processing operations represent the most significant contributors of lithium-bearing wastewater. Extraction from hard rock minerals (e.g., spodumene) and lithium-rich brines generate large volumes of process water containing residual lithium salts that are not fully recovered during the primary extraction stage. These waste streams, often with high lithium concentrations, present substantial opportunities for enhanced recovery and reduced environmental impact (Meshram et al. 2014; Flexer et al. 2018).

Industrial effluents also contribute notably to lithium inputs in wastewater. Sectors such as battery manufacturing, ceramics, glass, pharmaceuticals, and metallurgy discharge process water containing lithium salts, including lithium carbonate (Li_2CO_3), lithium hydroxide (LiOH), and lithium chloride (LiCl) (Choubey et al. 2017; Chen et al. 2017).

Recycling of spent lithium-ion batteries, especially via hydrometallurgical processes, produces leachates enriched with lithium and other critical metals such as cobalt, nickel, and manganese. Acid leaching is employed to dissolve these metals

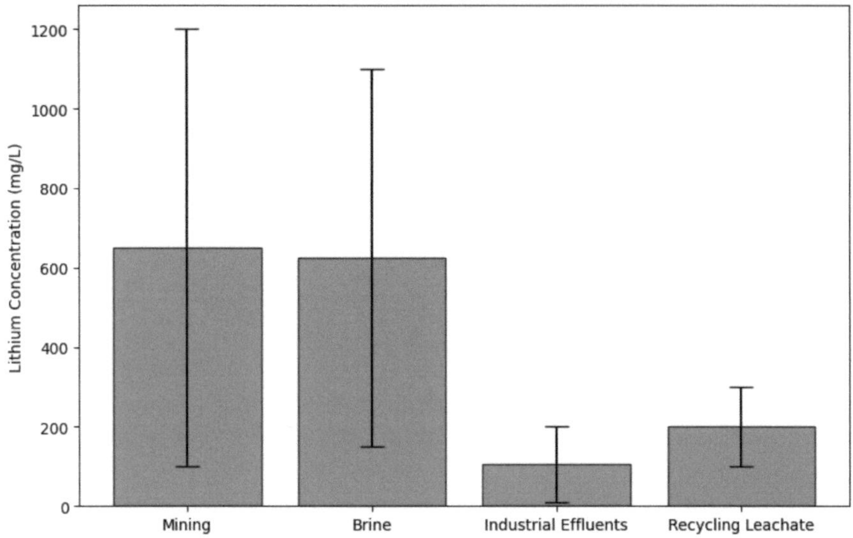

Fig. 3 Graphical representation of lithium in wastewater sources, showing variability through error bars (Meshram et al. 2014; Choubey et al. 2017; Chen et al. 2017; Zhang et al. 1998; Flexer et al. 2018)

into solution, and selective separation of lithium from such complex mixtures is essential for sustainable resource recovery (Zhang et al. 1998).

3 Importance of Lithium Recovery

Lithium recovery from wastewater is becoming increasingly critical due to its significant environmental and economic implications. As the demand for lithium rises due to the rapid growth of electric vehicles (EVs) and renewable energy storage systems, traditional sources such as mineral ores and brines are facing increasing pressure (Flexer et al. 2018; Swain 2017).

Recovering lithium from wastewater offers a sustainable alternative, helping to supplement the lithium supply chain while reducing the environmental impact caused by lithium discharge into water bodies. Moreover, this approach supports circular economy principles by enabling the reuse of valuable resources, lowering dependence on trad, and transforming waste streams into economically and environmentally beneficial raw materials (Vega-García et al. 2023). Integrating lithium recovery into industrial and municipal wastewater management systems can therefore contribute to resource efficiency, pollution reduction, and the development of more sustainable energy and manufacturing sectors.

4 Emerging Recovery Technologies

Several methods have been developed to recover lithium from wastewater, with a particular focus on membrane-based technologies due to their high selectivity, energy efficiency, and suitability for continuous operation (Li and Li 2019; Quist-Jensen et al. 2019; Luo et al. 2015; Xu et al. 2016). These advanced technologies are capable of efficiently extracting lithium even from complex and variable aqueous matrices, making them highly suitable for industrial and municipal wastewater treatment applications. The most important characteristics of the membrane separation technologies to recover a lithium resource from an aqueous environment are summarized in Table 1.

The supported liquid membrane, ion-imprinted membrane, and ion-sieve membrane immobilize highly selective adsorbents towards lithium, such as ionic liquids, ion-imprinted polymers, and ion sieves, onto the membrane carriers, respectively (Li and Li 2019; Quist-Jensen et al. 2019; Luo et al. 2015; Xu et al. 2016). Nanofiltration membranes can separate monovalent ions without chemicals, using Donnan exclusion, dielectric exclusion, and steric hindrance as the main mechanisms (Gong et al. 2018).

Membrane distillation crystallization enables the simultaneous production of freshwater and recovery of minerals from concentrated brine, utilizing low-grade heat (Pramanik et al. 2016). Electricity-driven membrane technologies, including selective electrodialysis and capacitive deionization, use monovalent-selective ion-exchange membranes as the separation medium. Under an applied electric field, these membranes allow efficient and selective transport of monovalent cations and anions, enabling effective separation from multivalent ions in aqueous solutions (Nie et al. 2017; Siekierka et al. 2018).

5 Conclusion

Recovering lithium from wastewater provides a sustainable solution to meet the growing global demand while reducing environmental pollution. Emerging technologies, including nanofiltration membranes, supported liquid membranes, ion-imprinted membranes, ion-sieve membranes, membrane distillation crystallization, selective electrodialysis, and capacitive deionization, have demonstrated strong potential for efficiently extracting lithium from complex wastewater streams.

Despite these advances, challenges such as low lithium concentrations, limited scalability, high operational costs, and issues like membrane fouling or low adsorption capacity remain. Future research should focus on developing cost-effective, scalable, and environmentally friendly processes. The integration of hybrid systems combining multiple technologies and the use of artificial intelligence (AI) to optimize process performance can further enhance recovery efficiency and reduce energy consumption. Collaboration between researchers, industry, and policymakers will

Table 1 Overview of membrane-based technologies for lithium recovery from aqueous environment

Method	Mechanism	Maturity	Extraction efficiency	Advantages	Limitations	References
Supported liquid membrane	Selective ion transport via solvent impregnated membrane	Lab scale	Recovery > 95%	Large adsorption capacity, high selectivity, low footprint	Organic solvent leakage, need for chemical reagent for desorption	Li and Li (2019); Quist-Jensen et al. (2019)
Ion-imprinted membrane	Selective adsorption by chelating sites	Lab scale	Li^+/Na^+, K^+, Ca^{2+}, Mg^{2+} separation factor: 4–51; adsorption: 4–50 mg/g	High selectivity	Low adsorption capacity, need for chemical reagent for desorption	Li and Li (2019); Luo et al. (2015)
Ion-sieve membrane	Selective adsorption by intercalation	Lab scale	Li^+/Na^+, K^+, Ca^{2+}, Mg^{2+} separation factor: 99–5312; adsorption: 10.3–27.8 mg/g	Large adsorption capacity, high selectivity, chemical stability	Particle leakage, need for chemical reagent for desorption	Li and Li (2019); Xu et al. (2016)
Nanofiltration membrane	Steric hindrance and Donnan exclusion	Full scale	Li^+/Mg^{2+} separation factor: 2.6–10.4 (up to 1815 with MOF-based membrane)	Low footprint	Membrane fouling, high investment and operating costs	Li and Li (2019); Gong et al. (2018)
Membrane distillation crystallization	Vapour pressure gradient across hydrophobic membrane	Pilot scale	Recovery > 73%	Produces fresh water and salts simultaneously, uses different energy sources	Membrane wetting, fouling, critical salt separation issues	Li and Li (2019); Pramanik et al. (2016)
Selective electrodialysis	Electrical potential drives ion movement	Pilot scale	Recovery > 95%	High selectivity for monovalent ions, eco-friendly	Membrane fouling, energy cost rises with salinity	Li and Li (2019); Nie et al. (2017)
Capacitive deionization	Electrostatic adsorption	Lab scale	Recovery > 83%	High efficiency, eco-friendly	Low desorption efficiency	Li and Li (2019); Siekierka et al. (2018)

be crucial to accelerate technology adoption and commercialization, ultimately supporting a circular economy and contributing to sustainable energy infrastructure.

References

Baudino, L., Santos, C., Pirri, C.F., La Mantia, F., Lamberti, A.: Recent advances in lithium recovery from water resources: from passive to electrochemical methods. J. Power. Sources **558**, 232585 (2023)

Bello, A.S., Zouari, N., Da'ana, D.A., Hahladakis, J.N., Al-Ghouti, M.A.: An overview of brine management: emerging desalination technologies, life cycle assessment, and metal recovery methodologies. J. Environ. Manage. **288**, 112358 (2021)

Chen, X., Ma, H., Luo, C., Zhou, T.: Recovery of valuable metals from waste cathode materials of spent lithium-ion batteries using mild phosphoric acid. J. Hazard. Mater. **326**, 77–86 (2017)

Choubey, P., Chung, K.S., Kim, M.S., Lee, J.C., Srivastava, R.: Advance review on the exploitation of the prominent energy-storage element lithium. Part II: from seawater and spent lithium-ion batteries (LIBs). Miner. Eng. **110**, 104–121 (2017)

Flexer, V., Baspineiro, C.F., Galli, C.I.: Lithium recovery from brines: a vital raw material for green energies with a potential environmental impact in its mining and processing. Sci. Total. Environ. **639**, 1188–1204 (2018)

Gong, L.Y., Ouyang, W., Li, Z.R., Han, J.: Direct numerical simulation of continuous lithium extraction from high Mg^{2+}/Li^{+} ratio brines using microfluidic channels with ion concentration polarization. J. Membr. Sci. **556**, 34–41 (2018)

Granata, G., Moscardini, E., Pagnanelli, F., Trabucco, F., Toro, L.: Product recovery from Li-ion battery wastes coming from an industrial pre-treatment plant: lab scale tests and process simulations. J. Power. Sources **206**, 393–401 (2012)

Huang, T.Y., Pérez-Cardona, J.R., Zhao, F., Sutherland, J.W., Paranthaman, M.P.: Life cycle and techno-economic assessment of lithium recovery from geothermal brine. ACS Sustain. Chem. Eng. **9**(19), 6551–6560 (2021)

International Energy Agency (IEA). Global EV Outlook 2019. IEA, Paris (2019)

Kim, S., Kim, J., Kim, S., Lee, J., Yoon, J.: Electrochemical lithium recovery and organic pollutant removal from industrial wastewater of a battery recycling plant. Environ. Sci.: Water Res. Technol. **4**(2), 175–182 (2018)

Li, X., Li, J.: Membrane-based technologies for lithium recovery from aqueous resources: a comprehensive review. J. Membr. Sci. **591**, 117317 (2019)

Luo, X., Guo, B., Luo, J., Deng, F., Zhang, S., Luo, S., Crittenden, J.: Recovery of lithium from wastewater using development of Li Ion-imprinted polymers. ACS Sustain. Chem. Eng. **3**(3), 460–467 (2015)

Manjong, N.B., Usai, L., Burheim, O.S., Strømman, A.H.: Life cycle modelling of extraction and processing of battery minerals—a parametric approach. Batteries **7**(3), 57 (2021)

Meng, F., McNeice, J., Zadeh, S.S., Ghahreman, A.: Review of lithium production and recovery from minerals, brines, and lithium ion batteries. Miner. Process. Extr. Metall. Rev. **40**(4), 231–259 (2019)

Meshram, P., Pandey, B.D., Mankhand, T.R.: Extraction of lithium from primary and secondary sources by pre-treatment, leaching and separation: a comprehensive review. Hydrometallurgy **150**, 192–208 (2014)

Narins, T.P.: The battery business: lithium availability and the growth of the global electric car industry. Extr. Ind. Soc. **4**(2), 321–328 (2017)

Nie, X.Y., Sun, S.Y., Song, X.F., Yu, J.G.: Ion-fractionation of lithium ions from magnesium ions by electrodialysis using monovalent selective ion-exchange membranes. Desalination **403**, 128–135 (2017)

Pramanik, B.K., Thangavadivel, K., Shu, L., Jegatheesan, V.: A critical review of membrane crystallization for the purification of water and recovery of minerals. Rev. Environ. Sci. Biotechnol. **15**(3), 411–439 (2016)

Quist-Jensen, C.A., Ali, A., Drioli, E., Macedonio, F.: Perspectives on mining from sea and other alternative strategies for minerals and water recovery—the development of novel membrane operations. J. Taiwan Inst. Chem. Eng. **94**, 129–134 (2019)

Siekierka, A., Tomaszewska, B., Bryjak, M.: Lithium capture from geothermal water by hybrid capacitive deionization. Desalination **436**, 8–14 (2018)

Swain, B.: Recovery and recycling of lithium: a review. Sep. Purif. Technol. **172**, 388–403 (2017)

U.S. Geological Survey. Mineral Commodity Summaries 2025 (ver. 1.2, March 2025). U.S. Geological Survey, p. 212 (2025)

Vega-García, L., Ho, Y.C., Thant, M.M.M., Han, D.S., Lim, J.W.: Lithium in a sustainable circular economy: a comprehensive review. Processes **11**(2), 418 (2023)

Xu, X., Chen, Y., Wan, P., Gasem, K., Wang, K., He, T., Adidharma, H., Fan, M.: Extraction of lithium with functionalized lithium ion-sieves. Prog. Mater. Sci. **84**, 276–313 (2016)

Zhang, P., Yokoyama, T., Itabashi, O., Suzuki, T.M., Inoue, K.: Hydrometallurgical process for recovery of metal values from spent lithium-ion secondary batteries. Hydrometallurgy **47**(2–3), 259–271 (1998)

Comparative Assessment of the Groundwater Occurrence and Depth (GOD) and Rainfall Travel Time (RTt) Groundwater Vulnerability Methods for Leachate Contamination: A Case Study of the Botshabelo Landfill Site, South Africa

Silent Ruzvidzo, Saheed Adeyinka Oke, Thandi Patricia Gumede, Moeketsi Sesing, and Innocent Mugudamani

Abstract Groundwater vulnerability to leachate contamination is a significant public health concern, particularly in developing countries where non-engineered landfill sites are common. These sites pose a huge risk of contaminating groundwater resources used for drinking. The study employed a comparative case study research design utilising the GOD and RTt vulnerability assessment methods to evaluate the risk of leachate contamination at the Botshabelo non-engineered landfill site. Field data, including depth to groundwater and nitrate concentration, were collected using a Solinst 101 water level meter, and groundwater samples were analysed at the University of the Free State Institute for Groundwater Studies. Results showed that the GOD method indicated a very low vulnerability with an index of 0.08, while the RTt method suggested a moderate vulnerability with an index of 53. Validation using nitrate concentration yielded a mean of 0.33 mg/ℓ, confirming very low contamination levels which aligned closely with the GOD model. The rise in values observed in RTt was attributed to the shallow depth of groundwater and slope sensitivity, while the relatively flat trend in GOD findings was linked to the dominance of confining lithological layers. The findings highlight the importance of selecting assessment methods that are well-adapted to local conditions such as soil

S. Ruzvidzo (✉) · T. P. Gumede · M. Sesing
Department of Life Science, Central University of Technology, Bloemfontein, Free State, South Africa
e-mail: Silentr14@gmail.com

S. A. Oke · I. Mugudamani
Faculty of Engineering Built Environment and Information Technology, Department of Civil Engineering, Centre for Sustainable Smart Cities, Central University of Technology, Bloemfontein, Free State, South Africa

H. Xu (ed.), *Proceedings of the 8th International Symposium on Water Resource and Environmental Management*, Environmental Science and Engineering,
https://doi.org/10.1007/978-3-032-19088-8_7

type, lithology, precipitation, and slope. Although the study indicated low contamination risk, ongoing monitoring and additional water quality testing are recommended to ensure the continued safety of groundwater for consumption.

Keywords Groundwater · Vulnerability · Contamination · GOD · RTt · Comparison

1 Introduction

Groundwater vulnerability to leachate contamination represents a growing environmental and public health concern, particularly in developing and underdeveloped countries where waste management infrastructure remains insufficient or entirely lacking. In many such regions, landfill sites are non-engineered, lacking proper lining systems, leachate collection mechanisms, and monitoring facilities (Lapworth et al. 2017). This increases the likelihood of contaminants percolating into underlying aquifers, threatening groundwater quality and availability (Mor et al. 2006; Kumar et al. 2015). Inadequate waste disposal practices exacerbate the risks to communities that rely on groundwater as their primary source of potable water (UN-Habitat 2010).

Intrinsic groundwater vulnerability assessment methods have emerged as vital tools in identifying areas at risk of contamination and in informing water resource protection strategies. These methods including (Aller et al. 1987; Foster et al. 2002; Civita 1994; Oke 2016), among others evaluate the natural susceptibility of groundwater systems to contamination based on physical and hydrogeological parameters such as depth to water table, recharge rates, aquifer media, topography, and soil properties. One of the main advantages of these methods is their relative simplicity and reliance on accessible data, making them particularly suitable in resource constrained settings. However, despite their practicality, these methods may not always adequately capture site-specific risks, particularly from point source pollution such as landfill leachate, due to their generalized nature (Vrba and Zaporozec 1994; Gogu and Dassargues 2000).

The effectiveness and applicability of vulnerability assessment tools often depend on local environmental and geological conditions, including soil and rock types, elevation, vegetation cover, and intensity of rainfall (Rahman 2008; Sorensen et al. 2006). Therefore, continuous testing and adaptation of these methods to different geographic and hydrogeological contexts is essential. Such adaptations can lead to more refined models that are better suited for guiding practical management decisions. Enhancing these tools through the integration of localized parameters or hybridizing them with risk-based models may improve their accuracy and relevance, offering water managers more reliable evidence for policy formulation and resource protection.

Leachate pollution has become one of the most pressing threats to groundwater quality in developing countries. The World Bank (2012) estimates that more than two billion tonnes of municipal solid waste are generated globally each year, with sub-Saharan Africa contributing approximately 62 million tonnes per year, much of which is disposed in the open and in poorly managed dump sites. These conditions facilitate the migration of leachate into subsurface water bodies. In many sub-Saharan African countries, groundwater is the dominant source of drinking water due to its relative accessibility, resilience to drought, and generally acceptable quality (Misi et al. 2018; MacDonald et al. 2012). In South Africa, particularly in rural and peri-urban areas such as Botshabelo, groundwater has long been a vital source of domestic water supply, especially where municipal water services are inadequate (Burke and Villholth 2007).

This study investigates and compares the effectiveness of two intrinsic groundwater vulnerability assessment methods, the GOD method and the RTt method, in detecting vulnerability to leachate contamination. The methods were chosen due to their popularity, simplicity and ease of application (Ruzvidzo 2019). It is thus imperative to determine their effectiveness in identifying groundwater vulnerability due to leachate contamination. The landfill, characterized by uncontrolled and unlined waste disposal, presents a representative case of informal landfill management common in urban peripheries of South Africa. The outcomes of this study aim to contribute to the refinement of vulnerability assessment methods and inform evidence-based strategies for groundwater protection in similar socio-environmental contexts.

2 Materials and Methods

2.1 Description of the Study Area

The Botshabelo landfill site is situated in the eastern periphery of Botshabelo Township in the Free State Province of South Africa. The landfill lies approximately 1.3 km from the nearest residential area, at geographical coordinates 29°14′46″ S and 26°45′03″ E (World Population Review 2025). Botshabelo is home to an estimated population of around 191,000 people (DWS 2012a). The township functions as a satellite settlement of Bloemfontein, the judicial capital of South Africa, and is integrated into the broader socio-economic and service delivery framework of the Mangaung Metropolitan Municipality.

Botshabelo is located in South Africa's summer rainfall region, within a sub-humid, warm climatic zone characterized by distinct seasonal variations in precipitation and temperature. The average annual rainfall is approximately 533 mm, with the bulk of rainfall occurring between October and March in the form of soft precipitation (DWS 2012a; SAWS (South African Weather Service) 2023). Temperature extremes are notable, with summer highs reaching up to 30 °C in January and winter lows dropping to about 1 °C in June. Evaporation rates also follow seasonal patterns, with peak monthly values of around 323 mm in December and lows of approximately 85 mm in July (SAWS (South African Weather Service) 2023).

The landfill is classified as a non-engineered waste disposal site, lacking essential containment structures such as liners, leachate collection systems, and gas extraction infrastructure (Mangaung Metropolitan Municipality 2016). The site is designated as a General, Medium, Class B (GMB) landfill, which permits the disposal of general (non-hazardous) waste generated by households, small businesses, and non-industrial activities (Mangaung Metropolitan Municipality 2016). Common waste types include household refuse, construction and demolition debris, discarded tyres, and green waste such as garden trimmings.

Geologically, the landfill site is underlain by sedimentary formations of the Karoo Supergroup, primarily consisting of mudstone, shale, and sandstone units. These sedimentary rocks are intersected by dolerite dykes and sills, which manifest as linear ridge like structures across the landscape (Johnson and Wolmarans 2025).

The landfill is equipped with four groundwater monitoring boreholes strategically positioned at the site. These boreholes serve as critical observation points for detecting potential leachate impacts on groundwater quality. Their spatial arrangement provides coverage for assessing both upgradient and downgradient water quality conditions relative to the landfill's waste body. Figure 1 shows the map of the Botshabelo landfill site.

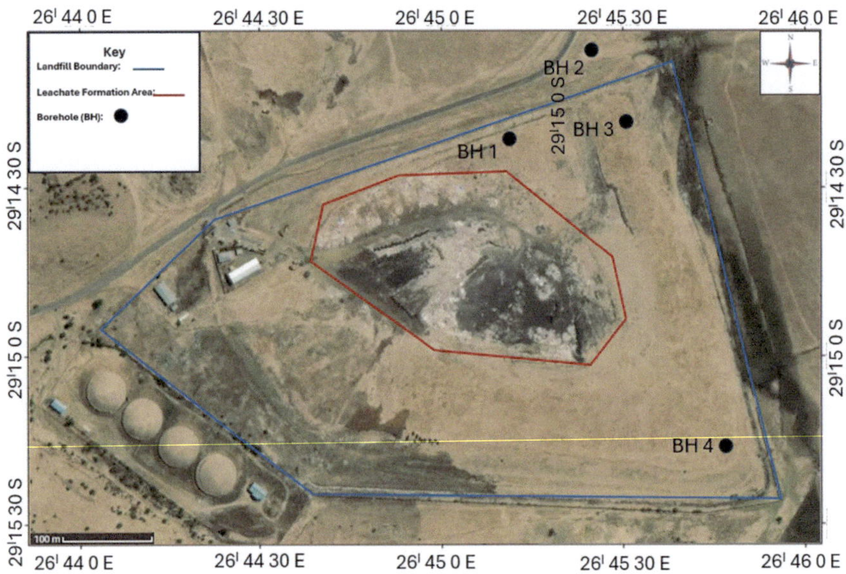

Fig. 1 Map of the Botshabelo non-engineered landfill site (Maps 2025)

2.2 Sample Collection and Analysis

Groundwater samples were collected from each of the four monitoring boreholes during two distinct seasonal periods in 2023 summer (February) and winter (July) to capture temporal variations in water quality associated with seasonal changes. The grab sampling technique was employed to sample groundwater using a plastic bailer, which was thoroughly rinsed with distilled water prior to use and between sampling events to avoid cross-contamination. From each borehole, groundwater was transferred into pre-sterilized 500 ml High-Density Polyethylene (HDPE) bottles, which were labeled with the borehole identification code. Care was taken to fill the containers fully, leaving minimal headspace to prevent oxidation or volatilization of certain parameters. The samples were transported at the Institute of Groundwater Services at the University of Free State where they were analysed using an Ultraviolet (UV) Spectrophotometer for nitrate analysis.

Application of the GOD method. The GOD method is a convectional groundwater vulnerability assessment method that have been used to assess the groundwater vulnerability in coal and gold mining areas (Foster et al. 2002; Sukmawati Rukmana et al. 2020). GOD stands for the following parameters: G = groundwater assurance; O = overall lithology, and D = depths to groundwater. The method produces 5 vulnerability classes ranging from Negligible (0–0.1), Low (0.1–0.3), Moderate (0.3–0.5), High (0.5–0.7), and Extreme (0.7–1.0). The overall GOD method is outlined in Eq. 1. The underlying rock and aquifer type was identified using data from the Geological Map of South Africa (Johnson and Wolmarans 2025). The depth to groundwater was measured seasonally at each borehole using a Solinst Model 101 Water Level Meter. The results were averaged to find the mean depth to groundwater.

$$\text{GOD index} = \text{Ca} \times \text{Cl} \times \text{Cd} \quad (1)$$

where: Ca = type of aquifer

Cl = type of lithology

Cd = depth to groundwater.

Application of the RTt method. The RTt method was developed by Oke (2016) to evaluate the vulnerability of shallow groundwater aquifers in the Dahomey basin in western Nigeria. The concept of the RTt method is mainly based on the hazard–pathway receptor model. The RTt is mainly a subjective vulnerability method that considers rainfall, slope, porosity, hydraulic conductivity, and depth to water table as the main parameters affecting groundwater vulnerability. The outline of the RTt method is shown in Eq. 2.

$$\text{RTt} = \text{R} \times 10 + (\text{D} \times \text{S})/(\text{Ksat}/\theta) \times 10 \quad (2)$$

where:

R = Rainfall rating

D = Depth to groundwater

S = Slope

Ksat = Hydraulic conductivity

θ = Porosity.

The RTt index is divided into 5 vulnerability classes ranging from very low (12–29), low (29–47), moderate (47–65), high (65–83), and very high (83–100). Hydraulic Conductivity (Ksat) was derived from the South Africa Council of Geoscience datasets and corroborated with regional pumping test records whilst porosity (θ) was estimated from lithological profiles of mudstones, shales, and sandstones, validated with published hydrogeological averages (Council for Geoscience 2021; Shapiro et al. 2017). The rainfall ratings were derived from the South African Weather Service data whilst the slope was calculated using GIS from the ArcGIS 10 maps.

Analysis and Validation of results. The groundwater vulnerability index maps for both the GOD and RTt method were produced using the Arc GIS 10 software. The outcomes of the GOD and RTt vulnerability methods were validated using nitrate concentration data. The use of nitrate for validation was due to its close association with household waste materials.

3 Results and Discussion

3.1 Vulnerability of Groundwater from Leachate Using the GOD and RTt Methods

The groundwater vulnerability assessment conducted using the GOD index method revealed an overall score of 0.08 across all four monitoring boreholes during both the summer and winter seasons. This score falls within the "very low vulnerability" category, indicating minimal susceptibility of the groundwater system to contamination. Although the general vulnerability was low, spatial analysis using the GIS-generated GOD index map (Fig. 2) revealed that boreholes 1, 2, and 3 displayed slightly elevated vulnerability levels relative to borehole 4. These differences, however, were not statistically significant to shift the overall vulnerability classification.

The low vulnerability values observed across the site can be attributed to several key hydrogeological factors. Notably, the presence of confining layers, such as impermeable mudstone formations and a protective clay cover, plays a crucial role in limiting the vertical movement of contaminants. These lithological features act as natural barriers that significantly reduce infiltration rates and limit contaminant transport pathways, thereby enhancing the intrinsic protection of the aquifer system.

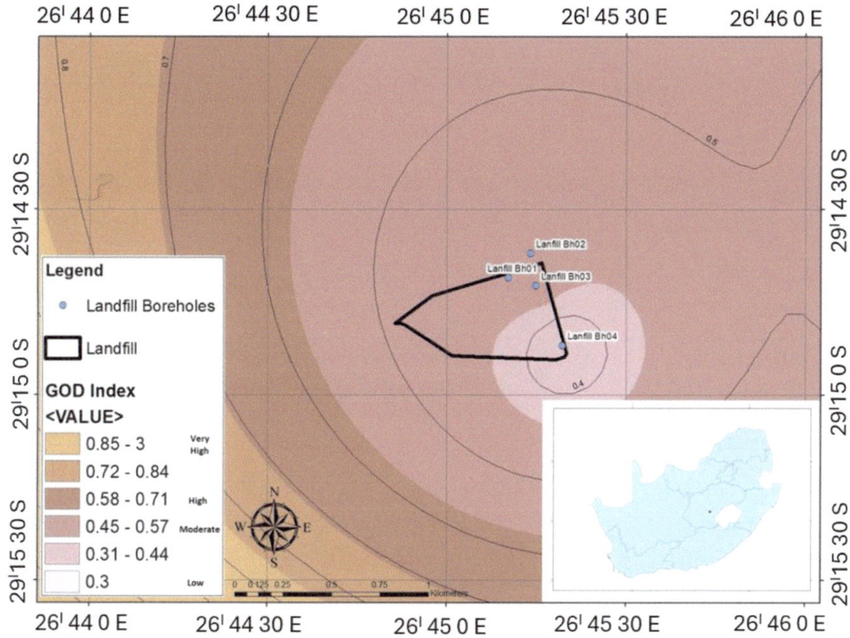

Fig. 2 GOD vulnerability index map for groundwater

In contrast to the outcome of the GOD method, the application of the RTt method yielded an average index value of 53 across the same boreholes and seasons. According to the RTt vulnerability classification, this falls within the "moderate vulnerability" range. Several environmental and hydrogeological parameters contributed to this assessment. Specifically, low annual rainfall, moderate slope gradients, low soil porosity, and moderate hydraulic conductivity were identified as key factors influencing the vulnerability score. These parameters collectively reduce percolation and contaminant mobility, thus moderating the overall vulnerability risk.

Despite the overall moderate vulnerability classification under the RTt method, spatial differentiation again highlighted that boreholes 1, 2, and 3 exhibited higher vulnerability levels, ranging between moderate to high. A likely explanation for this trend is the shallow groundwater table recorded at these boreholes: 4.8 m (BH1), 4.1 m (BH2), and 3.1 m (BH3), compared to a significantly deeper water table of 10.37 m at borehole 4. Shallow groundwater depths are commonly associated with increased vulnerability as they reduce the buffer zone between surface contaminants and the saturated zone, thereby facilitating more rapid contaminant transport. Figure 3 shows the groundwater vulnerability map for the RTt method.

The disparity in vulnerability outcomes between the two methods underscores the importance of a multi-method approach in vulnerability assessments. While the GOD method focuses on detailed aquifer characteristics, the RTt method incorporates

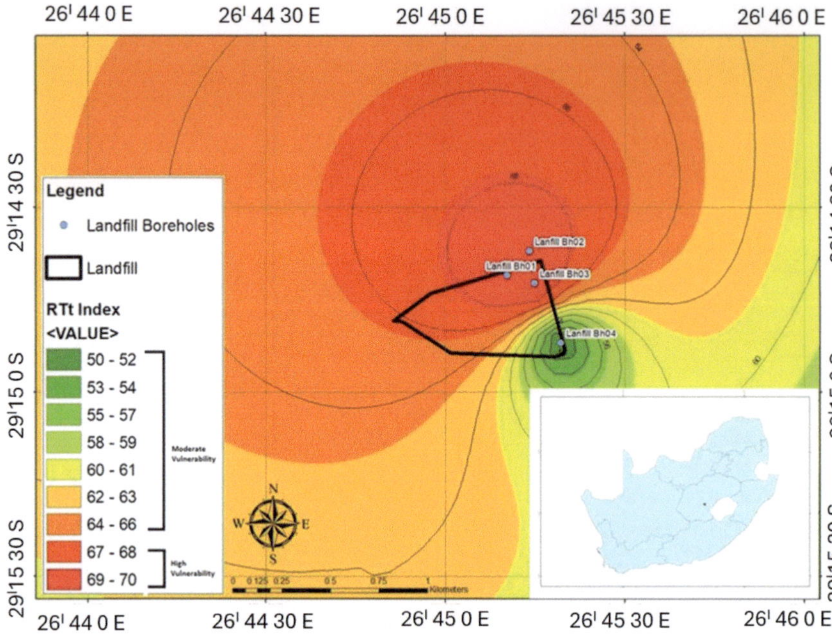

Fig. 3 RTt vulnerability index map for groundwater

surface and climatic variables that reflect recharge potential and contaminant transport mechanisms. This complementarity allows for a more nuanced understanding of site-specific vulnerability patterns and provides stronger evidence for guiding monitoring and remediation strategies.

3.2 Validation of Groundwater Vulnerability Indexes with Nitrate

The linear regression analysis presented in Fig. 4 revealed a positive correlation between nitrate concentrations and the GOD vulnerability index across the study site. The measured mean nitrate concentration was 0.33 mg/ℓ during both the wet and dry seasons well below the South African and WHO drinking water guideline of 50 mg/ℓ. These consistently low nitrate levels were in alignment with the very low vulnerability indices obtained using the GOD method. This strong correspondence suggests that the GOD model reliably reflects the intrinsic vulnerability of the aquifer system based on its hydrogeological conditions.

The GOD method's predictive accuracy in this context can be attributed to its focus on intrinsic aquifer properties, such as depth to groundwater, presence of confining layers, and unsaturated zone lithology. These factors are directly related

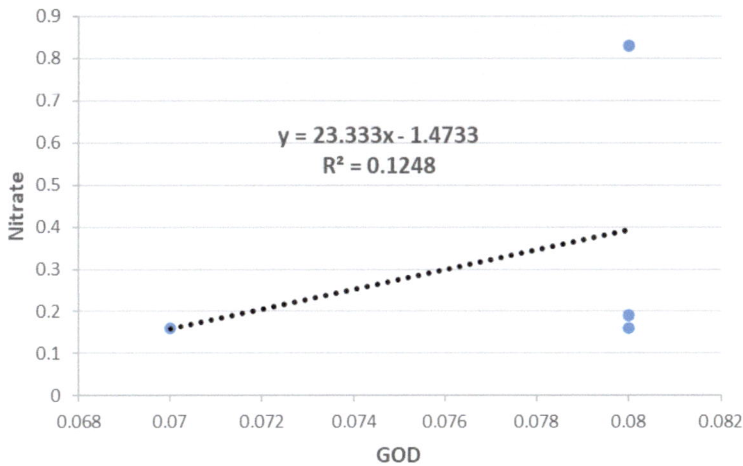

Fig. 4 Linear regression plot for nitrate validation of the GOD method

to the aquifer's capacity to buffer and filter out contaminants like nitrates. Hence, the positive correlation observed provides empirical support for the suitability of the GOD index in assessing leachate-induced contamination risk in landfill settings underlain by confined or semi-confined aquifers.

In contrast, the RTt method displayed an inverse relationship with nitrate concentration, as illustrated in Fig. 5. For instance, areas where the RTt index suggested moderate to high vulnerability exhibited low nitrate levels, while areas with lower RTt scores did not necessarily show higher nitrate concentrations. This mismatch between model output and water quality data raises questions about the appropriateness of the RTt method for this specific site context.

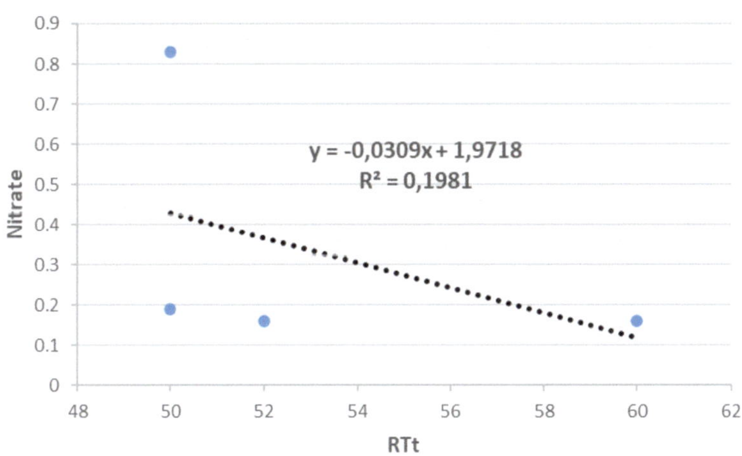

Fig. 5 Linear regression plots for nitrate validation of the RTt method

Table 1 Literature review table

Author and year	Method applied	Study area	Reported accuracy	Reported precision	Key observations
Foster et al. (2002)	GOD	Global	High for confined aquifers	Moderate	Effective where lithology dominates protection
Oke (2016)	RTt	Dahomey Basin, Nigeria	Moderate	High	Captures recharge effects well
Rukmana et al. (2020)	GOD	Indonesia	High	High	GOD effective in mining areas
Rahman (2008)	DRASTIC	India	Moderate	Moderate	Sensitive to recharge variability
Gogu and Dassargues (2000)	Overlay and index	Europe	Variable	Variable	Highlights model limitations

A likely explanation lies in the parameter composition of the RTt model, which emphasizes surface-related variables such as rainfall, slope, soil type, and hydraulic conductivity. While these are important for assessing potential recharge and contaminant migration at a broader scale, they may not sufficiently account for site-specific subsurface controls, such as low-permeability layers and depth to water table, which are critical in a landfill hydrogeological setting like Botshabelo. Additionally, the low and relatively uniform rainfall patterns across the site could have diminished the influence of this parameter in predicting nitrate mobility.

These contrasting outcomes between the two models highlight the need for context-specific vulnerability assessments and underscore the importance of validating model outputs with actual water quality data. While the RTt method may be more suitable in unconfined or highly dynamic recharge zones, the GOD method proved more aligned with observed contaminant behavior in this study, making it a more reliable tool for assessing intrinsic groundwater vulnerability under the prevailing conditions at the Botshabelo landfill site.

Table 1 summarizes comparative studies, emphasizing accuracy and precision as benchmarks for evaluating GOD and RTt outcomes.

4 Conclusion

This study underscores the critical importance of selecting appropriate groundwater vulnerability assessment methods when evaluating the impact of leachate contamination to groundwater in landfill sites. The choice of method should not be based solely on the expected outcome but must also consider site-specific hydrogeological conditions and contextual relevance. In this assessment, the GOD method classified the groundwater vulnerability as very low, while the RTt method indicated a moderate level of vulnerability. Validation using nitrate concentration levels which remained low across all boreholes aligned more closely with the GOD model's predictions, suggesting that the aquifer system is currently well protected from leachate infiltration, likely due to confining geological formations and low permeability soils. However, it is important to note that low nitrate levels do not necessarily confirm groundwater safety for human consumption. Comprehensive water quality analyses, including microbial and heavy metal testing, are essential to determine the potability of the groundwater. Future research should focus on adapting and refining existing models such as GOD and RTt to improve their applicability to diverse landfill settings, particularly those with complex subsurface conditions. Additionally, long-term monitoring across all seasons is recommended to capture temporal variations in vulnerability and contamination risks, thereby enhancing the robustness and reliability of the assessment.

References

Aller, L., Bennett, T., Lehr, J.H., Petty, R.J., Hackett, G.: DRASTIC: A Standardized System for Evaluating Groundwater Pollution Potential Using Hydrogeologic Settings. Environmental Protection Agency, U.S (1987)

Burke, J.J., Villholth, K.G.: Groundwater for Emergency Situations: A Methodical Approach to Planning and Management. UNESCO (2007)

Civita, M.: Le carte della vulnerabilità degli acquiferi all'inquinamento: Teoria & Pratica. Pitagora Editrice (1994)

Council for Geoscience: Catalogue of Geoscience Data and Information at the Council for Geoscience (Document No. KIMS-CAT-001 rev1). Council for Geoscience (2021)

DWS (Department of Water and Sanitation): Aquifer Classification of South Africa (2012a). https://www.dws.gov.za/Groundwater/documents/Aquifer%20Classification.pdf

Foster, S., Hirata, R., Andreo, B.: The GOD Method: A Groundwater Pollution Potential Mapping Methodology. British Geological Survey (2002)

Gogu, R.C., Dassargues, A.: Current trends and future challenges in groundwater vulnerability assessment using overlay and index methods. Environ. Geol. **39**(6), 549–559 (2000)

Google Maps. https://www.google.com/maps/@-29.2461614,26.7515816,723m/data=!3m1!1e3?entry=ttu&g_ep=EgoyMDI1MDczMC4wIKXMDSoASAFQAw%3D%3D. Accessed 02 June 2025

Johnson, M.R., Wolmarans, L.G.: Geology Map of South Africa. Council for Geoscience. https://maps.geoscience.org.za/portal/apps/sites/#/council-for-geoscience-interactive-web-map1/apps/f17eb56408b34add9c43878de5ccd4c2/explore. Accessed 07 Apr 2025

Kumar, A., Bhowmick, S., Chakraborty, S.: Assessment of groundwater quality near a municipal solid waste landfill site in Lucknow, India: a case study. Environ. Monit. Assess. **187**(7), 1–15 (2015)

Lapworth, D.J., Nkhuwa, D.C.W., Okotto-Okotto, J., Pedley, S., Stuart, M.E., Tijani, M.N., Wright, J.: Urban groundwater quality in sub-Saharan Africa: current status and implications for water security and public health. Hydrogeol. J. **25**(4), 1093–1116 (2017)

MacDonald, A.M., Bonsor, H.C., Dochartaigh, B.É.Ó., Taylor, R.G.: Quantitative maps of groundwater resources in Africa. Environ. Res. Lett. **7**(2), 024009 (2012)

Mangaung Metropolitan Municipality: Integrated Waste Management Plan: Review of Integrated Waste Management Plan; Mangaung Metropolitan Municipality: Bloemfontein, South Africa (2016)

Misi, S.N., Gumindoga, W., Dube, T.: Groundwater potential mapping using remote sensing and GIS techniques: a case study of Bulawayo, Zimbabwe. South Afr. Geogr. J. **100**(1), 1–20 (2018)

Mor, S., Ravindra, K., Dahiya, R.P., Chandra, A.: Leachate characterization and assessment of groundwater pollution near municipal solid waste landfill site. Environ. Monit. Assess. **118**(1–3), 435–456 (2006)

Oke, S.: Evaluation of the Vulnerability of Selected Aquifer Systems in the Eastern Dahomey Basin, South Western Nigeria. Institute for Groundwater Studies, Bloemfontein (2016). https://www.scribd.com/document/561886200/Oke-2016

Rahman, A.: A GIS based DRASTIC model for assessing groundwater vulnerability in shallow aquifer in Aligarh, India. Appl. Geogr. **28**(1), 32–53 (2008)

Ruzvidzo, S.: Effects of Evapotranspiration on Water Quality and Quantity Along the Modder River Catchment. Central University of Technology (2019). Available at: http://ir.cut.ac.za/bitstream/handle/11462/2380/Ruzvidzo%2C%20Silent.pdf

SAWS (South African Weather Service): Annual state of the climate of South Africa Pretoria, South Africa (2023). https://www.weathersa.co.za/Documents/Corporate/Annual%20State%20of%20the%20Climate%202023_02042024155932.pdf (2022)

Shapiro, A.M., Evans, C.E., Hayes, E.C.: Porosity and pore size distribution in a sedimentary rock: Implications for the distribution of chlorinated solvents. J. Contam. Hydrol. **203**, 70–84 (2017)

Sorensen, J.P.R., Lapworth, D.J., Marchant, B.P., Nkhuwa, D.C.W., Stuart, M.E., Pedley, S., Bell, R.A.: Emerging contaminants in African groundwater: evidence and impacts. Environ. Sci. Technol. **49**(9), 5529–5536 (2006)

Sukmawati Rukmana, B.T., Bargawa, W.S., Cahyadi, T.A.: Assessment of groundwater vulnerability using GOD method. IOP Conf. Ser. Earth Environ. Sci. **477**, 012020 (2020)

UN-Habitat: Solid Waste Management in the World's Cities: Water and Sanitation in the World's Cities 2010. Earthscan (2010).

Vrba, J., Zaporozec, A.: Guidebook on mapping groundwater vulnerability. In: IAH International Contributions to Hydrogeology, vol. 16 (1994)

World Bank. What a Waste: A Global Review of Solid Waste Management. In: Urban Development Series Knowledge Papers, No. 15. World Bank, Washington, DC (2012)

World Population Review. Botshabelo Population 2025 (2025). Retrieved from: https://worldpopulationreview.com/world-cities/botshabelo-population

Satellite-Based Monitoring of Soil Moisture and Land Degradation in Iraq's Major River Systems

Zahraa Khaled Al-Salihi, Diaa F. Hassan, Mustafa Qais Hamid, and Sajjad Hussain

Abstract Iraq faces critical environmental challenges, particularly land degradation and water resource management in its arid climate. Remote sensing (RS) technologies can monitor soil moisture and vegetation dynamics to address these issues, providing valuable insights into land cover changes and soil conditions. This study utilized Landsat, Sentinel-2, MODIS, and SMAP L4 satellite datasets to analyze soil moisture and vegetation cover in the Babylon, Basra, and Mosul governorates. The normalized difference vegetation index (NDVI) assessed vegetation changes, while geographic information system (GIS) platforms enhanced data analysis and visualization. Significant variability in soil moisture was observed across the regions, influenced by climatic and environmental factors. Babylon experienced fluctuations in soil moisture, Basra showed higher moisture levels likely due to irrigation, and Mosul exhibited high variability and consistently low vegetation cover. NDVI data indicated increased vegetation in Babylon due to improved agricultural practices but decreased cover in Mosul, highlighting distinct hydrological and ecological conditions across the regions. RS and GIS technologies play a crucial role in monitoring soil moisture and degradation in Iraq, providing valuable insights into land and water management strategies. The findings emphasize the need for effective water resource management and agricultural planning in Iraq's arid climate. Future research should focus on enhancing data availability and developing multi-scale analysis methods to further improve the monitoring and management of soil and water resources in Iraq.

Z. K. Al-Salihi
Department of Soil and Water Resources Sciences, College of Agricultural, University of Wasit, Wasit, Iraq

D. F. Hassan (✉)
College of Engineering, Al-Qasim Green University, Babylon, Iraq
e-mail: diaafliah@wrec.uoqasim.edu.iq

M. Q. Hamid
Department of Soil Science and Water Resources, College of Agriculture, University of Al-Qadisiyah, Al-Diwaniyah, Iraq

S. Hussain
Department of Environmental Sciences, COMSATS University Islamabad, Islamabad, Pakistan

H. Xu (ed.), *Proceedings of the 8th International Symposium on Water Resource and Environmental Management*, Environmental Science and Engineering,
https://doi.org/10.1007/978-3-032-19088-8_8

Keywords Remote sensing · Soil moisture · Land degradation · Vegetation dynamics · Iraq river systems

1 Introduction

Remote sensing (RS) techniques have been widely adopted for detecting and measuring land degradation in Iraq, particularly in the alluvial plain region. Other studies have also employed the various types of satellite imagery (Landsat, Sentinel-2) for land cover changes (Dalhel et al. 1973; Ziboona et al. 2022). These have illustrated the quantity of land degradation enhanced heavy handed until 1976, retreated after 1996 and then re-worsened (Ziboona et al. 2022). Research has also been focussed on specific regions, such as southern Baghdad where they calculate agriculture degradations based on VEGETATION and water indices (Dhamin et al. 2020). Application of RS and GIS methods allowed researchers to assess and map various environmental problems in Iraq such as soil salinity, land degradation, and vegetation cover change (Al-Quraishi and Negm 2020). These findings demonstrate the usefulness of geomatics techniques for monitoring and modeling land degradation and land use dynamics in Iraq over a longer time period. Monitoring of soil moisture and estimation of its degradation are important subjects for sustainable land management as well as food security. The remote sensing technique has been established as an effective approach in the monitoring of soil properties and its degradation processes (Wang and Zhang 2005; Wang et al. 2023). These approaches have benefits in contrast to traditional methods, allowing for big volume evaluation and rapid data gathering. Remote sensing could be utilised to assess land degradation using indicators like mineral composition, organic matter, and vegetation status (Wang et al. 2023). However, it is still difficult to effectively obtain in situ soil moisture value using only RS technique without any surface measurement (Wang and Zhang 2005). Combining several techniques such as geographical information systems (GIS), global positioning system (GPS), remote sensing and geostatistics, may lead to a more accurate assessment and simulation of soil moisture (Zhenjian 2004).

To cope with the complexity of detecting land degradation, scientists recommend numerical modeling work and the use of integrative techniques (hyper spectral, multi-sensorial and multi temporal) (Shoshany et al. 2013). Iraq's two largest river bodies, the Tigris and Euphrates rivers are experiencing a critical water shortage that is heavily polluted. These historic watercourses, integral to the civilization and irrigation in the region for millennia, are threatened due to upstream dams, climate change and pollution (Hussain 2040; Al-Ansari et al. 2019). Decreased water flow has also resulted in seaborne intrusion of marine water into the Shatt Al-Arab estuary, which has affected agriculture and local fauna (Mahdi and Fawzi 2014). The quality of the rivers water has been degraded by inadequate management and raising amounts of direct dumping not treated waste in the absence, which becomes hazardous for health and environmental damage (Mahdi and Fawzi 2014). The arid climatic conditions of the Middle East would aggravate this situation given that its annual precipitation ranging from 166 mm/year (Al-Ansari et al. 2019). To tackle such challenges, water management plans based on diversification of non-conventional water sources, efficiency in irrigation units and social participation have been suggested by specialists (Hassan et al. 2021). Remote sensing is increasingly playing an important role in monitoring LD since it provides efficient and broad-scale observations (Wang et al. 2023). Dalhel et al. (1973) used different satellite sensors, such as Landsat and Sentinel-2, to assess land degradation in Iraq's alluvial plain. Direct indicators such as soil mineral composition and moisture content and indirect proxies like vegetation status have been identified by researchers for assessing land degradation (Wang et al. 2023). Although there are difficulties in the development of strong relationships between remote sensing measurements and slight to moderate degradation levels, successful approaches include multi-annual vegetation condition observations as well as integrating optical and radar data (Lobell 2010). In Iraq, remote sensing (RS) and geographic information systems (GIS) have been widely used for modeling and mapping soil properties, including soil salinity, soil moisture, texture, and land degradation, supporting agricultural planning and sustainable water and land resource management under arid and semi-arid conditions (Al-Quraishi and Negm 2019). Directions for future research include establishing multi-scale analysis methods, strengthening data availability and making a land degradation dataset shared (Wang et al. 2023).

The study area of the current study is the main Iraws rivers (Tigris and Euphrates). The two rivers are the lifeblood of the country, providing water for agriculture, industry and drinking. The Tigris River, which flows through Iraq is sourced in eastern Turkey And like the Euphrates River (also sourced in Turkey and flowing through Syria) it irrigate farmlands all along its path. Particular sites within these river systems will be identified according to several criteria, including the extent of agricultural lands, ease of access for field surveys and the availability of historical RS data. The research objectives are to

(1) Elaborate the spatial and time distribution of soil moisture in the considered sites on the basis of RS and GIS methods;
(2) Map the scale of land degradation, and
(3) Investigate the correlation between soil moisture and land degradation, and
(4) To assess the performance of RS techniques for monitoring these water environmental variables in Iraq river systems.

2 Materials and Methods

2.1 Study Area

The study area of this research for RS application in monitoring soil moistures and degradation includes important locations situated along the main rivers of Iraq. The Euphrates, Tigris and the river Shatt al-Arab are vital for Iraq's water resources and agricultural needs. It is hoped that using remote sensing methods this study will capture and analyze soil moisture dynamics, to identify areas suffering degradation. The Euphrates River plays vital role in local agriculture and communities of Babylon Governorate. In this area, the region of the clayed overland flow covers about 150 km^2 in ten Km away to the river and very sensible to changes in soil moisture and degradation. This closeness implies that it is directly influenced by the river and so, it is a good geographical region to be monitored using advanced remote sensing. The complex interrelation between the dynamics of the river and the soil properties requires regular measurements to manage piece of land effectively and reduce degradation.

It continues in a southerly direction to flow through the Basra Governorate of Iraq, where it joins the Tigris and Euphrates rivers et al-Fawwwar to form the Shatt al-Arab waterway, which empties into the Persian Gulf. This work focuses on a 200 km according to the São Francisco River. The Shatt al-Arab region is sensitive to salinity and changing moisture regimes influenced by tides and human interventions. It is important to study the soil moisture and degradation over this region because they have negative impact on agriculture and livelihoods in the local, thus deserve a complete evaluation using remote sensing. Description Situated in northern Iraq, the Tigris River passes through the Mosul Governorate as it cuts across a region of great historical relevance and agricultural importance. The study is carry out in an 170 km^2 region situated approximately. Since about the year 1000 A.D. more, less 11 km decurrent of time are required for the Tigris river (Table 1). This proximity of spatial relations also emphasizes the frequent influence of fluvial processes on development of soil, showing that it is necessary to keep constant vigil in order to detect and offset eventual degradation trends. A more thorough understanding of Tigris River influence on local soil moisture variation will be helpful to properly manage land use in a sustainable manner and to ensure that regional agricultural productivity is sustained. Together, these sample regions of Iraq's main rivers serve as representative examples of river-based soil conditions. Conclusions: Through the application of advanced remote sensing it is hoped that this study will provide important information about patterns of moisture and degradation by which to manage resources in a sustainable manner and conserve these essential ecosystems. The study highlights the need for using remote sensing data to keep a track of variations in environmental conditions and conservation of ecological balance in these critical inland riverine ecosystems.

Table 1 Soil areas exposed to moisture and degradation in major river regions of Iraq, based on RS data

Major River	Region	Soil area exposed to moisture and degradation ((km^2)	Approximate distance from river (km)
The Euphrates River	Babylon Governorate	150	10
The Shatt al-Arab	Basra Governorate	200	15
The Tigris River	Mosul Governorate	170	12

2.2 Satellite Data Sources

For monitoring soil moisture and degradation in the headwaters of Iraq major river systems, satellite data sources are much needed to be utilized RS effectively. Through the power of Google Earth Engine, researchers are able to access and analyse data from 2015 to 2024. This is carried out through JavaScript code in the home application of the platform, and enables several CSV and TIFF data readouts to be quickly acquired (Fig. 1). These formats help facilitate working with large sets of data, and have flexibility in terms of building visualization or analyzing the data. The SPL4SMGP is one of the major satellite data sources used in this study. 007 SMAP L4 Global product. This dataset has 3-hourly measurements at the 9-km grid scale, and it is useful for surface and root zone soil moisture information. This high temporal monitoring identifies fine-scale details of soil moisture evolution along the river systems of Iraq, which is key information to interpret the interactions between soil regimes and environmental fluctuations. By using SMAP data, researchers can record the short and long term changes in soil moisture, enabling them to pin down the exact pattern of degradation.

Additionally, the MOD44B. 006 Terra Vegetation Continuous Fields dataset provides insight into the vegetation properties of the study area. At global 250 m resolution, this data set provides information on distribution and density of GuMG(c) vegetation cover, a key parameter required in land degradation studies. The new nature of degradation in the study area led to changes in soil and ecosystem over time that could be assessed jointly with vegetation, allowing trends in the soil stability and ecosystem health following disturbances. The MOD44B land cover product is a key input for following land use change at larger scales and potential effects on soil and water resources. Furthermore, the MOD11A1. Terra Land Surface Temperature and emissivity (Collection 6.1) provides daily global 1 km observations which enabled this analysis to be conducted. This temperature information can have critical impacts on soil moisture content and plant health in riverine systems. Research has the capability to combine LST data to determine thermal thresholds that lead to deterioration of soil and vegetation and subsequently, improved identification of environmental stressors. Collectively, these satellite-derived sources of data represent a useful series for investigators wishing to develop a comprehensive understanding of the changing quality of soil moisture across Iraq's major river systems. When integrating various

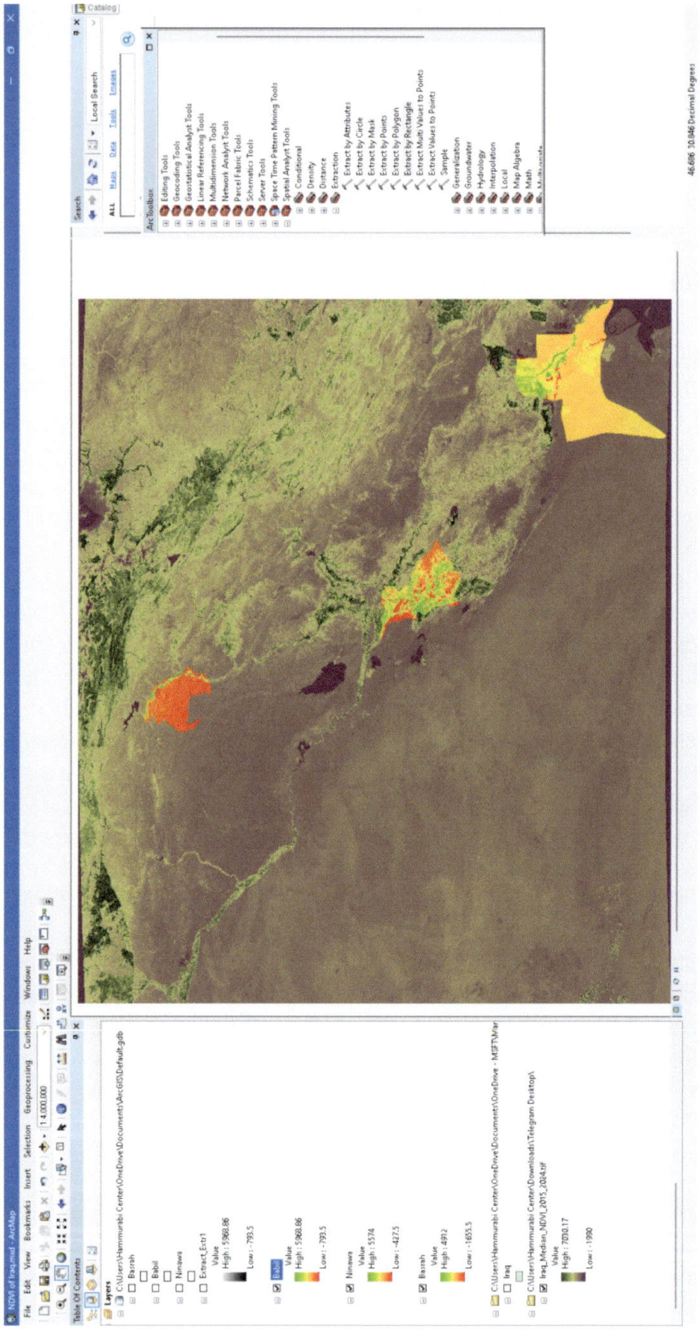

Fig. 1 Framework of Arc GIS of field of study

datasets in the Google Earth Engine, it enables a multi-angle assessment of soil and vegetation status with respect to both space and time. Consequently, the application of high-resolution remote sensing data offers a sound foundation for addressing environment-related problems in the region and assist in planning long-term optimal strategies aimed at sustainable land–water resource use.

2.3 *Soil Moisture Estimation and Degradation Assessment*

Soil moisture estimation and degradation assessment data were obtained from Google Earth Engine through its scalable cloud based platform which facilitates efficient processing and analysis of satellite data. Data were fetched and manipulated during filtering via JavaScript in the Earth Engine Code Editor from respective satellite missions (e.g., SMAP for soil moisture or moda24h for vegetation) or other MODIS products (temperature). The process usually started with the definition of spatial extent (e.g. Iraq major river systems) and temporal range (for example, 2015–2024). After the data had been filtered to the appropriate criteria, actions were carried out—calculating indices of soil moisture and land degradation. These data were further viewed in the platform to check layers and results providing a more comprehensive evaluation of the environmental condition of the study area.

For download output data from Google Earth Engine, analysis results were exported into file formats such as CSV or TIFF for further processing outside the platform. To export data, which as soil moisture over time or spatial patterns of degradation, use the export table to drive function was used to export the CSV (for time series and tabular data analyses). Meanwhile, the export image to drive function allowed the export of spatial data in TIFF, retaining geospatial information and specifying export parameters such as scale and region. These files were then downloaded for further analysis into GIS or statistical databases. This was the salient requirement for transferring knowledge from remote sensing data to stakeholders and achieving efficient water resources management and land conservation schemes of the Iraqi river systems. Accordingly, the Google Earth Engine greatly streamlined data retrieval and processing, opening up an easy gateway to accessing large-scale satellite-based data for environmental monitoring applications.

2.4 *Integration with Geographical Information Systems (GIS)*

Integration with GIS was an important part of incorporating remote sensing data from satellite sources such as MODIS, SPL4 and Sentinel-2 to monitor soil moisture and degradation in main river systems in Iraq. GIS as the useful tool for spatial data processing, analysing, and representation which was beneficial to study complex interactions in river systems. Based on GIS and remote sensing overlays, we compiled a set of interdisciplinary maps depicting soil moistures, vegetative coverage and hotspots of degradation. This aggregation also allowed precise spatial analysis, necessary for monitoring trends and fluctuations over time that make it possible to adapt management as it changes the environment. The process of data from the satellite mission to GIS applications took several stages. Datasets from MODIS, SPL4 and Sentinel-2 were processed in Google Earth Engine (GEE) as GeoTIFF adjacent to their geospatial information and easy import into a Geographic Information System. This processing was essential to collect advanced spatial analysis (area statistics, running geostatistical model to assess the spatiality of land degradation). The extracted data was integrated with some ancillary GIS information, like land use maps or topographic layer, to better deal with the river systems' environmental dynamics. Through facilitating multiple dimensions of analysis, GIS integration allowed monitoring to be undertaken on a broad scale and more than that supported the development of interventions focused on reducing degradation and conserving water resources in Iraq's critical alluvial landscapes (Fig. 1).

3 Results

3.1 Annual Soil Moisture

Soil Moisture data for years 2015–2024 analyzed is originated from the Babylon, Basra and Mosul regions of Iraq using SPL4SMGP. 007 SMAP L4 data, shows disparate patterns and variability in these regions. In Babylon, the average soil moisture had high variability in different years, which varied from 0.0528 (2018) to 0.1835 (2016). The substantial range of standard deviations from 0.022 to 0.0451, suggests dynamic soil moisture environment affected by annual climatic and environmental conditions. Particularly, the highest soil moisture content values were persistently higher in 2018 and 2021 despite they had quite moderate mean values indicating at episodic events of water availability or heavy rain periods. Comparatively, Basra exhibited generally higher mean soil moisture values than Babylon, indicating wetter soil conditions. The mean values ranged from 0.0543 in 2015 to peaks of 0.1963 in 2021 and 0.1927 in 2019, with corresponding high maximum values, particularly 0.31214 in 2019 (Fig. 2). The standard deviations in Basra were relatively high across the years, particularly in 2017 and 2018, suggesting more variability in soil moisture, possibly due to fluctuations in water availability from tidal influences or irrigation practices which are prevalent in this region due to its geographical proximity to the Shatt al-Arab waterway.

Mosul showed the highest variability in soil moisture trends. The mean soil moisture values in Mosul ranged notably higher, especially in 2019, with a significant peak of 0.3403, reflecting a potentially anomalous year with abundant moisture availability (Table 2). This was accompanied by the highest maximum soil moisture levels

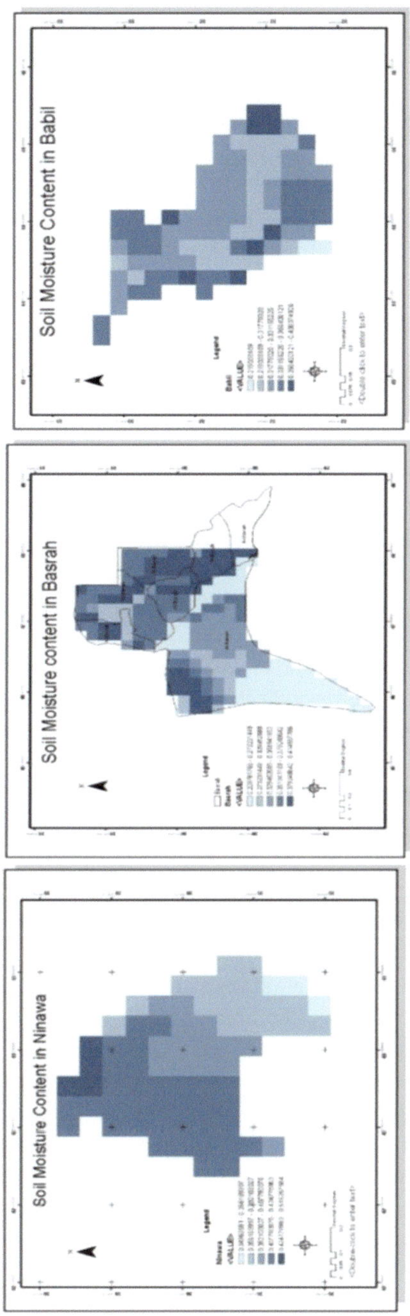

Fig. 2 Shapefiles of soil moisture of Iraqi environment

observed at 0.4203 in that year. The sharp increase in 2019 stands in contrast with other years and regions, highlighting unusual climatic conditions or water management strategies that could have affected the readings (Fig. 3). The substantially higher peaks contrasted with the others showing off a different hydrological or environmental status in Mosul; such could be due to its special geographical or climatic localities. Although with weak comparisons, these findings point to the existence of contrasting patterns in soil moisture behavior mainly associated with local climatic conditions and geographical settings possibly weakened by human interventions in water management and land use. These findings offer important implications for water resources and agricultural planning within the main river basins of Iraq.

3.2 *Normalized Difference Vegetation Index (NDVI)*

The FAPAR MOD44B (NDVI) results for 2015–2024. 006 Terra Vegetation Continuous Fields, Mongolia The large-field continuous fields products display degrees of vegetation development in the Babylon-Basra-Mosul regions of Iraq. In Babylon, NDVI values indicate a relatively steady but low vegetation index condition with a mean of between 0.03 and 0.06 (Fig. 4). Overall, the standard deviation was low in all years with small variation of vegetation cover. Particularly interesting is the stable mean in 2019 (mean = 0.06) and 2023 (mean = 0.06) while the mean value decreased during about a magnitude step to a mean of 0.03 between 2018 with higher variability might indicate the presence of some environmental stress including practices related to land use significantly compromising vegetation health that year.

On the other hand, NDVI values of Basra indicate a different vegetation pattern, with a higher average value as compared to Babylon mean value ranging constantly from 0.40 and 0.88 (table3). Maximum values remained constant (near 0.70) during this time, indicating high vegetation density areas. The standard deviation in Basra remained at 0.16 for all years, illustrating great variability of the vegetation density across the regions studied. Especially, in 2020, the average NDVI increased to 0.88 either because of favorable environment or effective plants control. This high variability and index are in line with dynamic vegetation patterns that might be affected by Basra's ecological position along the Shatt al-Arab waterway, which may bring different water sources for plants.

For the two regions of Mosul, NDVI shows a sharp curve where all three parameters minimum, maximum and mean fluctuates around − 0.98 to − 0.99 throughout years possibly there is very less or no vegetation over the considered time span (Fig. 5). This static and uniformly low NDVI values indicates a harshly barrenscape, potentially with contributions from land degradation under drought or anthropogenic influences, clearly in need of careful monitoring of its ecological state. This comparative analysis also illustrates the variability in vegetation condition and health between Iraq's regions and describes specific environmental conditions that determine different NDVI trends for the various regions.

Table 2 Annual soil moisture statistics from SPL4SMGP.007 SMAP L4 for regions in Iraq (2015–2024)

Region	The year	Min	Max	Mean	S.D.
Babylon	2015	0.0177	0.13	0.0545	0.022
	2016	0.0335	0.2377	0.1835	0.0451
	2017	0.0333	0.1844	0.1069	0.0268
	2018	0.0109	0.2595	0.0528	0.0393
	2019	0.0396	0.2526	0.1104	0.0393
	2020	0.0533	0.1656	0.0925	0.0225
	2021	0.0372	0.2576	0.0833	0.0392
	2022	0.0210	0.1415	0.0533	0.0235
	2023	0.0192	0.1886	0.0708	0.0352
	2024	0.0142	0.2088	0.0845	0.0292
Basra	2015	0.0809	0.1710	0.0543	0.0216
	2016	0.0673	0.2668	0.1519	0.0358
	2017	0.011	0.2717	0.0863	0.0465
	2018	0.0020	0.2662	0.0779	0.051
	2019	0.0796	0.31214	0.1927	0.0462
	2020	0.0816	0.2726	0.1786	0.0361
	2021	0.0767	0.24137	0.1963	0.0420
	2022	0.0445	0.2652	0.155	0.0490
	2023	0.11682	0.11682	0.1861	0.0336
	2024	0.0427	0.3234	0.1168	0.0387
Mosul	2015	0.0658	0.2266	0.0876	0.0278
	2016	0.05871	0.3029	0.1704	0.0330
	2017	0.08477	0.35192	0.19546	0.04136
	2018	0.07618	0.3308	0.1681	0.05264
	2019	0.08236	0.4203	0.3403	0.04511
	2020	0.12390	0.4002	0.2018	0.4002
	2021	0.0457	0.33435	0.1082	0.0451
	2022	0.0705	0.32112	0.15752	0.0465
	2023	0.09909	0.27383	0.1763	0.0401
	2024	0.0837	0.3709	0.16450	0.0542

4 Discussion

Soil moisture varies due to a lot of factors, climate, topography and make-up < 0 these 22 soil. Research has indicated that the spatio-temporal variability of soil moisture can differ considerably depending on region and time scale (Ahmed and Hadi 2019; Fajardo 2021). In areas of flat topography, soil moisture may display clear

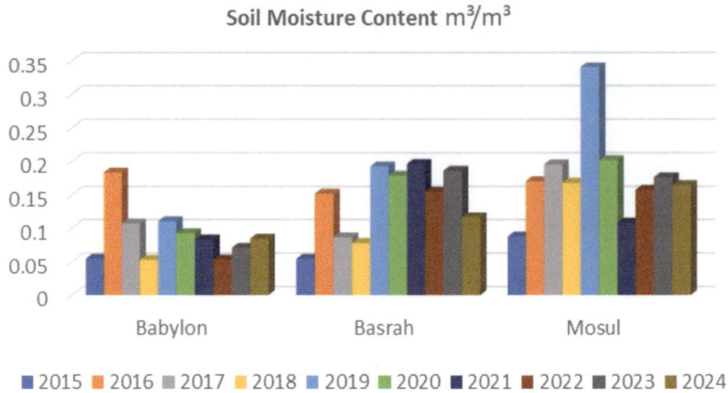

Fig. 3 Soil moisture content in Iraq

spatial patterns in response to wet conditions that are diluted during dry conditions (Fajardo 2021). Soil moisture variability can be related to the mean water content, and is mainly controlled by soil hydraulic properties and spatial heterogeneity in soil properties (Vereecken 2007). Inter-annual and intra-annual climate variabilities have strong effects on soil moisture temporal dynamics, thus shaping the annual pattern of soil moisture cycle and distribution of wet versus dry soil moisture status (Mahmoud et al. 2025). It is essential for hydrological studies to be aware of such trends, as they can have implications on water availability, evaporation process, agriculture and ecosystem in different locations.

AMT Soil moisture variability is driven by vegetation, soil character, topography and climate (Teuling and Troch 2005). Iraq has been studied and diverse aspects of soil moisture dynamic variation have been demonstrated. Mosul had the highest and most variable as well as mean value in soil moisture, Basrah had the lowest (Piles et al. 2019). A significant negative relationship between soil moisture and temperature has been observed for the entire group of Iraq stations (Piles et al. 2019). Internationally, the SMOS satellite mission has also demonstrated valuable information on spatiotemporal variability of soil moisture at multiple scales (i.e., seasonal and interannual) (Al-Taai and Nassif 1530). Patterns of coherent variability and long-term linear trends consistent with ENSO teleconnections were identified in the data. The knowledge and description of these soil moisture dynamics is vital for climate monitoring, weather forecasting, and hydrological modeling (Al-Taai and Nassif 1530; Hamid et al. 2025). Assimilating satellite-derived soil moisture with in situ and model data can also help to improve our knowledge about soil water content variability.

Normalized Difference Vegetation Index (NDVI) has been applied for monitoring vegetation condition in different parts of Iraq. In Babylon, a rise in vegetation cover has been reported from between 2000 and 2021 (Al-Turaihi et al. 2022), as well as that observed between 2015 and 2020 due to better agricultural practices. However, the NDVI values had some ups and downs in Mosul between 2014 and 2018 (general decreasing). An investigation of climatic modulation in an NDVI response to climatic

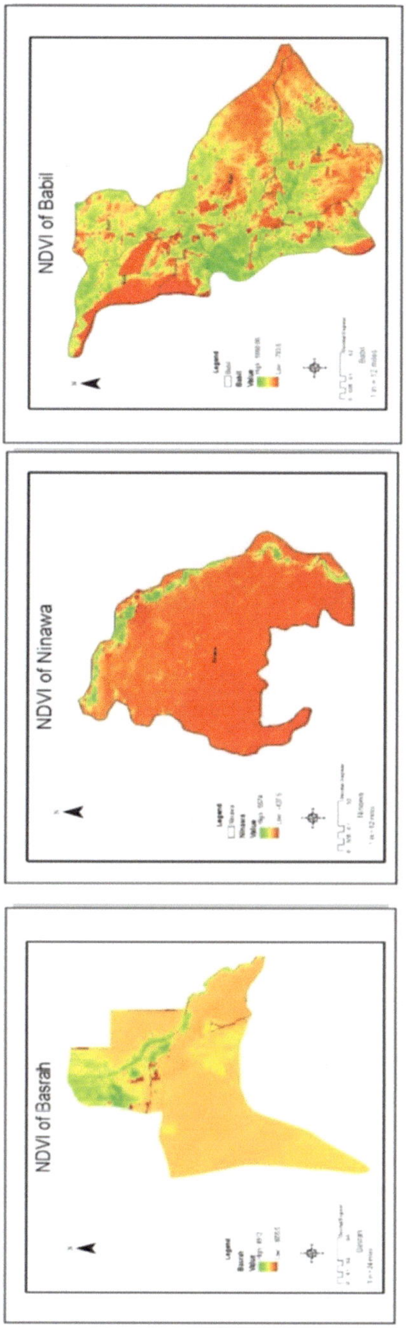

Fig. 4 Shapefiles of NDVI of Iraqi environment

Table 3 NDVI statistics for Iraq's regions using MOD44B.006 Terra vegetation continuous fields (2015–2024)

Region	Year	Min	Max	Mean	S.D.
Babylon	2015	0.05	0.08	0.06	0.01
	2016	0.06	0.06	0.06	0.00
	2017	0.01	0.06	0.04	0.02
	2018	0.04	0.09	0.03	0.04
	2019	0.03	0.12	0.06	0.03
	2020	− 0.02	0.11	0.06	0.05
	2021	0.01	0.07	0.05	0.02
	2022	0.04	0.07	0.06	0.01
	2023	0.03	0.08	0.06	0.01
Basra	2015	0.03	0.70	0.52	0.16
	2016	0.03	0.70	0.65	0.16
	2017	0.03	0.70	0.70	0.16
	2018	0.03	0.70	0.40	0.16
	2019	0.03	0.70	0.54	0.16
	2020	0.03	0.70	0.88	0.16
	2021	0.03	0.70	0.52	0.16
	2022	0.03	0.70	0.65	0.16
	2023	0.03	0.70	0.70	0.16
	2024	0.03	0.70	0.40	0.16
Mosul	2015	− 0.98	− 0.98	− 0.98	0.00
	2016	− 0.99	− 0.99	− 0.99	0.00
	2017	− 0.99	− 0.99	− 0.99	0.00
	2018	− 0.99	− 0.99	− 0.99	0.00
	2019	− 0.98	− 0.98	− 0.98	0.00
	2020	− 0.98	− 0.98	− 0.98	0.00
	2021	− 0.99	− 0.99	− 0.99	0.00
	2022	− 0.99	− 0.99	− 0.99	0.00
	2023	− 0.99	− 0.99	− 0.99	0.00
	2024	− 0.99	− 0.99	− 0.99	0.00

factors across various Iraqi regions showed that there is a highly positive relationship between NDVI and precipitation with 4 months time lag, but partially negative association with air temperature (Daham et al. 2018). These investigations reveal the suitability of remote sensing and GIS methodologies to monitor the impact of vegetation cover, as well in terms of its growth behavior under changing climatic conditions due to precipitation, which is a fundamental factor influencing vegetation growth in Iraq's semi-arid terrain.

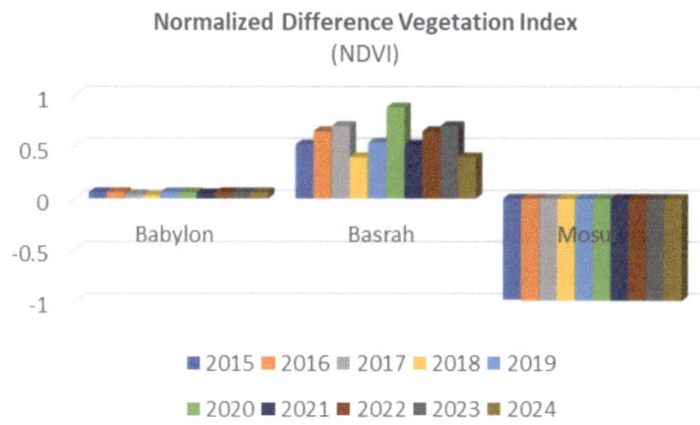

Fig. 5 Normalized difference vegetation index in Iraq

NDVI is an important index used to monitor the vegetation conditions in Iraq. Research has found that NDVI has a close correlation with precipitation but lags by 4 months in certain places (Daham et al. 2018). In Mesopotamian marshes, NDVI calculated from several satellite images (Landsat, MODIS, AVHRR) resulted in a strong relationship and there were no significant reduction in vegetation over the course of 17 years (Albarakat and Lakshmi 2019). Groundwater depth also has a strong effect on the distribution of vegetation, and areas with shallow groundwater are found to have high NDVI values (Seeyan et al. 2014). In urban type cities as Baghdad, the NDVI patterns are associated with seasonal climate drivers such as air temperature and precipitation (Naif et al. 2019). Such studies together illustrate the potential of NDVI to monitor Vegetation dynamics and its association with different environmental variables in various parts of Iraq.

The analyses focus on NDVI and rainfall data to study vegetation cover and drought conditions across the north of Iraq. Nassiriya, Mosul, Kirkuk and Salah al-Din had different NDVI values with the lowest being in Mosul (Muhaimeed 2013; Akol et al. 2024). Vegetation cover in Mosul declined between 2014 and 2017, with a minimal increase observed in 2018 (Dahham et al. 2025; Hassan et al. 2020). The Kurdistan region showed spatial variance in the distribution of vegetation with the northwest and southeast having a higher level compared to the west (Mustafa 2020). Drought effects were notable, especially in 2007–2008 that impacted on all the governorates studied (Al-hedny and Muhaimeed 2019). NDVI was strongly related to rainfall, and thus can be used as an effective drought monitoring platform in the area. Importantly, the results herein emphasized the necessity of integrating remote sensing and GIS methods for monitoring vegetation cover changes and drought conditions in northern Iraq (Dahham et al. 2025; Hamid 2025).

The role of soil moisture to land degradation is an important theme in relation the health of ecosystems, especially areas such as Iraq where water shortage and land degradation are major concerns. The soil-moisture relationship is essentially a direct function of several land-degradation processes (erosion, salinization and the loss of organic matter) affecting the soil's water-holding capacity. For example, soil erosion decreases the soil's water holding capacity through topsoil which is rich of moisture storage and nutrients eroded (Hundi et al. 2025; Mohammed et al. 2019). RS has many advantages such as wide and inaccessible area coverage, rapid data availability, and the ability to observe the temporal and spatial changes in environmental conditions. Nonetheless, the RS data still need to be validated with ground truth observations in order to prove that the RS methods are effective. Through pairing these statistical tools, we could be able to show conclusively how RSMs can in deed capture soil moisture and land degradation patterns with an enhancement and become more usable for environmental assessment and decision making in that region (Al-Hamandi et al. 2019; Aljumaily et al. 2022).

5 Conclusion

The research highlights the potential of RS and GIS technologies to monitor soil water content and desertification throughout Iraq, particularly in terms of the main river systems; Tigris and Euphrates. Using satellite data (Landsat, Sentinel-2, MODIS and SMAP L4) the study properly estimates land cover changes, vegetation dynamics and temporal soil moisture variability.

- The results also indicate that Babylon is affected by climate variability and farming practices on soil moisture and vegetation variation, Basra regardless of extreme agriculture intervention have increased moist conditions, while in Mosul there was high fluctuations as well as below normal values for both moist and vegetation.
- The incorporation of RS data in GIS is advantageous for spatial and temporal assessment, serving as valuable inputs to sustainable land and water resource management practices.
- Identification of the presence and extent of land degradation was achieved by the use of combined vegetation health, soil moisture patterns and land cover change analysis with RS effectively able to identify which regions have been affected most by degradation e.g. those areas associated with soil erosion and salinization along riversides.
- The work sheds light on the potential of these technologies to contribute to addressing environmental challenges in Iraq and tentatively its semi-arid climate, and water resource management.
- The connection between land degradation and soil moisture was investigated by studying the effect of degradation processes, such as erosion and salinity, on the soil's who moisture retention. They found a direct association between the tendency of landslides and low values in soil moisture content (the level at which it tends to hold) Read more: Choking Australian cities should look to the skies under climate change For every percentage point that soil moisture content fell below its site-specific threshold, landslide hazards rose by four times.
- It is recommended that future research be directed toward increasing the availability of data and developing multi-scale procedures for improved monitoring and management of soil–water resources in Iraq.

- Future research may benefit from the application of Multi-Criteria Decision Making (MCDM) methods to take into account a set of environmental indices, including soil moisture, vegetation health and land degradation as well as socio-economic variables, in order to enrich the analysis process. With such an approach, the effort will enable to priorities the areas more in need of intervention and develop a stand for and comprehensive frame for land and water management on Iraq's river systems'.

Acknowledgements Acknowledgements The authors are grateful to the organizations and individuals that provided support for this research. We acknowledge the satellite data providers including NASA and ESA for providing MODIS, Sentinel-2, and SMAP L4 datasets used in our study. We also thank the developers of Google Earth Engine, which is a powerful platform that will be ideal for processing and analyzing large-scale remote sensing data. Authors would like to thank our coworkers and collaborators for their input and comments throughout the course of this project.

Conflict of Interests The authors declare no conflict of interest.

References

Ahmed, O.T., Hadi, S.H.: Analysis of the monthly and annual change of soil moisture and evaporation in Iraq. Al-Mustansiriyah J. Sci. (2019)
Akol, A.M., Hassan, D.F., Mohammed, R.J., Al Janaby, Z.A.A., Abed, M.A.K., Hussain, S., Razzaq, H.K.: Optimizing wheat yield and water use efficiency using AquaCrop model calibration and validation in various irrigation and tillage systems under climate change. Soil Sci. Annual **75**(3), 195823 (2024)
Al-Ansari, N., Adamo, N., Sissakian, V.K.: Hydrological Characteristics of the Tigris and Euphrates Rivers (2019)
Albarakat, R., Lakshmi, V.: Comparison of normalized difference vegetation index derived from Landsat, MODIS, and AVHRR for the Mesopotamian Marshes between 2002 and 2018. Remote Sens. **11**, 1245 (2019)
Al-Hamandi, H.M., Al-Obaidi, M.A., Aljumaily, M.M.: Study on quantity and intensity of potassium in the alluvial soils in Baghdad. Plant Archives **19**(2), 123–130 (2019)
Al-hedny, S., Muhaimeed, A.S.: Drought monitoring for northern part of Iraq using temporal NDVI and rainfall indices. Environ. Rem. Sens. GIS Iraq (2019)
Aljumaily, M.M., Al-Hamandi, H.M., Farhan, M.J., Kareem, H.A.: Relationship between Zn and Cd in soil and plant. J. Agric. Sci. **1**(XXXIII), 33–42 (2022)
Al-Quraishi, A.M., Negm, A.M.: Updates, conclusions, and recommendations for environmental remote sensing and GIS in Iraq. Environ. Remote Sens. GIS Iraq (2019)
Al-Quraishi, A.M., Negm, A.M.: Environmental Remote Sensing and GIS in Iraq. Environ. Rem. Sens. GIS Iraq (2020)
Al-Taai, O.T., Nassif, W.G.: The relationship between soil temperature and volumetric soil water content in Iraq using ECMWF data for the period 1980–2016. J. Phys. Conf. Ser. **1530** (2020)
Al-Turaihi, A.S., et al.: Vegetation cover variations in central Iraq by using GIS and remote sensing data. Iraqi Geol. J. (2022)
Daham, A., et al.: Analysis of NDVI variability in response to precipitation and air temperature in different regions of Iraq using MODIS vegetation indices. Environ. Earth Sci. **77**, 1–24 (2018)

Dahham, I.T., Kareem, H.A., Khair, A.M., Hamid, M.Q.: Evaluation of hydrological properties of gypsiferous soils cultivated with wheat under varying gypsum content. Plant Sci. Today **12**(4), 1–8 (2025)

Dalhel, F.T., Albayati, M.M., Ziboon, A.R.: Investigation soil degradation in Iraq by using geomatics techniques. J. Phys. Conf. Ser. **1973** (2021)

Dhamin, T.A., Khanjer, E.F., Mashee, F.K.: Detection Agriculture Degradation for the South of Baghdad City Using Remote Sensing Data for Years 2010–2019 (2020)

Fajardo, L., et al.: Surface soil moisture variability in a sector of a humid basin characterized by extremely flat relief. Ecohydrology **15** (2021)

Hamid, M.Q.: Response of physical properties of sandy soil treated with different levels of natural soil conditioners zeolite and perlite. Sarhad J. Agric. **41**(2), 591–599 (2025)

Hamid, M.Q., Abd, E.H., Al-Salihi, Z.K., Muhammed, R.J., Hassan, D.F.: Effect of organic conditioners on the physical properties of sandy soil under drip irrigation conditions. Sarhad J. Agric. **41**(3), 1133–1142 (2025)

Hassan, D.F., Ati, A.S., Neima, A.S.: Effect of irrigation uniformity and efficiency on water consumption, yield of maize using different irrigation and cultivation methods. Int. J. Agric. Stat. Sci. **17**(1), 1441–1450 (2021)

Hassan, D., Thamer, T., Mohammed, R., Almaeini, A., Nassif, N.: Calibration and evaluation of AquaCrop model under different irrigation methods for maize (*Zea mays* L.) in central region of Iraq. In: Kallel, A., et al. (eds.) Selected Studies in Environmental Geosciences and Hydrogeosciences. CAJG 2020. Advances in Science, Technology & Innovation. Springer, Cham (2023)

Hundi, H.K., Hamid, M.Q., Noori, A.A.M.: Role of Ochrobactrum bacteria and organic matter in plant growth and the content of N, P, and K under soil salinity stress. J. Environ. Earth Sci. **7**(5), 130–139 (2025)

Hussain, Z.N.: No water by 2040: the crisis engulfing Iraq and its historic river flow. BMJ **380** (2023)

Lobell, D.: Remote sensing of soil degradation: introduction. J. Environ. Qual. **39**(1), 1–4 (2010)

Mahdi, B.A., Fawzi, N.A.: Iraq's Inland Water Quality and Their Impact on the North-Western Arabian Gulf (2014)

Mahmoud, S., Hundi, H., Razzaq, R., Hamid, M.: Effect of zinc and potassium humate spraying on growth and yield of tomato (*Solanum lycopersicum* L.). Plant Sci. Today **12**(4) (2025). https://doi.org/10.14719/pst.11104

Mohammed, R.J., Abdulkadhim, K.A., Hassan, D.F., Kadhim, T.F.: Effect of wheat straw as organic matter and different water quality on some chemical soil properties and growth of pepper (*Capsicum annuum*). IOP Conf. Ser. Earth Environ. Sci. **344**(1), 012034 (2019)

Muhaimeed, A.S.: Evaluation of long-term vegetation trends for northeastern Iraq: Mosul, Kirkuk and Salah al-Din (2013)

Mustafa, Y.T.: Spatiotemporal Analysis of Vegetation Cover in Kurdistan Region–Iraq Using MODIS Image Data (2020)

Naif, S.S., et al.: Seasonal normalized difference vegetation index responses to air temperature and precipitation in Baghdad. Open Agric. **5**, 631–637 (2019)

Piles, M., et al.: Dominant features of global surface soil moisture variability observed by the SMOS satellite. Remote Sens. **11**, 95 (2019)

Seeyan, S., et al.: Investigation of the relationship between groundwater level fluctuation and vegetation cover by using NDVI for Shaqlawa Basin, Kurdistan Region-Iraq. J. Geogr. Geol. **6**, 187 (2014)

Shoshany, M., Goldshleger, N., Chudnovsky, A.: Monitoring of agricultural soil degradation by remote-sensing methods: a review. Int. J. Remote Sens. **34**, 6152–6181 (2013)

Teuling, A.J., Troch, P.A.: Improved understanding of soil moisture variability dynamics. Geophys. Res. Lett. **32** (2005)

Vereecken, H., et al.: Explaining soil moisture variability as a function of mean soil moisture: a stochastic unsaturated flow perspective. Geophys. Res. Lett. **34** (2007)

Wang, X., Zhang, Z.: A review: theories, methods and development of soil moisture monitoring by remote sensing. IEEE Int. Geosci. Rem. Sens. Symp. (2005)
Wang, J., Zhen, J., Hu, W., Chen, S., Lizaga, I., Zeraatpisheh, M., Yang, X.: Remote sensing of soil degradation: Progress and perspective. Int. Soil Water Conserv. Res. (2023)
Zhenjian, X.: Progress in monitoring and simulation of soil moisture. Progr. Geogr. (2004)
Ziboona, A.T., Abdul-Husseinlb, S.A., Albayatic, M.M., Turkey Dakheld, S.F.: Monitoring of land degradation in Alluvial Plain in Iraq by using geomatics techniques. IOP Conf. Ser. Earth Environ. Sci. **961** (2022)

Assessment of Ecological Impacts and Effectiveness of Protection Measures in the 1st-Phase Hanjiang-Weihe Water Diversion Project on Aquatic Ecological Sensitive Areas North of Qinling Mountains

Lijun Wang, Songsong Zhang, Xiaomei Kou, Le Niu, and Fan Wan

Abstract The First-phase Project of the Hanjiang-to-Weihe River Water Diversion Project (1st-phase Hanjiang-Weihe WDP) is characterized by its complex construction, lengthy duration, and extensive environmental impacts, particularly in the northern foothills of the Qinling Mountains. Some engineering facilities intersect with nationally or locally designated aquatic ecologically sensitive areas, directly or indirectly causing habitat degradation and biodiversity loss in these regions. This paper analyzes and identifies the ecological impacts of the project on aquatic ecologically sensitive zones, reviews the implementation of mitigation measures, and evaluates their effectiveness. By assessing the efficacy of these measures, this study aims to determine the project's environmental sustainability and provide insights for ecological restoration in future similar projects.

Keywords Aquatic ecological sensitive area (AESA) · Ecological restoration measures · Restoration effect

L. Wang (✉) · X. Kou · L. Niu · F. Wan
Power China Northwest Engineering Corporation Limited, Xi'an, China
e-mail: wanglij@nwh.cn

S. Zhang
China Railway First Survey and Design Institute Group Co., Ltd., Xi'an, China

X. Kou · F. Wan
Shaanxi Union Research Center, University and Enterprise for River and Lake Ecosystems Protection and Restoration, Xi'an, China

H. Xu (ed.), *Proceedings of the 8th International Symposium on Water Resource and Environmental Management*, Environmental Science and Engineering,
https://doi.org/10.1007/978-3-032-19088-8_9

1 Project Overview

1.1 Background Introduction

The Hanjiang-Weihe WDP plays a vital role in the economic and social development of Shaanxi Province (Zhou 2011). The 1st-phase Hanjiang-Weihe WDP comprises three major components: the Huangjinxia Hydropower Hub, the Sanhekou Hydropower Hub, and the Qinling Water Conveyance Tunnel. Its primary objective (Liu 2015) is to divert water from the Hanjiang River in southern Shaanxi to the Weihe River Basin in the Guanzhong region. This project features large-scale construction, complex project components, a long construction period, and a wide scope and degree of environmental impact (Liu 2012). Especially in the northern foothills of the Qinling Mountains, construction facilities of some spoil disposal sites and temporary camps intersect with aquatic ecologically sensitive areas (AESAs) designated by national or local authorities, which may directly or indirectly degrade habitats and reduce biodiversity in these regions.

The affected AESAs include the Shaanxi Heihe Rare Aquatic Wildlife National Nature Reserve ("SHRAWNNR") and Heihe National Aquatic Germplasm Resources Conservation Area for *Onychostoma macrolepis* ("HNAGRC-O. macrolepis"). They Zhao et al. (2017), Gao et al. (2020) are all located in Zhouzhi County, Shaanxi, and SHRAWNNR protects species such as the *Brachymystax lenok tsinlingensis*, *Andrias davidianus*, *Onychostoma macrolepis*, and *Batrachuperus tibetanus*. HNAGRC-O. macrolepis mainly protects *Onychostoma macrolepis*, alongside other species like the otter.

While the main infrastructure avoids these AESAs, ancillary construction activities, such as the 5#, 6#, and 7# branch tunnels in the northern Qinling foothills—directly occupy or indirectly disrupt habitats through spoil disposal and camp operations. These disturbances threaten protected aquatic species and riverine ecological functions. This study conducts a follow-up survey on the ecological protection measures proposed in the project design, objectively evaluates the impacts of the project construction process on the aquatic ecological sensitive areas and the restoration effects after the implementation of the measures, and provides reference experiences for other projects (Fig. 1).

1.2 The Positional Relationship Between the Project Layout and AESAs

The TBM construction method is adopted for the No. 5 branch tunnel, while the drill-and-blast method is used for the No. 6 and No. 7 branch tunnels. The positional relationships between the above-mentioned engineering construction activities and the aquatic ecological sensitive areas are shown in Table 1.

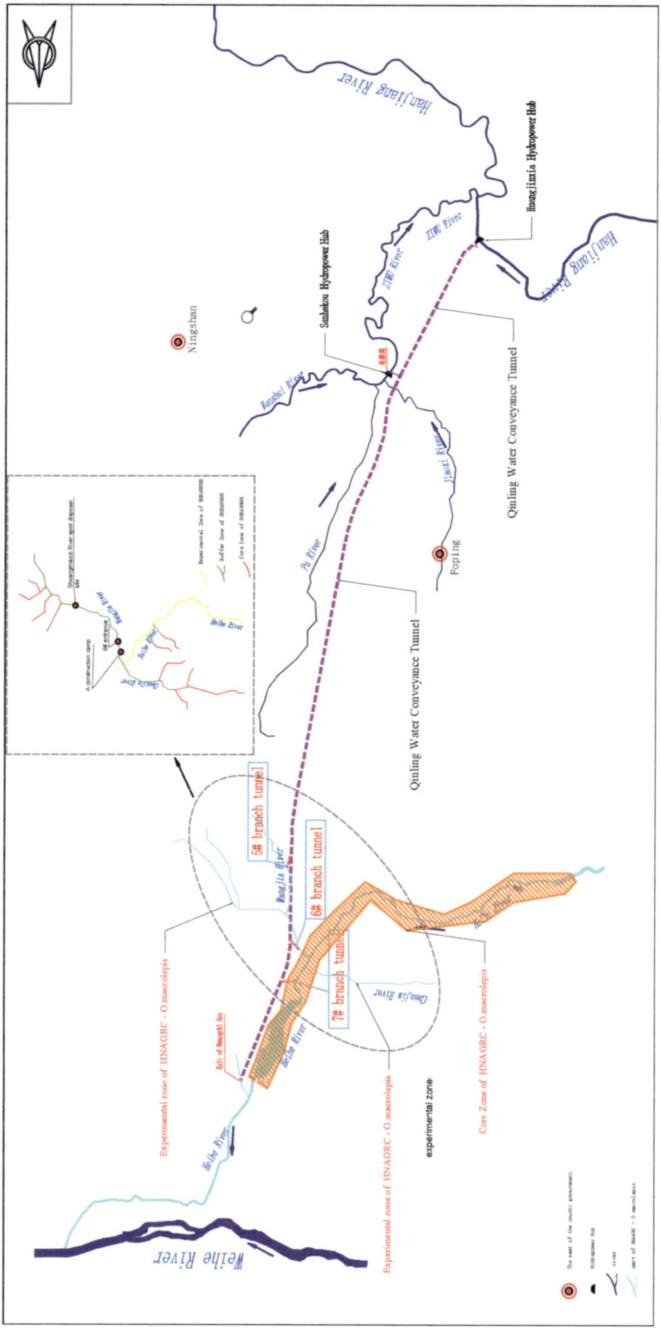

Fig. 1 Schematic diagram of the geographical location of the project and AESA

Table 1 List of AESAs occupied by the Hanjiang-to-Weihe River water diversion project

No.	Construction activity	Occupation of aquatic ecological sensitive areas
1	A construction camp	The camp is located on the left-bank of the Wangjia River. It occupies an area of the river channel with a length of about 350 m and an average width of 35–40 m, involving the experimental area of the SHRAWNNR and HMSSSNAGS
2	No. 5 branch tunnel of the Qinling Tunnel	Located in Xigou, a tributary of Wangjia River, occupying a river section of 500 m in length and 30–35 m in width. Involves the experimental zone of the SHRAWNNR
3	No. 6 branch tunnel and its ancillary projects	Located on both banks of Wangjia River, covering three sections with a total length of 250 m and area of 8978 m^2. Overlaps the experimental zones of the SHRAWNNR and the core/experimental zones of the HMSSSNAGS
4	No. 7 branch tunnel and its ancillary projects	Located on the left bank of the Heihe River main channel, occupying a river section of 560 m in length and 40–45 m in average width. Situated within the core zone of the HMSSSNAGS, with the Heihe Bridge crossing the river to connect the construction site
5	Shuangmiaozi River spoil disposal site	Located on the right-bank floodplain of Donghe (a tributary of Wangjia River): Spoil Site 1: Occupies 1.2 km of river length (30–50 m width); Spoil Site 2: Occupies 970 m of river length (40 m width). Both involve the experimental zones of the SHRAWNNR and the core/experimental zones of the HMSSSNAGS
6	Huangchigou spoil disposal site and project flood -discharge facilities	Located in Huangchigou, occupying a river section of 625 m in length and 30–100 m in average width. Overlaps the core/experimental zones of the HMSSSNAGS
7	Huangshiban'gou spoil disposal site	Located in the Xigou tributary of Wangjia River, occupying 600 m of natural river channel with a total area of 46,458 m^2. Involves the core/experimental zones of the HMSSSNAGS
1	A construction camp	The camp is located on the left-bank of the Wangjia River. It occupies an area of the river channel with a length of about 350 m and an average width of 35–40 m, involving the experimental area of the SHRAWNNR and HMSSSNAGS
2	No. 5 branch tunnel of the Qinling Tunnel	Located in Xigou, a tributary of Wangjia River, occupying a river section of 500 m in length and 30–35 m in width. Involves the experimental zone of the SHRAWNNR

2 Identification and Analysis of Environmental Impact

Since the two ASEAs were established during the construction of the 1st-phase Hanjiang–Weihe WDP, the EIA report (Changjiang Water Resources Protection Institute 2014) did not identify and analyze their impacts, let alone propose specific protection measures. Referring to the analysis of conventional projects, the environmental impacts of the construction are significantly phased, mainly manifested during the construction phase. The main impacts are the effects of construction land occupation and construction activities on the surrounding environment. There are basically no environmental impacts during the operation phase. The major potential impacts on natural and aquatic ecosystems are summarized in Table 2.

From the table, the primary impacts on aquatic ecological sensitive areas are surface water quality, terrestrial ecology, and aquatic ecosystems. Secondary impacts include noise, air quality, slag disposal, and groundwater. According to the construction-period data (Shaanxi Province Yinhanjiwei Survey and Design Research Institute Co., Ltd. 2022), the main pollution factors are also basically the same as those mentioned above. For example, the records show that:

Construction activities wastewater includes wastewater from tunnel operations, tunnel water inflow, concrete mixing stations, and domestic sewage. For example, the daily water inflow in the no.5, no.6, and no.7 adits was about 2000 m^3, and can even reach 2800, with COD concentrations of 25 mg/L and petroleum hydrocarbons at 0.08 mg/L.

The amount of waste slag generated by the excavation of the No. 5, No. 6, and No. 7 branch tunnels exceeded 2 million m^3. Except for part of it being used for the construction of living camps and flood-control retaining walls, all the rest is transported to the waste slag disposal site of Huangshiban or Shuangmiaozi.

The project mainly involves temporary land occupation (85 acres), mostly shrubland and grassland, with only 5.1 acres of permanent land use outside the No. 6 tunnel. Permanent land use will lead to a slight reduction in forest/grassland area after hardening, while temporary land will have vegetation restored post-construction.

Table 2 Environmental impact factor identification matrix

Environmental factor	Impact factor	Font size and style				
		Land occupation	Earth works	Mechanical operations	Waste water	Slag disposal
Natural environment	Surface water		2-R		3-R	2-R
	Groundwater		1-R		1-R	
	Acoustic quality		2-R	2-R		1-R
	Air quality		2-R	1-R		1-R
	Soil erosion		2-R	1-R		
Ecological environment	Terrestrial ecology	2-R	2-L			2-L
	Aquatic ecology	2-R	2-L		2-R	2-L

Note 3: Major impact; 2: Moderate impact; 1: Minor impact; +: Positive; −: Negative; R: Reversible or short-term; L: Irreversible or long-term

The original river in Huangchigou has been transformed into a box culvert. The water flows through the box culvert to the downstream section. A spoil disposal site, a concrete mixing station, and temporary construction sites are set up on the box culvert, completely changing the original flow direction and habitat of the gully. The Huangshiban spoil disposal site occupies the meandering section of the original tributary of the Wangjiahe West Stream. The meandering channel was straightened and a culvert was built for water passage. The above construction activities have an adverse impact on the water environment of the river and the aquatic organisms. Meanwhile, they have changed the original morphology of the river, which has a certain impact on the structure of the aquatic ecosystem in this area.

3 Ecological Restoration Measures and Their Effects

During the implementation process, relevant measures were proposed with reference to the main environmental protection objectives in the EIA. According to the collected supervision reports and the completion environmental protection acceptance report, in the construction activities, with reference to the protection objectives in the EIA of the main project, the following main habitat restoration measures were taken.

3.1 River Water Quality Restoration Measures

During the construction period, treatment measures were implemented for production and domestic sewage to ensure comprehensive utilization or reuse of treated wastewater. Higher-standard measures were taken for large-inflow cavern water gushing to protect the project area's water environment. The measures are as follows:

Concrete mixing wastewater. Sedimentation tanks (12 m^3, 8 m^3) were built at branch-tunnel concrete mixing stations. Wastewater entered the tanks via diversion channels for natural sedimentation. Supernatant was reused for cleaning, and sediment was sent to the spoil disposal site.

Sand and gravel washing wastewater. Wastewater from the Shuangmiaozi sand and gravel yard was treated by a horizontal flow sedimentation tank and reused for production, with river discharge prohibited.

Domestic sewage. A 100 m^3 collection tank was built for main-camp canteen, washing, and bath sewage. An 80 m^3 septic tank was built for camp toilets, and the sewage was used for orchard fertilization. Some camps had integrated treatment facilities, e.g., the Huangcaopo checkpoint of the No. 6 tunnel used a specific process. The No. 7 and No. 5 branch-tunnel camps had equipment with a 16.8 m^3/d capacity, and treated sewage was reused for construction.

Tunnel water inflow. For the No.6 test tunnel, a temporary station with "automatic dosing + flocculation reaction + secondary sedimentation tank" (2000 m^3/d capacity) was used. For the No.6 main tunnel, the designed Qinling Tunnel No.6 water-gushing treatment station with "high-efficiency sedimentation tank + rapid filter + activated carbon filter" (16,000 m^3/d capacity) was used (Ren 2018). Treated water met Class II surface-water standards and was discharged into the Wangjia River.

Wastewater and gushing water from the No.7 branch tunnel were pumped through four-stage stations to an external treatment station. Due to site restrictions, treatment evolved from simple to secondary sedimentation, then to "automatic dosing + flocculation reaction + secondary sedimentation tank". Finally, "high-efficiency sedimentation tank + rapid filter + activated carbon filter" (8000 m^3/d capacity) was adopted. Sludge was dewatered by a centrifugal dehydrator and transported out.

3.2 Riverbank Ecological Restoration Measures

During the project implementation, the main slope ecological restoration measures cover soil–water conservation for slag yards and vegetation restoration in construction-occupied areas, with details as follows:

Soil and water conservation measures. The vegetation at the cave entrance top and construction camp area remained intact. After construction, wire mesh protected exposed rock surfaces, and humus-rich soil was covered. The cave entrance camp was hardened with conservation facilities, and slope-protection grid vegetation started to recover.

For spoil dumps, the principle of "upper drainage, lower interception, soil covering and improvement, reclamation and greening" was adopted. Topsoil from some spoil-dump areas was stripped for later greening.

Shuangmiaozi Spoil Dump: A geogrid slag-retaining embankment was built along the Donghe River floodplain, with a cut-off ditch at the slag-hillside junction. Post-construction, pre-stripped topsoil was backfilled.

Huangchigou Spoil Dump: The retaining dam was an impervious rock-slag dam with a composite geomembrane core wall, using Qinling Tunnel excavation materials.

Huangshiban Spoil Dump: A C30 concrete flood-discharge channel was built to prevent gully-flood scouring. After slag piling, the land was improved and greened with *Amorpha fruticosa* and *Lolium perenne* (Fig. 2).

Vegetation restoration measures in the construction- occupied areas. During construction: Construction was limited to the expropriated area, with retaining walls for isolation. Cave-entrance excavation height was controlled within 10 m, and wire mesh protected cave-entrance slopes.

Huangshiban'gou spoil disposal site   Huangchigou spoil disposal site

Fig. 2 The spoil disposal site during the construction period and after restoration

After construction: Landform was restored promptly, especially in areas with severely damaged terrestrial vegetation. Temporary branch-tunnel land was restored, and parts of the Wangjia River roadbed and riverbanks were renovated.

Rare-plant treatment: Encountered rare plants were transplanted to the owner's camp botanical garden. Cave entrances were greened with climbing plants like *Sabia japonica* for disguise and to comfort animals.

Other management measures. The construction unit hung environmental-protection signs around the site and camp and trained employees. Through inspections and meetings, staff were required to avoid random vegetation felling and wildlife hunting. No such violations occurred outside the construction area during the project.

3.3 River Ecological Restoration Measures

In response to rectification measures for project spoil disposal sites and their impact on the *Varicorhinus macrolepis* habitat in the Heihe River near Huangchigou, habitat restoration was carried out. The habitat, about 1.5 km upstream of the Huangchigou water distribution hub, had identification signs. It was in good condition, with clear surface water, gentle riverbank slopes, rich vegetation, a gravel-bottomed side-beach, small back-water bays and a small deep pool.

The main measure for this good-quality habitat was to protect it from human damage. For river sections with spoil-covered and damaged riverbeds, ecological restoration of riverbank slopes and artificial restoration of the riverbed substrate were conducted. First, the spoil was removed, then the riverbed's stability was ensured and its structural diversity increased. Boulders were placed in the riverbed to mimic mountain-stream water flow.

As the slopes on both sides of the river section were flat, boulders were added to the riverbed and slopes, and slope-protection structures like rip-rap were built. The repair width was varied to restore the natural uneven width of the Huangchi River's riverbed.

3.4 Retrospective Evaluation of the Restoration Measures

According to the water quality monitoring results of the cross-section 500 m downstream of Wangjia River around the project area during the wet, normal, and dry periods in 2016 and 2020, the water quality met the requirements of Class II water quality. It can be seen that the water treatment measures during the construction period effectively ensured that the river water quality was not affected by the construction, indicating that these measures were effective.

No. 5#, 6#, 7# branch tunnel: Construction and restoration are completed. The nearby habitat has recovered well, with clear water in the upstream and tunnel-entrance rivers and rich benthic organisms. There is no construction, vehicle activity, or pollution. Construction facilities in the area are sorted and waiting for removal.

The slag disposal operations at all disposal sites have been completed. Reinforcement, drainage, soil conservation, soil covering, and vegetation restoration are completed. Nearby rivers are clear with rich aquatic life, and there is no pollution.

4 Conclusions and Recommendations

Based on a comparative analysis of construction-phase video records and supervision logs with the ecological status during the post-restoration investigation period, it is evident that the Hanjiang-to-Weihe River Water Diversion Project did not cause significant irreversible adverse impacts on water ecological sensitive areas. Meng et al. (2024) evaluated the ecological health status of water ecological sensitive areas to reflect the impacts of large-scale water conservancy projects, and the conclusion is consistent with that of this paper. This outcome highlights the effectiveness of the implemented water quality restoration measures and riverbank/river habitat rehabilitation measures during the project's execution. It is recommended to conduct research on fish-passing facilities for the fish activities at the exit of the Huangshiban Tunnel discovered during the investigation, and to demolish and rehabilitate the uneven sites to further mitigate the environmental impacts of the project on sensitive areas.

References

Changjiang Water Resources Protection Scientific Research Institute: Environmental Impact Report of the Hanjiang-to-Weihe River Water Diversion Project in Shaanxi Province (2014)

Gao, N.N., Bian, K., et al.: Primary research of feeding habits of onychostoma macrolepis in Qinling Bashan Mountains. Acta Agriculturae Boreali-Occidentalis Sinica **29**(8), 1141–1147 (2020)

Liu, H.X.: Discussion on the feasibility of the Hanjiang-to-Weihe river water diversion project. Henan Wat. Resour. South-to-North Water Divers. **8**, 95–96 (2012)

Liu, B.: Overall layout of Haniiang-to-Weihe river water diversion project. China Wat. Resour. **14**, 74–76 (2015)

Meng, X.Z., Li, J.H., et al.: Effect evaluation of ecological protection measures for streams at the northern foot of Qinling Mountains. J. Hydroecol. **46**(4), 218–228 (2024)
Ren, J.T.: Application of high sink double filter of water treatment technology in the leading Han-Ji-Wei Project. Shaanxi Wat. Resour. **4**(July), 54–58 (2018)
Shaanxi Province Yinhanjiwei Survey and Design Research Institute Co., Ltd.: Completion Environmental Protection Acceptance Inspection Report for Qinling Tunnel No. 6 Exploration and Test Tunnel Project. Shaanxi Hanjiang–Weihe River Water Diversion Survey, Shanxi (2022)
Shaanxi Province Yinhanjiwei Survey and Design Research Institute Co., Ltd.: Completion Environmental Protection Acceptance Inspection Report for Qinling Tunnel No.7 Exploration and Test Tunnel Project (2022)
Zhao, H., Zhang, H.X., et al.: Analysis of variation in water temperature of rare aquatic wildlife nature reserve of Heihe River. Genom. Appl. Biol. **36**(4), 1505–1510 (2017)
Zhou, W.: A Preliminary Analysis of the Status and Role of the Hanjiang-to-Weihe River Water Diversion Project in the Economic and Social Development of Shaanxi Province. Shaanxi Shuili, vol. 3, pp. 20–22 (2011)

Detailed Investigation and Comprehensive Prevention Strategy Research into Geological Disasters in Fuyuan County, Yunnan Province

Hui Geng, Wei Wang, Jianting Liu, Xinke Wang, Siyu Su, Kang Gao, and David Benson

Abstract This study is based on the framework of the "Research progress on special series for the construction of geohazard prevention and control system in Yunnan Province, China", and a systematic investigation and comprehensive evaluation of geological hazards in Fuyuan County, China. Using remote sensing interpretation, engineering geological mapping, drilling and monitoring techniques, the study comprehensively analyzed the types, distribution patterns and formation mechanisms of geological hazards in Fuyuan County. The survey results show that there is a total of 331 geological hazard sites in the county, including 219 landslides (66%), 66 ground collapses (20%), 39 collapses (12%), and 7 debris flows (2%). The study reveals that the development of geological hazards is influenced by the combined effects of topography and landform, rock and structural characteristics, rainfall, and anthropogenic engineering activities (especially mining activities), among which the collapse of mined-out areas accounts for 87.9% of all collapse disasters. This study identifies geological hazard-prone areas and hazardous areas based on detailed survey data and then proposes a comprehensive prevention system comprised of an "engineering governance + relocation avoidance + monitoring and early warning"

H. Geng
Institute of International Rivers and Eco-Security, Yunnan University, Kunming, People's Republic of China

W. Wang (✉) · X. Wang · K. Gao
Xiaolangdi Multipurpose Dam Project Management Center, Ministry of Water Resources, Zhengzhou, People's Republic of China
e-mail: w.wei@exeter.ac.uk

W. Wang · D. Benson
Environment and Sustainability Institute, University of Exeter, Penryn Campus, Cornwall, UK

J. Liu · S. Su
College of Mathematics and Statistics, North China University of Water Resources and Electric Power, Zhengzhou, People's Republic of China

H. Xu (ed.), *Proceedings of the 8th International Symposium on Water Resource and Environmental Management*, Environmental Science and Engineering,
https://doi.org/10.1007/978-3-032-19088-8_10

approach, providing a scientific basis for the future prevention and control of geological hazards and land use planning in Fuyuan County. The research results also have important reference value for disaster prevention in similar geological conditions.

Keywords Geological disasters · Risk assessment · Prevention and control strategies · Fuyuan County

1 Introduction

Geological disasters, as a major type of natural disaster, pose a significant threat to human life, property safety, and regional economic and social development (Fathani et al. 2016; Bui et al. 2019). Yunnan Province, as one of the provinces most affected by geological hazards in China, has complex geological conditions and widespread anthropogenic engineering activities, which contribute to the prevalence and severity of geological hazards (Fathani et al. 2023; Dobrev et al. 2013). Fuyuan County, located in the eastern part of Yunnan Province at the junction of the Dian Dong Plateau and the Qi Xi Plateau, has a complex geological structure, highly dissected terrain, and frequent mining activities. It is an important coal production centre in southern China and is also a high-risk area for geological hazards (Rahardjo et al. 2003; Faris and Fathani 2012).

In recent years, with the intensification of climate change and increased intensity of anthropogenic engineering activities, geological hazards in Fuyuan County have occurred more frequently (Fathani et al. 2016; Peranić et al. 2019). Following the findings of the 2022 Flood Control and Disaster Reduction Work Conference, Fuyuan County adheres to the principles of "prevention is better than relief" and "better to be overly cautious than to be negligent," and has comprehensively strengthened its geological hazard prevention and control (Fathani et al. 2023). The latest monitoring data for 2025 show that all 226 geological hazard points in the county are under effective supervision, achieving the goal of a "zero-incident record" for geological hazard events (Fathani et al. 2023; Faris and Fathani 2012).

This study is based on the key contents of the "Research progress on special series for the construction of geohazard prevention and control system in Yunnan Province, China" and systematically investigates the development characteristics and distribution patterns of geological hazards in Fuyuan County (Kang et al. 2023). It also analyzes the formation mechanisms and inducing factors of geological hazards, conducts geological hazard risk assessment and zoning, and ultimately proposes targeted comprehensive prevention strategies (Bui et al. 2019; Dobrev et al. 2013; Rahardjo et al. 2003). The results of this study not only provide direct guidance for geological hazard prevention in Fuyuan County but also offer an important scientific reference for regions with similar geological and environmental conditions (Faris and Fathani 2012; Peranić et al. 2019; Muñoz-Carpena et al. 2002).

2 Research Area and Methods

2.1 A Subsection Sample

Fuyuan County is located between 103°58′ E and 104°49′ E and 25°00′ N and 25°58′ N. It covers a total area of 3348 km^2 and administers 9 towns, 1 township, and 2 sub-district offices, with 161 village (neighborhood) committees. The terrain in the county is highly undulating, with the highest point being the main peak of Menghong Town's Mengpan Mountain (2748.9 m above sea level), and the lowest point being the Teshu Canyon in the southeast (1110 m above sea level). Fuyuan County belongs to the south temperate—north subtropical mountainous monsoon semi-humid climate, with an annual average temperature of 13.8 °C and an annual average rainfall of 1100 mm. The rainfall is unevenly distributed in time, with the majority concentrated in June to August; accounting for 56% of the annual rainfall.

Geologically, Fuyuan County is located at the junction of the Yanzhou quasi-plateau Diandong platform fold belt and the South China fold system Dian Southeast fold belt. The Xinhua Xiaoxi fold belt in this area is the controlling fault. Various tectonic systems such as the mountain-shaped structure, lotus-shaped structure, north-eastward structure, north-westward structure and arc-shaped structure are developed in the area. The strata are exposed from the Precambrian to the Quaternary, with the Mesozoic Triassic, Paleozoic Permian and Carboniferous rock series being the main ones.

Fuyuan County also experiences severe geological hazards (see Fig. 1). Based on the findings of this survey, the areal distribution of disaster sites within the region shows a similarity to the linear distribution of rivers, presenting an overall linear pattern. The density of disaster sites along these lines is closely related to the incision depth of the rivers and the types of valleys. From a spatial perspective, it is evident that areas with relatively high densities of disaster sites are mainly distributed along the Zhawai River in the Huangni River basin, Fucun, Yingshang, and the Dahe section in the Kuaize River basin, as well as along the Bumu River. These rivers or river sections are all characterized by deep incision, V-shaped valleys, valley slopes steeper than 30°, and in some places cliff-like slopes, which are high-prone areas for geological disasters.

Fuyuan County is an important coal production centre in southern China. It covers an area of 826 km^2, accounting for 25% of the county's total land area. The geological reserves amount to 141.02 billion tons, with the proven reserves reaching 64.57 billion tons. Coal mining activities are extremely frequent, which has had a significant impact on the development of geological hazards. To identify these hazards, this study employed a comprehensive investigation method using multiple technical means:

Field detailed investigation: Using a 1:50,000 scale topographic map as the base map, a combination of point, line, and area-based geological hazard investigation was conducted. Each geological hazard area with basic disaster-causing conditions was inspected individually. Field investigations were conducted in residential areas

Fig. 1 Photographs of typical geological hazards in the study area

in mountainous areas and around the basin margins, and large-scale mapping was selected for unstable or severely hazardous hazard sites.

Remote sensing interpretation: Utilizing high-resolution remote sensing images combined with drone aerial photography technology, the terrain and landforms of geological hazard points were identified and analyzed, using especially detailed interpretation of areas that are difficult for humans to reach.

Engineering geological mapping: Special mapping work was carried out for different engineering geological rock groups by dividing them into 6 types: cohesive soil, gravel, sand and silt multi-layer soil group; thin to medium-thick layered soft mudstone, sandstone, gravel, coal seam rock group; thin to medium-thick layered hard shale, mudstone, sandstone, slate rock group; medium to thick layered strong karstified hard limestone, mudstone, dolomitic limestone, limestone with calcite rock group; medium to thick layered strong karstified hard limestone, limestone, dolomitic limestone, dolomitic limestone rock group; and almond-shaped hard basalt rock group.

Drilling and geophysical exploration: For severely collapsed areas, a stability evaluation was carried out using the "geophysical exploration + drilling + deep hole displacement monitoring" technology, providing a basis for disaster control.

Data analysis and integration: A geological hazard information system for Fuyuan County was established by integrating geological environmental conditions, disaster distribution, and monitoring data for geological hazard zoning evaluation and risk prediction.

3 Results

3.1 Temporal and Spatial Distribution of Geological Hazards

The survey revealed that a total of 331 hazard sites in Fuyuan County, including 219 landslides (194 hazard points), 66 collapses (64 hazard points), 39 collapses (28 hazard points), and 7 debris flows (1 hazard point). From the regional distribution perspective, the hazard sites in Fuyuan Town were the most prevalent (54), while the hazard sites in Shengjing Sub-district Office, Zhong'an Sub-district Office, and Guegang Township were relatively fewer. The survey results of disaster types are shown in Table 1.

Landslides are the most common geological hazard type in this area. They can be further classified based on the nature of the sliding mass, the depth of the sliding surface, and their scale. According to the nature of the rock and soil mass, there are 202 occurrences of soil landslides (92%) and 17 occurrences of rock landslides (8%). According to the relationship between the sliding surface and the rock layers of the slope, there are 71 occurrences of parallel-layered landslides (32%), 56 occurrences of cutting-layered landslides (26%), 49 occurrences of reverse-layered landslides (22%), 23 occurrences of horizontal-layered landslides (11%), and 20 occurrences of similar soil landslides (9%). According to the thickness of the landslide mass, there are 210 occurrences of shallow landslides (96%) and 9 occurrences of middle-layer landslides (4%). According to their volume, there are 197 occurrences of small landslides (90%), 18 occurrences of medium-sized landslides (8%), and 4 occurrences of large landslides (2%).

Ground subsidence mainly includes two types: subsidence in mined-out areas and karst subsidence. Among them, there are 58 occurrences of subsidence in mined-out areas (accounting for 87.9% of the total), mainly occurring in the strata of the Quaternary period, the Mesozoic Triassic Lower Series, and the Upper Paleozoic Permian Series; there are 8 occurrences of cave subsidence and soil cave subsidence

Table 1 Statistics of geological disaster types in Fuyuan County

Disaster type	Cases	Threatened population	Affected population	Main locations
Landslide	219	66	194	Fucun Town, Dahe Town, Mohong Town
Ground collapse	66	20	64	Housuo Town, Dahe Town, Laochang Town
Rockfall	39	12	28	Huangnihe Town, Shibalianshan Town
Debris flow	7	2	1	Pide River, Kuaize River Basin
Total	331	100	287	–

(accounting for 12.1%), mainly occurring in the strata of the Quaternary period and the Upper Paleozoic Permian Series.

The collapse disasters mainly consist of two types: collapse of the fourth-order soil and collapse of the bedrock. The scale of soil collapse is generally small, while the collapse of the bedrock is mainly distributed in the major river basins and their tributary banks, as well as the steep cliffs within mining rights and the road cutting slopes, with the scale mainly being medium-sized and small-sized.

Debris flows can be classified into two types: valley-type debris flows and slope-type debris flows. Valley-type debris flows are the main type of debris flows in the working area, mainly distributed in the river valley banks of the Pi De River, Kuai Ze River and the first-order tributaries of the Da River. Slope-type debris flows mainly refer to the flow of solid substances on the slope triggered by heavy rain and are relatively rare in the working area.

The distribution of geological hazards in Fuyuan County is characterized as "numerous and widely distributed, with uneven distribution". Among the 12 townships in the county, except for Zhong'an Sub-district, Shengjing Sub-district and Guegan Township which have fewer disaster sites, the rest of the townships have more disaster sites. Particularly, Fuyun Town, Mo Hong Town and Dahe Town have a total of 138 disaster sites, accounting for 48% of the total number in the county. The disaster results of each township are shown in Table 2.

From the perspective of time distribution, the majority of landslides detected in Fuyuan County are modern landslides. Their occurrence time is relatively concentrated between 1997 and 2015, with a small portion occurring between 1980 and 1988. According to the investigation and interviews, the occurrence time of the landslides is mostly concentrated between June to September, and heavy rain is the dominant triggering factor for their formation.

From the perspective of their relationship with geological conditions, approximately 50% of these landslides are distributed along or near faults and at the contact zones of strata. Among the 7 debris flow gullies, 6 are located on the gully faults. In terms of rock types, landslides, debris flows, and collapses mainly occur in mudstone, siltstone, silt-sandy mudstone, mudstone limestone, and basalt strata.

Human engineering activities, especially mining activities, have a significant impact on the development of geological hazards. Fuyuan County is rich in coal resources. In recent years, frequent mining activities have occurred and the resultant collapse of mining tunnels, mined-out areas and the drainage of mines have triggered surface landslides, subsidence, and collapses. According to statistics, 87.8% of ground collapses, 30% of landslides, and 42% of collapses are caused by mining activities.

Table 2 Statistics of geological disaster distribution in townships of Fuyuan County

No.	Township/subdistrict	Area (km^2)	Population	Population density (persons/km^2)	Number of village committees	Number of geological disaster sites
1	Housuo Town	430.00	67,689	157.4	12	35
2	Shengjing Subdistrict	193.58	50,600	261.4	10	4
3	Zhong'an Subdistrict	327.00	90,700	277.4	15	9
4	Dahe Town	247.00	84,900	343.7	17	42
5	Mohong Town	495.50	62,551	126.2	16	42
6	Zhuyuan Town	164.05	52,700	321.2	11	22
7	Yingshang Town	152.30	83,100	545.6	16	36
8	Fucun Town	330.53	101,400	306.8	21	54
9	Laochang Town	236.00	48,100	203.8	9	22
10	Huangnihe Town	263.27	65,600	249.2	14	33
11	Gugan Township	82.60	16,400	198.5	3	8
12	Shibalianshan Town	334.00	71,200	213.2	17	24
Total	–	3348.00	794,940	237.4	161	331

3.2 *Analysis of Disaster-Causing Mechanisms and Influencing Factors in Coal Mine Goafs*

Mining is a process in which the equilibrium state of a geologic body is disrupted due to disturbance, followed by stress redistribution until a new equilibrium state is achieved. In this process, the shear stress of surrounding rock forms a stress-balanced arch as the working face advances. Stress concentration causes the failure of overlying rock, with displacement and detachment occurring sequentially from bottom to top, starting with the immediate roof strata of the coal seam. Once a goaf (or mined-out area) is formed, rock strata move until the stress stops. Based on the degree of damage, the affected rock strata are roughly divided into three zones: the caving zone, the fractured zone, and the bending deformation zone. During mining, the greater the advance distance of the working face, the larger the subsidence, and the scope of the three zones expands accordingly. When the maximum subsidence value of the caving zone exceeds the critical value, roof caving occurs, and the three zones extend upward. The roof rock adjacent to the goaf undergoes the most significant change in stress state, directly losing support and being prone to collapse. Photographs of land subsidence in the study area are shown in Fig. 2.

Fig. 2 Photographs of ground subsidence in the study area

Due to the large mining area, long mining history, and advanced coal mining mechanization technology employed in Fuyuan County, coal mining-induced disasters are wide ranging, concentrated in distribution, and cause relatively severe surface movement, deformation, and damage.

During coal seam mining, the support for surrounding rock in roadways and goafs during the extraction process is temporary. After the goaf is formed, its roof strata can bend and move downward under gravity and the pressure of the overlying rock-soil mass. When the tensile stress formed inside the roof strata exceeds the ultimate tensile strength of the rock, the goaf roof will fracture, collapse, and cave in. Subsequently, three distinct zones (the caving zone, the fractured zone, and the bending deformation zone) are formed in the overlying rock-soil mass of the goaf (Fig. 3a). As the scope of the goaf expands, the affected moving rock strata also continues to expand, thereby forming surface subsidence or a subsidence basin much larger than the goaf in scope on the ground surface (Fig. 3b).

According to collected data, the Jike Coal Mine has a mining area of 6.58 km^2, with an elevation range of 1800–1350 m. The shallowest coal seam is within the Changxing Formation, with a thickness of 5.66 m. The mine is designed to extract six coal seams, with seam dip angles generally ranging between 5° and 15°. According to the "Specifications for Coal Pillar Retention and Under-Water, Railway, Building and Shaft Extraction" (State Administration of Work Safety et al. 2017), the calculated caving height indicates that areas where the mining depth of coal seams are less than 28.28 m are classified as high-prone zones for mining-induced geological disasters.

Based on the deformation areas within the survey region, which are located at elevations between 1840 and 1950 m, the back-calculated coal seam burial depth ranges from 1820 to 1930 m. Coal extraction forms underground goafs of considerable scale, which alter the subsurface stress state. Under the combined effects of the overburden weight, stress release, and groundwater, the roofs of the goafs deform and fail, eventually collapsing and inducing large-scale surface deformation.

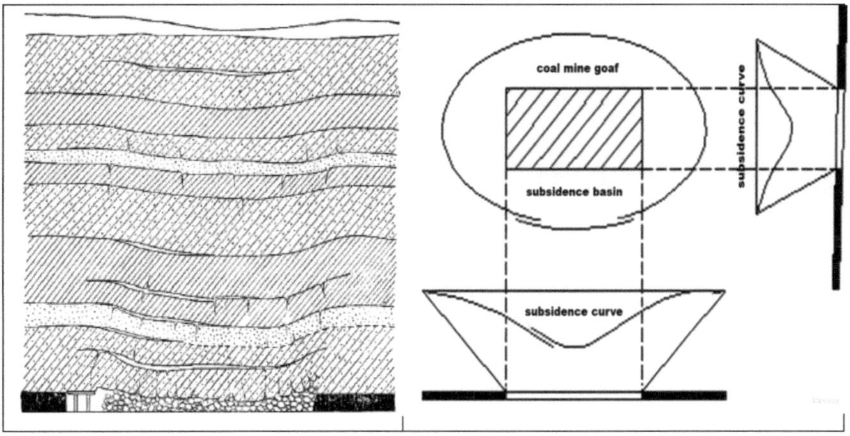

Fig. 3 **a** Overlying rock movement in Goaf. **b** Surface subsidence basin

4 Results

Based on the detailed survey results, this study classifies the geological hazards in Fuyuan County into different prone areas and hazardous areas: *High-prone areas* are mainly distributed in regions such as Huangni River–Shibalian Mountain, Fucun–Laochang, Dahe–Yingshang, and Housuo–Mohong, characterized by significant topographic relief, concentrated coal mines, and intensive engineering activities; *medium-prone areas* primarily cover the Gungan–Yingshang and Zhuyuan–Fule areas; while *low-prone areas* are located in relatively flat regions such as Zhong'an and Shengjing. Correspondingly, high-hazard areas mainly include goafs and steep slope zones, medium-hazard areas cover general mountainous areas and mining areas with moderate intensity, and low-hazard areas refer to flat areas and gentle slope lands.

In terms of prevention and control strategies, the study proposes a comprehensive system of "engineering governance + relocation & avoidance + monitoring & early warning". For engineering governance, targeted measures can be adopted for landslides, collapses, ground subsidence, and debris flows, including slope cutting for load reduction, retaining walls, anti-sliding piles, anchoring, grouting, rockfill backfilling, and construction of slag retaining dams and diversion channels. For relocation & avoidance, centralized relocation and resettlement can be implemented for hazard points that cannot be fundamentally controlled through engineering measures, such as the relocation projects that have been completed or are in progress in Heichong Village (Dahe Town), Xinhua Village (Housuo Town), and Dashanjiao Village (Laochang Town). For monitoring and early warning, a "human–machine integrated monitoring" system can be established, with video surveillance, displacement meters, and rainfall stations deployed to form a three-level mass monitoring and prevention network (county, township, and village), full-time monitors assigned, and an inter-departmental information sharing mechanism set up to realize real-time transmission and early warning of hazard information. For emergency capacity building, emergency plans should be improved, regular drills conducted, and the public's ability to identify hazards, avoid risks, and conduct self-rescue and mutual rescue should be enhanced through brochures, posters, and door-to-door talks with residents. Overall, Fuyuan County has initially formed a systematic and scientific approach for geological hazard prevention and control. It is recommended that a rainfall–disaster triggering model (e.g., single-day rainfall > 50 mm triggering high risk) is established and embedded into CMIP6 climate models to simulate changes in rainfall patterns under a 1.5 °C warming scenario by 2050, thereby quantifying the relationship between extreme rainfall frequency and landslide/subsidence mechanisms. Human engineering activities account for 87.6% of disasters: in the past five years, coal mining areas have expanded by 12%, significantly increasing the subsidence risk of mined-out zones with burial depths < 28.28 m; from 2020 to 2023, the surface hardening rate increased by 15%, reducing rainfall infiltration capacity and exacerbating landslide risk. Mining plans and land development schedules should therefore be incorporated into the assessment system to predict high-risk zones of human-induced disturbances over the next five years (e.g., deeply incised V-shaped valleys). In the Huangni River and Kuaize River basins, an IoT sensor network should be deployed to monitor rainfall, groundwater levels, and rock mass displacement in real time. Combined with a GIS platform, risk zoning can be dynamically updated, upgrading traditional community-based monitoring and prevention into an integrated "meteorology–geology–human activity" intelligent early-warning system.

Acknowledgements This work was supported by the Special Geological Disaster Prevention Fund of Yunnan Province. The authors would like to thank Kunming Survey and Design Institute of Southwest Nonferrous Metals Co., Ltd. for technical support during field investigations and data collection, and express gratitude to the Fuyuan County Natural Resources Bureau and the people's governments of various towns for their assistance with fieldwork.

References

Bui, T.A., Fathani, T.F., Wahyu, W.: Landslide risk assessment for designing monitoring and early warning system. J. Appl. Geol. **4**(1), 1–8 (2019)

Dobrev, N., et al.: Landslide problems in Bulgaria: factors, distribution and countermeasures. Landslide Sci. Pract. **7**, 187–193 (2013)

Faris, F., Fathani, F.: A coupled hydrology/slope kinematics model for developing early warning criteria in the Kalitlaga Landslide, Banjarnegara, Indonesia. Prog. Geo-Disaster Mitig. Technol. Asia **1**, 453–467 (2012)

Fathani, T.F., Dwikorita, K., Wahyu, W.: An integrated methodology to develop a standard for landslide early warning systems. Nat. Hazard. **16**(9), 2123–2135 (2016)

Fathani, T.F., et al.: Strengthening the resilience by implementing a standard for landslide early warning system. Prog. Landslide Res. Technol. **1**(1), 277–284 (2023)

Kang, X.B., et al.: Research progress on special series for the construction of geohazard prevention and control system in Yunnan Province, China. Chinese J. Geol. Hazard Control **34**(6), 146–157 (2023)

Muñoz-Carpena, R., et al.: Field evaluation of the new Philip-Dunne permeameter for measuring saturated hydraulic conductivity. Soil Sci. **167**(1), 9–24 (2002)

Peranić, J., Jagodnik, V., Arbanas, Ž.: Rainfall infiltration and stability analysis of an unsaturated slope in residual soil from flysch rock mass. In: The XVII European Conference on Soil Mechanics and Geotechnical Engineering, vol. 1 (2019)

Rahardjo, H., et al.: Effectiveness of horizontal drains for slope stability. Eng. Geol. **69**(3–4), 295–308 (2003)

State Administration of Work Safety, et al.: Notice on issuing the "specifications for coal pillar retention and under-water, -railway, -building and -shaft extraction". Bulletin No. Anjian Zong Mei Zhuang **2017**(66) (2017)

Application of Improved Flap Filter in a Water Purification Plant in Zhejiang Province

Ling Zhu, Yue Jiang, Bo Zhang, and ZhiHan Jiang

Abstract This paper focuses on turbidity treatment technology in a Zhejiang water purification plant with a daily capacity of 80,000 tons, using Taihu Reservoir as its water source. The raw water exhibits characteristics including low annual turbidity, seasonal algal blooms, excessive manganese content, and low pH in summer, while the plant must meet stringent standards—effluent turbidity ≤ 0.3 NTU per Zhejiang's modern water plant assessment, stricter than the national 1 NTU limit. To this end, it adopts the process: "pre-ozone contact + flocculation and sedimentation + V-type sand filtration + ozone contact + granular activated carbon adsorption + disinfection". Aiming at flaws of traditional flap filters (e.g., increased head loss from embedded pipes, no manholes for maintenance, large footprint, algal growth), improvements were made: optimizing layout, adding manholes, adjusting overflow design, and adopting enclosed structures. The improved filters use coal-based activated carbon (8×30 mesh, thickness 2200 mm) and follow a three-stage backwash (air, air–water, water). Since commissioning in 2021, effluent turbidity (monitored Apr 2023–Mar 2024) has consistently outperformed standards, verifying the technology's effectiveness in turbidity control.

Keywords Flap filter · Granular active carbon filter · Backwash

1 Introduction

With the rapid development of social economy in China, people's living standards have been continuously improved. The requirements for urban drinking water quality have become increasingly strict, and the water supply guarantee rate and water supply safety have also been constantly enhanced. In particular, the turbidity index, which

L. Zhu · Y. Jiang (✉)
Anhui Zhongban Engineering Design Group Co., Ltd., Anhui, China
e-mail: 527240316@qq.com

B. Zhang · Z. Jiang
Southeast University, Nanjing, China

H. Xu (ed.), *Proceedings of the 8th International Symposium on Water Resource and Environmental Management*, Environmental Science and Engineering,
https://doi.org/10.1007/978-3-032-19088-8_11

reflects the comprehensive conditions such as sensory properties, general chemical indicators and microbial content, has become more and more stringent. In addition to adopting conventional treatment processes, newly-built water purification plants have all added advanced treatment processes. Among them, the advanced treatment process is mostly dominated by the "ozone oxidation + activated carbon adsorption" process. This process has low and stable effluent turbidity, can effectively remove microorganisms in water, and greatly improve the biological safety of drinking water (Staehelin and Hoigne 1985). Considering that activated carbon is a light filter medium, in order to reduce the loss of activated carbon filter medium, the "flap filter" type is mostly adopted. A water purification plant in Zhejiang Province is a modern water plant that uses the improved flap filter type as the granular active carbon filter. The water source of the plant is Taihu Reservoir, the engineering treatment capacity is 80,000 tons per day, and the water purification adopts the process of "pre-ozone contact + flocculation and sedimentation + sand filtration + ozone contact + activated carbon adsorption + disinfection".

2 Characteristics of Granular Active Carbon Filters

With the increasing pollution of water sources and the improvement of drinking water quality standards, in order to overcome the shortcomings of conventional treatment processes, further advanced purification of the effluent from conventional treatment processes has become one of the options for water purification plants. The ozone-biologically activated carbon (O_3-BAC) filter process is currently the main advanced treatment process for water purification (Kawamur 1999; Liu et al. 2020). The main difference between granular active carbon filters and ordinary filters lies in that the filter medium used in the former is activated carbon.

Activated carbon is a black solid material with carbon as its framework structure, which is made with well-developed pores through gasification (carbonization or activation) (Hassan and Khong 2023). The well-developed pores of activated carbon result in a large specific surface area, endowing it with excellent adsorption properties. The adsorption of activated carbon refers to the process in which pollutants in water are enriched or concentrated on the surface of activated carbon. The reason for this adsorption is the existence of intermolecular and intramolecular bonding forces—these forces are relatively weak interactions between molecules in the aggregated state of substances.

3 Characteristics of Flap Filters

Flap filters, also known as Sulzer filters, are a research achievement of the Technology and Engineering Department (CTE Company) under Sulzer AG of Switzerland. The term "flap" comes from the fact that the backwash drainage flap valve flips 0–90° to open and close during operation (Peng 2016). This type of filter has made new technological breakthroughs in the backwash system, drainage system and filter medium selection, thus featuring significantly improved effluent quality, less backwash water consumption, shorter backwash time, longer backwash interval, lower infrastructure investment, lower operating costs, simple construction and short construction period (Crittenden et al. 2012; Zhou and Jiang 2006).

Working Principle of Flap Filters: During filtration, the inlet water flows evenly into the filter through the weir, and filtration is carried out from top to bottom in a constant water level filtration mode. During backwashing, the water inlet valve is first closed; when the residual water level in the filter drops to 20–30 cm above the filter medium surface, the water outlet valve is closed, and backwashing is performed in three stages: air washing, air–water combined washing, and water washing. After backwashing, the drainage flap valve is opened to discharge the backwash wastewater (see Fig. 1).

Due to the adoption of valve-closed backwashing, flap filters can achieve high-intensity expansion washing of the filter medium layer, resulting in thorough and clean washing while preventing the loss of filter medium (Wang et al. 2023). This basic feature, which distinguishes flap filters from other types of filters, allows for diverse selection of filter media for flap filters, such as double-layer filter media, activated carbon filter media, or sand filter media. In addition, no drainage occurs during the washing process, which can reduce backwash water consumption.

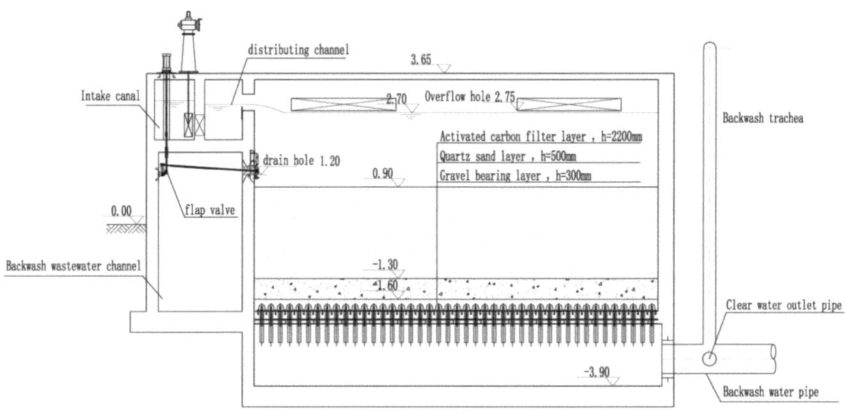

Fig. 1 Conventional flap filter type

4 Comparison and Selection Between Improved Flap Filters and Traditional Flap Filters

The water source of a water purification plant project in Zhejiang Province is a certain reservoir. According to the raw water quality test results from the Taizhou Monitoring Station of the National Urban Water Supply Quality Monitoring Network, the raw water is generally classified as Class II when measured against the *Environmental Quality Standards for Surface Water* (GB 3838-2002). In addition, in recent years, the raw water from Taihu Reservoir has shown characteristics such as relatively low turbidity throughout the year, seasonal algal blooms, seasonal excessive manganese content, and low pH value in summer. The project has high requirements for effluent quality: in addition to meeting the national *Standards for Drinking Water Quality* (GB 5749-2006), the effluent must also comply with the local *Evaluation Standards for Modern Urban Water Supply Plants in Zhejiang Province* (2013 Edition), with the turbidity of the finished water not exceeding 0.3 NTU.

Considering multiple factors such as raw water quality, effluent quality, land use conditions, and the owner's requirements, the water purification process of "pre-treatment + conventional treatment + advanced treatment" is adopted, specifically as follows:

Raw water → Pre-ozone contact tank → Flocculation and sedimentation tank → V-type sand filter → Ozone contact tank → Granular active carbon filter → Clear water tank → Water supply pump house → Effluent.

Among them, the advanced treatment adopts the "ozone contact + activated carbon adsorption" process. Compared with other filter types, flap filters not only have lower water consumption but also avoid filter medium loss. Therefore, it is determined that flap filters will be used as the filter type for the "granular active carbon filters" in the project * MERGEFORMAT * MERGEFORMAT (* MERGEFORMAT (* MERGEFORMAT (Yang 2015).

Through the investigation of the current domestic flap filter process in various water purification plants, it is found that the traditional activated carbon flap filter has the following shortcomings * MERGEFORMAT * MERGEFORMAT (* MERGEFORMAT (* MERGEFORMAT (Li et al. 2006) (see Fig. 1):

(1) The flap valve is arranged on the same side as the water inlet, and a flap valve connecting rod needs to be installed in the main water inlet channel of the filter. Therefore, it is necessary to pre-embed a steel pipe (with the flap valve connecting rod placed inside) in the main water inlet channel. When the pre-embedded steel pipe is installed in the water inlet channel, the cross-sectional area of the flow in the water inlet channel is suddenly reduced, the flow state is changed, and the head loss is increased.
(2) According to Article 3.0.15 of Code for Design of Filter Tanks (CECS 321: 2012), the outlet channel of a single filter cell shall be provided with a manhole for easy installation and maintenance. The air and water distribution channel (also serving as the outlet channel) at the bottom of the traditional flap filter is not equipped with a manhole, which fails to meet the requirements of the

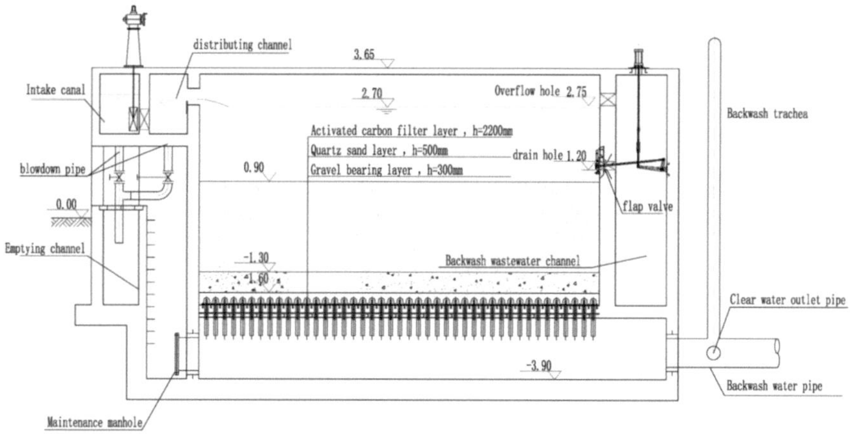

Fig. 2 Improved flap filter configuration

relevant code. During the construction of the filter, it is difficult to disassemble the formwork and clean up; after the filter is completed, the inspection and maintenance are complicated.

(3) The overflow hole is arranged on one side of the filter length. When multiple groups of activated carbon filters are arranged in parallel, an overflow channel needs to be added between the two groups of filters to discharge the overflow water, which increases the floor area of the filters.

(4) The filter media consists of a supporting layer and an activated carbon layer from bottom to top. Most filters adopt an open design. Due to sufficient sunlight and dissolved oxygen, algae in the filter are prone to growth and may penetrate the carbon filter layer into the clear water tank, affecting the effluent quality.

In response to the above issues, the project design incorporated improvements to the traditional flap filter pool type by considering the layout, process flow, and land area. Specific improvements are as follows (see Fig. 2).

5 Design of the Process Scheme

The project at a water purification plant in Zhejiang employs the "Ozonation Contact + Activated Carbon Filter" process, with a designed capacity of 80,000 tons per day. The specific design is as follows:

1. Ozone Contact Tank

A total of two ozone contact tanks are provided, with an effective contact time of 15 min. The structure is fully enclosed. The plan dimensions of a single ozone contact tank are 12.8 m × 10.8 m. The structural depth of the tank is 6.15 m, with an average effective water depth of 5.5 m.

The maximum ozone dosage is 2.5 mg/L, and the average dosage is 1.5 mg/L. The ozone contact tank is equipped with specialized ozone diffusion discs. Ozone is injected at three points per tank, with the ratio of gas injection quantities being 50%, 25%, and 25%.

2. Activated Carbon Filter

   (1) Design Parameters

   The activated carbon filter adopts a modified valve filter type. One filter unit is provided, internally divided into six cells arranged in two rows (three cells per row). A pipe gallery is installed between the two rows of filter cells.

   Design filtration rate: 10.0 m/h, Forced filtration rate: 12.0 m/h.

   Overall plan dimensions: 38.1 m × 19.6 m, Net dimensions per filter cell: 10 m × 6 m, Structural depth of filter cell: 5.75 m, Water depth above filter media surface: 1.8 m, Pipe gallery dimensions: 8.7 m × 19.6 m, Clear height of pipe gallery: 9.35 m.

   (2) Major Equipment and Materials

   Filter media: Columnar coal-based activated carbon is used, with a particle size of 8 × 30 mesh (equivalent to 2.38 × 0.6 mm); bulk density: 0.35–0.55 g/cm^3; uniformity coefficient $K_{80} = 1.9–2.0$; media thickness: 2200 mm; methylene blue value > 200 mg/g; iodine value > 900 mg/g; specific surface area ≥ 900 m^2/g. It shall meet the requirements of CJ/T 345-2010 "Coal-Based Activated Carbon for Drinking Water Treatment".

   Supporting layer: Homogeneous quartz sand is used, with an effective particle size $d_{10} = 0.7 \pm 0.03$ mm, uniformity coefficient $K_{80} < 1.4$, and thickness 500 mm.

   Underdrain layer: Gravel is used, with a thickness of 300 mm. The particle size distribution from top to bottom is: 2–4 mm (h = 50 mm), 4–8 mm (h = 100 mm), 8–16 mm (h = 100 mm), 16–32 mm (h = 50 mm).

   Backwashing equipment: A total of 4 backwash pumps are provided for the activated carbon filter, using horizontal centrifugal pumps, all equipped with variable frequency speed control devices; operating mode: 3 duty, 1 standby; pump parameters: Q = 333 L/s, H = 12 m, N = 55 kW. A total of 3 backwash blowers are provided for the activated carbon filter, using screw-type blowers, all equipped with variable frequency speed control devices; operating mode: 2 duty, 1 standby, shared with the V-type sand filters; blower parameters: Q = 1815 m^3/h, H = 5 m, N = 37 kW. The backwash water for the activated carbon filter comes from the filtered water of the V-type sand filters. The backwash pumps and blowers are installed separately in the backwash pump house.

   (3) Filtration Process

   Raw water is distributed from the main inlet pipe to both side inlet channels, enters each filter cell through the distribution channel of each cell, continues to flow downward through the activated carbon layer and the water collection system, passes through the air–water distribution channel (also functioning as the outlet channel), and is then conveyed via the outlet pipe to the main outlet channel before proceeding to the subsequent clear water tank.

(4) Backwashing Process

During filter backwashing, the inlet gate is first closed. Then, the corresponding valves (plates) are operated in three stages: air washing, combined air–water washing, and water washing. Typically, after repeating twice, the drainage valve (plate) is closed, the inlet gate is opened, and the filter returns to normal filtration operation. The specific steps are as follows:

a. When the opening degree of the outlet valve reaches maximum and the head loss reaches maximum (2.0 m), the inlet gate is closed, and filtration continues;
b. When the water level in the filter drops to 0.2 m above the top of the filter media layer, the outlet valve is closed;
c. The backwash air inlet valve is opened for air washing alone, lasting 4 min;
d. The backwash water inlet gate is opened for combined air–water washing (with low water flow rate), lasting 4 min;
e. The backwash air inlet valve is closed, and the water washing intensity is increased for water washing alone (with high water flow rate), lasting 2 min;
f. The backwash water inlet valve is closed. After an idle period of about 20 s, the flap valve is opened for drainage, which lasts about 1.5 min;
g. The flap valve is closed. Water washing alone (with high water flow rate) is performed for 2 min, followed by an idle period of 20 s. The flap valve is then opened for drainage, which lasts about 1.5 min;
h. The flap valve is closed. Water washing alone (with high water flow rate) is performed for 1 min. The inlet gate and the initial filtered water valve are opened, and the drainage period lasts about 20 min;
i. The initial filtered water valve is closed, and the outlet valve is opened to restart filtration.

The duration of each step may be appropriately adjusted based on actual field conditions after the filter has been in operation for a period. The entire backwashing process takes approximately 35–40 min. When the flap valve is opened for drainage, it is first slowly opened to 45° for draining. When the water level drops to 0.5 m above the surface of the carbon layer, the valve is fully opened for drainage.

6 Operation Status

The project was put into operation in 2021, and the overall effluent quality has been excellent (see Fig. 3). According to China's *Sanitary Standard for Drinking Water* (GB 5749-2022), the turbidity limit for finished water is 1 NTU, while the *Assessment Standard for Modern Water Plants in Urban Water Supply of Zhejiang Province* sets a more stringent turbidity limit of 0.3 NTU. After two years of operation, from

April 2023 to March 2024, the effluent turbidity over the one-year period has been significantly better than the standard requirements (see Fig. 4).

Fig. 3 Field application diagram

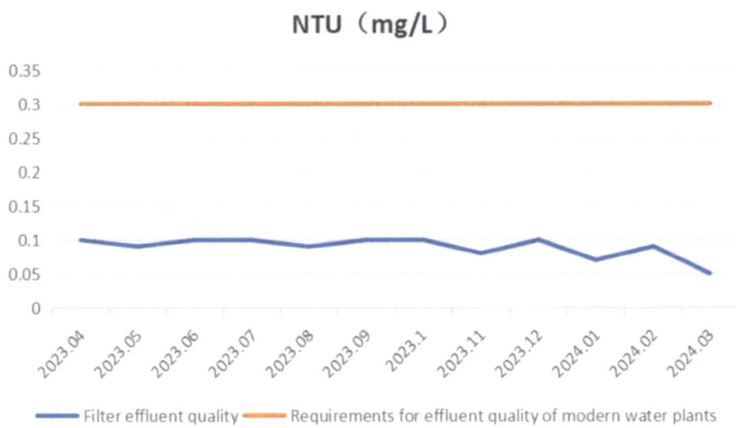

Fig. 4 Effluent turbidity

7 Conclusion

(1) In order to meet the increasingly high standards for effluent water quality in water purification plants, the "ozone contact + activated carbon filter" process is necessary in the water treatment of a certain plant in Zhejiang Province.
(2) The ozone-activated carbon filter adopts a new type of flap filter. Without changing the treatment efficiency, improvements have been made in the drainage valve, maintenance access, overflow channel, and filter media, as follows: (1) Change the installation position of the flap valve to reduce the head loss of the inlet channel; (2) Provide maintenance manholes in the water and air distribution channel to facilitate operation and maintenance; (3) Change the position of the overflow hole to reduce the footprint of the tank; (4) Add a sand filter layer to ensure effluent water quality.
(3) The experience of applying the new flap filter in domestic water purification plants is still relatively limited. Therefore, it is necessary to continuously summarize experience in practical engineering applications, promptly analyze and solve existing problems, and further optimize the design.

References

Crittenden, J.C., Trussell, R.R., Hand, D.W., et al.: Water Treatment Principles and Design. Wiley, USA (2012)
Hassan, S.S., Khong, H.Y.: Facile preparation of scalable bamboo-derived activated carbon for efficient wastewater treatment. BioResources **18**(3), 5292–5308 (2023)
Kawamur, S.: Design and operation of high-rate filter. AWWA **91**(12), 77–90 (1999)
Li, R., Dai, X., Chen, Y.: Discussion on design of shutter-type filter in practical projects. China Water Wastewater **22**(18), 48–51 (2006)
Liu, D., Xie, Q., Huang, X., et al.: Backwashing behavior and hydrodynamic performances of granular activated carbon blends. Environ. Res. **184**(prepublish), 109302 (2020)
Peng, L.: Application of reverse shutter filter to reconstruction and expansion project of a water treatment plant. China Water Wastewater **13**(24), 87–89 (2016)
Staehelin, J., Hoigne, J.: Decomposition of ozone in water in the presence of organic solutes acting as promoters and inhibitors of radical chain reactions. Environ. Sci. Technol. **19**(12), 1206–1213 (1985)
Wang, T., Zhang, J., Wang, P.: Engineering case based on the combination process of high-efficiency flocculation sedimentation tank and shutter filter. Chinese J. Environ. Eng. **17**(3), 1043–1050 (2023)
Yang, Y.: Considerations in design and construction of shutter filter. China Water Wastewater **31**(10), 61–63 (2015)
Zhou, J., Jiang, Q.: Application of shutter filter in Kunming for drinking water purification. Water Wastewater Eng. **33**(3), 37–39 (2006)

Water Supply–Demand Imbalance and Sustainable Management Strategies for the Yurungkash River Basin in Xinjiang

Yinyin Zhu, Lei Chai, and Tian Shao

Abstract As an important inland river in the Hotan region, the Yurungkash River holds strategic significance for regional sustainable development and ecological security. This study analyzes the contradiction between water supply and demand in the basin and its intensifying trend. The results indicate that although recent runoff has shown an increasing trend due to glacial melt, the natural endowment of "abundance in summer and scarcity in winter," coupled with insufficient regulatory capacity of water conservancy projects, leads to prominent spatiotemporal unevenness. On the demand side, structural imbalances are evident, with agricultural water use accounting for over 90%, and inefficient water use severely encroaches on ecological and domestic water supplies, resulting in the degradation of downstream green corridors and groundwater over-extraction. It is projected that by 2030, the supply–demand gap in dry years will reach 155 million m^3. The study proposes a four-pronged regulatory strategy system centered on "water conservation priority," encompassing "conservation-development-management-optimization." This system includes comprehensive measures such as agricultural water efficiency, construction of water conservancy hubs, improvement of water rights markets, and implementation of ecological flood pulse management, providing a decision-making basis for ensuring water security in the basin.

Keywords Yurungkash River · Contradiction between water supply and demand · Structural imbalance · Four-pronged regulatory strategy · Water security

Y. Zhu (✉) · T. Shao
Power China Northwest Engineering Corporation Limited, Xi'an, China
e-mail: xby13562@powerchina.cn

L. Chai
Shaanxi Water Environment Engineering Survey and Design Institute, Xi'an, China

H. Xu (ed.), *Proceedings of the 8th International Symposium on Water Resource and Environmental Management*, Environmental Science and Engineering,
https://doi.org/10.1007/978-3-032-19088-8_12

1 Introduction

Water is the source of life, the key to production, and the foundation of ecology. For Xinjiang, located in the hinterland of the Eurasian continent, water resources are a strategic asset that sustains the oasis economy, social stability, and ecological security. As one of the most important rivers in the Hotan region of Xinjiang and the southern edge of the Tarim Basin, the Yurungkash River's water resources are directly linked to the development of the southern hub of the core area of the "Silk Road Economic Belt." At both national and regional strategic levels, ensuring water security in Southern Xinjiang has been elevated to an unprecedented height, with the Yurungkash River Basin being a key area for achieving this strategic objective.

Against the backdrop of climate change, the accelerated melting of glaciers and snow in the source areas of the basin may temporarily increase runoff in the short term. However, in the long run, this will lead to the decline of the "solid reservoir" function, threatening the sustainability of water resources. Meanwhile, with population growth, accelerated urbanization, and economic development, human activities are increasingly interfering with water resources. The supply–demand relationship between the water resource formation areas in the upper mountainous regions and the water consumption areas in the middle and lower reaches is becoming increasingly strained. The growing imbalance between supply and demand has emerged as a core bottleneck restricting the sustainable development of the basin.

2 River Basin Profile

The Yurungkash River originates from the northern slopes of the Kunlun Mountains. It flows from south to north through the Hotan region, eventually converging with the Karakash River at Koshilax to form the Hotan River, which then discharges into the Tarim River. The total length of the river is approximately 504 km (Liu 2019; Yu et al. 2021). The topography of the basin is higher in the south and lower in the north, encompassing four geomorphological types: mid-high mountain areas, low mountain and hilly areas, alluvial-proluvial plains, and plain desert areas. The multi-year average runoff volume is 2.44 billion cubic meters. Located within the Taklimakan Desert region, the basin experiences a warm temperate, extremely arid desert climate (Gao and Li 2010). At the Tongguziluoke hydrological station at the river's outlet from the mountains, the multi-year average precipitation is about 94.2 mm (Yin 2002), while the evaporation is approximately 3093 mm, indicating an arid climate with scarce precipitation and intense evaporation (Huang et al. 2021). The surface runoff of the Yurungkash River primarily relies on meltwater from glaciers and snow, supplemented by some high-altitude precipitation (Luo et al. 2022). The intra-annual distribution of runoff within the basin is highly uneven, with runoff from March to May (spring) accounting for only 6.1% of the annual total, while the period from June to August constitutes 79.1% of the annual runoff volume.

Administratively, the basin spans Qira County, Hotan County, Hotan City, and Lop County. The population is predominantly concentrated in the plain oases of Hotan County and Hotan City, with the Uyghur ethnic group being the majority. The basin's regional production and development model is primarily based on irrigated agriculture and oasis economy. The industrial structure is dominated by agriculture, with characteristic forestry fruits (such as red dates and walnuts) and crop cultivation (such as cotton) serving as pillars of the economy. However, agricultural water use efficiency is relatively low, and the industrial foundation remains comparatively weak.

3 The Current Situation of the Water Resources Supply–demand Contradiction and Its Intensifying Trend

3.1 Analysis of the Water Resources Supply System

(1) Total Supply and Its Spatio-temporal Variation Characteristics

The multi-year average runoff of the Yurungkash River is approximately 2.44 billion m^3. According to statistics, influenced by climate change, the average annual runoff from 1960 to 2000 was 2.195 billion m^3. In the periods 2001–2010 (2.438 billion m^3) and 2011–2020 (2.688 billion m^3), the average annual runoff increased by 11.09% and 22.48%, respectively, compared to the 1960–2000 baseline (Zhou et al. 2024). Since 2001, the runoff has generally shown an increasing trend; however, the issue of uneven intra-annual distribution persists. Currently, insufficient engineering regulation and storage capacity further exacerbate the challenges of water resource utilization.

(2) Analysis of Supply Structure

Statistics show that in Xinjiang, surface water (including transferred water) accounts for 44.74 billion m^3, representing 81.39% of the total water supply, while groundwater supply is 10.06 billion m^3, accounting for only 18.30% (Yin and Zhang 2025). The water supply structure in the Yurungkash River Basin aligns with the overall pattern in Xinjiang, meaning the basin relies primarily on surface water, which constitutes over 90% of the supply. Groundwater is the secondary source, and localized over-extraction has already occurred in the middle and lower reaches of the basin. The utilization of recycled water is still in its infancy, with a very low proportion in the supply mix. The supply structure is singular, and the system lacks resilience.

Currently, a headworks is located about 19 km downstream of the river's exit from the mountains, responsible for supplying water to Lop County, the Yuhe Irrigation District in Hotan County, and the irrigation district of Hotan City. Additionally, the Yurungkash Water Control Project is under construction on the midstream section of the river, approximately 85 km upstream of the Yurungkash Headworks. This project is scheduled for completion in 2028. Furthermore, due to the low anti-seepage lining rate of the existing canal system, significant losses occur through seepage and evaporation during water conveyance. Coupled with the aging and poor maintenance of some water diversion structures, the overall water supply efficiency remains low.

(3) Assessment of Supply Reliability and Vulnerability

Climate change-induced glacier retreat and alterations in precipitation patterns cast doubt on the long-term reliability of runoff primarily recharged by glacial meltwater. At the same time, the frequency and intensity of extreme hydrological events (such as droughts and floods) are likely to increase, further heightening the vulnerability of the supply system.

3.2 Analysis of the Water Resources Supply System

(1) Analysis of Demand Structure

Agricultural irrigation is the dominant water user in the Yurungkash River Basin. The irrigated areas currently include the Yuhe Irrigation District in Hotan County, the Hotan City Irrigation District, and the Lop County Irrigation District, with a total irrigated area of approximately 127,000 hectares. The agricultural sector consistently accounts for over 90% of total water use. Spring is a critical period for crop growth; however, during this season, low temperatures result in reduced meltwater recharge from glaciers, leading to severe "spring droughts." The extreme spatial and temporal unevenness of water availability is a natural bottleneck for water resources management in the basin.

Regarding crop patterns, water-intensive crops such as cotton and corn occupy a significant proportion of the cultivated area. Traditional irrigation methods like flood irrigation are still widely practiced. The water efficiency coefficient for irrigation is substantially lower than advanced levels found elsewhere in China. The issue of "high water use with low output" is particularly pronounced.

The Green Corridor of the Yurungkash River is distributed along approximately 40 km of the riverbanks upstream of the confluence point at Koshilax, covering an area of about 215.74 km^2. It comprises dozens of plant species, including grey poplar, Populus euphratica, tamarisk, reed, and licorice. This corridor serves as a vital ecological pathway running north–south through the Taklimakan Desert within the Tarim Basin. Populus euphratica, as the dominant species in the desert vegetation, holds significant ecological value for improving and sustaining the regional environment (Ye et al. 2011; Zhang et al. 2017). Maintaining the ecological water requirement to sustain the vegetation of this Green Corridor is a fundamental baseline for ensuring regional ecological security.

However, with population growth and the expansion of irrigated areas in the basin, socioeconomic water use continuously encroaches upon ecological water allocations. As a result, the river channel downstream of the Yurungkash Headworks carries water only from June to September, remaining dry during the rest of the year. This has led to the degradation of natural forest and grassland vegetation within the Green Corridor, which relies on groundwater and floodwater for recharge, making the ecological crisis increasingly apparent.

(2) Future Demand Projections

With the advancement of the Belt and Road Initiative, urbanization and industrialization are expected to accelerate, exerting sustained pressure on water resources. Based on the socioeconomic development plans for the Hotan Prefecture, projections indicate that, under current engineering capacity and water use efficiency levels, the total water demand in the basin during relatively dry years may exceed the available water supply by 155 million m^3 by 2030. The supply–demand gap is expected to widen further.

3.3 Detrimental Impacts Stemming from the Supply–Demand Imbalance

The Yurungkash River Basin faces limited water resources. With ongoing socioeconomic development, rigid water demand will continue to grow, leading to total demand exceeding total supply. The agricultural sector dominates water use, and inefficient water consumption encroaches upon the shares allocated for ecological needs as well as future domestic and industrial water requirements for socioeconomic development, resulting in an irrational water use structure. Furthermore, the seasonal flow pattern of the Yurungkash River—characterized by abundance in summer and scarcity in winter—seriously misaligns with the water demand patterns of agricultural spring irrigation and year-round ecological needs. Spatially, water resources are generated in the upstream areas but consumed mainly in the middle and lower reaches, creating a mismatch between spatial distribution and demand patterns.

The above contradictions will trigger a series of severe negative effects, such as: degradation of the downstream Green Corridor and riverine ecosystems, along with loss of biodiversity; over-extraction of groundwater in the middle and lower reaches, causing a continuous decline in groundwater levels; meanwhile, intensifying competition for water resources increases the risk of water disputes among different regions and sectors, posing a potential threat to social stability.

4 Sustainable Management Strategies for Basin Water Resources

4.1 Regulation Principles and Objectives

Adhering to the fundamental guideline of "determining urban development, land use, population size, and production activities based on water availability," the carrying capacity of water resources shall serve as a rigid constraint on regional socioeconomic development. Guided by the water governance principle of "prioritizing water conservation, achieving spatial balance, adopting systematic governance, and leveraging both government and market forces," water conservation shall be placed at the forefront. This involves coordinating water use for domestic, production, and ecological purposes, harnessing the synergistic role of the government and the market, safeguarding ecological security, and supporting economic development.

4.2 A Quadripartite Strategic Framework for Regulation, Integrating Conservation, Development, Management, and Optimization

4.2.1 Strengthen Water Conservation to Enhance Efficiency

(1) Agricultural Water Conservation

Vigorously develop high-efficiency water-saving agricultural irrigation and continuously strengthen water conservation efforts in irrigation districts. Strictly control the irrigated area within the basin and actively develop water-saving irrigation agriculture. Promote agricultural industrialization to increase output per unit area and per unit of water used, thereby transitioning from extensive traditional irrigation agriculture to intensive modern irrigation agriculture. Implement scientific water allocation and management according to plans, and rigorously practice water conservation. In recent years, with strong promotion by the XPCC and local authorities in Xinjiang, the "drip irrigation under plastic film" technique has demonstrated significant benefits in increasing crop yield and farmer income year after year. The basin should, based on its actual conditions, vigorously develop efficient water-saving irrigation technologies such as drip irrigation, micro-irrigation, and drip irrigation under plastic film. Through rational planning, focus on promoting improved surface irrigation water-saving measures like "small border irrigation, shortening long borders, small-stream furrow irrigation, and furrow planting with furrow irrigation," as well as "on-film irrigation, side-film irrigation, and film-furrow irrigation," to vigorously develop water-saving irrigation and achieve both water savings and yield increases. Increase efforts in technological transformation for high-efficiency water-saving in the basin's irrigation districts, reduce irrigation quotas, and improve water resource use efficiency.

Implement canal lining for water conservation. Based on factors such as topography, climate, and soil conditions in the irrigation districts, rationally select lining types and materials. Simultaneously focus on increasing supply and reducing losses ("open source and reduce outflow"), improve canal system Supporting Facilities and carry out lining, reinforcement, and repair for water conservation, thereby increasing the comprehensive utilization coefficient of canal water.

(2) Urban Water Conservation

Reduce water loss from supply network leaks. Systematically replace old pipelines that have been in service for over 50 years or are made of outdated materials, using corrosion-resistant, long-lasting pipes such as PE pipes and ductile iron pipes.

Promote the use of water-saving appliances. Promote water-saving sanitary ware. Mandate the installation of sensor-activated faucets, delayed self-closing valves, automatic sensor flushometers, etc., in public places across counties and cities within the basin. Promote waterless or micro-water eco-friendly public toilets.

(3) Industrial Water Conservation

Implement water recycling and develop low-water-consumption industries. Actively promote recirculating cooling technology to increase the reuse rate of circulating water and reduce the intake of fresh water. Introduce and develop water-saving technologies and new processes, retrofit traditional production lines, and reduce water consumption per unit product. Strictly restrict or shut down high water-consumption and high-pollution enterprises. Vigorously develop low-water-consumption industries such as deep processing of characteristic forestry fruits, medicinal materials processing, and beverage manufacturing.

4.2.2 Science-Based Resource Development to Secure Water Supply

Enhance the utilization of unconventional water resources. Construct upgraded urban wastewater treatment plants and reclaimed water reuse pipeline networks. Treat wastewater to higher standards for use in urban landscaping, road cleaning, industrial cooling, and other purposes, thereby replacing fresh tap water.

Accelerate the construction of the Yurungkash Water Control Project. The Yurungkash Water Control Project is a pivotal regulatory structure on the Yurungkash River. Upon completion, it will enhance the basin's capacity to regulate water resources across time and space, improve water use conditions in the basin's irrigation districts, alleviate spring droughts, raise the basin's flood control standards, and promote sustainable socioeconomic development in ethnic minority border regions.

4.2.3 Implement Strict Control and Strengthen Supervision

Strictly implement the indicators of the "Three Red Lines" water resources management system. Strengthen the unified management of water resources in the basin, strictly control the total amount of socioeconomic water use, and ensure necessary and reasonable ecological water use. Adhere to determining land use based on water availability, avoid large-scale development activities in the basin, and guarantee that the total socioeconomic water use does not exceed the requirements of the "Three Red Lines" indicators. Effectively strengthen the management of water withdrawal in the basin's irrigation districts, take robust measures to enhance the management of water intakes at all diversion points, and strictly prohibit excessive water diversion. Establish a red line for water use efficiency control and resolutely curb water waste. Accelerate the establishment of water use efficiency indicator systems for various sectors within the basin, and strengthen water quota and planning management.

Establish and improve a water rights trading market. Clearly define the initial water rights for various counties, cities, sectors, and water users within the basin, and complete their registration. Issue water rights trading management measures to clarify the trading entities, scope, procedures, price formation mechanisms, and dispute resolution methods, ensuring that all transactions follow established rules. Establish a water rights trading center or platform to provide one-stop services for buyers and sellers, including information release, trade matching, and fund settlement, thereby reducing transaction costs.

Establish a responsibility and assessment system for water resources management. The head of the basin authority bears overall responsibility for the management and protection of water resources within the basin.

4.2.4 System-Wide Optimization and Coordinated Allocation

Groundwater recharge and replenishment from flood overflow during the wet season are key factors for the formation, growth, and reproduction of vegetation in the basin's Green Corridor (Aikeremu et al. 2022). Specifically, groundwater serves as the primary water source for the consumption of forest and grassland vegetation in the lower reaches of the Yurungkash River's Green Corridor. During the flood season, flooding plays a dual role: on one hand, it provides supplemental irrigation for vegetation in near-river areas; on the other hand, flood overflow creates essential conditions for the "seed dispersal" of the riparian forest dominant species, grey poplar and Populus euphratica, playing a vital role in their reproduction and regeneration. Utilizing the regulatory capacity of the Yurungkash Water Control Project, an artificial flood event with a return period of two years can be released from the Yurungkash Reservoir in August each year. This managed flood is designed to meet the growth and reproductive needs of the riparian forest and grassland vegetation in the Green Corridor downstream of the project, thereby safeguarding ecological security.

5 Conclusion

The core issue of water resources in the Yurungkash River Basin is a complex challenge characterized by insufficient total volume, structural imbalance, and temporal-spatial mismatch. Inefficient agricultural water use lies at the epicenter of this contradiction, while severe encroachment upon ecological water needs is its prominent consequence. Driven by the dual pressures of climate change and socioeconomic development, the future supply–demand gap for water resources is expected to widen further, posing long-term challenges to water security in the basin. This study provides a systematic analytical framework and viable implementation pathways for the sustainable management of water resources in the Yurungkash River Basin, offering valuable reference for similar inland river basins.

References

Aikeremu, A., Dong, Z., Aishajiang, A., et al.: Response of vegetation distribution and physiological indicators to moisture conditions in the lower reaches of the Hotan River. Forestry Environ. Sci. **38**(5), 77–87 (2022)

Gao, W., Li, W.: Analysis of hydrological characteristics in the Pishan River Basin, Xinjiang. Jilin Water Resources **30**(9), 44–47 (2010)

Huang, X., Chen, F., Zhao, Q., et al.: Assessment and combination analysis of high and low runoff in the Hotan River, Xinjiang. Arid Zone Res. **38**(6), 1570–1578 (2021)

Liu, X.: Simulation of Hydrological Effects of Climate and Land Use Change in the Upper Reaches of the Hotan River Based on SWAT Model. Qufu Normal University, Qufu (2019)

Luo, C., Chen, F., He, C., et al.: Applicability of CMADS for runoff simulation in the Yurungkash River. Arid Zone Res. **39**(4), 1090–1101 (2022)

Ye, M., Xu, H., Gong, J., et al.: Reasonable ecological water level for radial growth of *Populus euphratica* with different diameters at breast height. Scientia Geographica Sinica **31**(2), 172–177 (2011)

Yin, R.: Research on Water Resources Planning of the Hotan River Basin. Xi'an University of Technology, Xi'an (2002)

Yin, F., Zhang, F.: Strategies and suggestions for agricultural water saving and efficient water resource utilization in Xinjiang. J. Water Resour. Water Eng. **36**(2), 1–8 (2025)

Yu, Q., Zhang, J., Bai, Y., et al.: Runoff evolution characteristics of the headstreams of the Hotan River from 1957 to 2018. Arid Zone Res. **38**(2), 494–503 (2021)

Zhang, W., Liu, P., Feng, Q., et al.: Spatiotemporal response and reasons of growing season of *Populus euphratica* in Chinese oases to global warming during 1960–2015. Acta Geogr. Sin. **72**(7), 1151–1162 (2017)

Zhou, H., Yang, Y., Zhu, C., et al.: A preliminary study on the development and utilization potential of surface water resources at the county level on the northern slope of the Kunlun Mountains from the perspective of supply-demand balance. Arid Land Geogr. **47**(7), 1106–1114 (2024)

Study on the Determination of Ecological Flow for Small and Medium Rivers in Seasonally Water-Deficient Regions: A Case Study of the Qingjian River, a Primary Tributary of the Yellow River

Tian Shao, Hao Chen, Yinyin Zhu, Lijun Wang, Chaoyan Zhou, and Fan Wan

Abstract When formulating ecological flow standards, water authorities often overrely on hydrological methods and the 90% assurance rate standard, which frequently deviates from the actual ecological water demands of rivers. This issue is particularly pronounced in seasonally water-deficient small and medium rivers in northern China. Characterized by low flow volumes and significant differences between wet and dry periods, these rivers exhibit more sensitive and fragile aquatic ecosystems. The ecologically required flow based on current assessment standards often fails to meet the basic needs of these sensitive aquatic ecosystems, thereby exacerbating river ecological degradation. Taking the Qingjian River, a primary tributary of the Yellow River, as a case study, this paper integrates Hydrological Methods, the Tennant Method, and the Wetted Perimeter Method. Based on identifying key ecological requirements such as fish habitat and wetland water demand, the ecological flow for the main stem was recalculated. The results show that the recommended values determined by the comprehensive method are significantly higher than the current assessment standards and can better secure water supply during critical ecological stages. This study concludes and recommends that the determination of ecological flow for seasonally water-deficient rivers should abandon singular, rigid assessment standards and shift towards a dynamic strategy based on ecological integrity, considering the characteristics of both mainstream and tributaries, as well as operational feasibility. The findings provide both theoretical and practical foundations for optimizing the current ecological flow management system.

T. Shao (✉) · H. Chen · Y. Zhu · L. Wang · C. Zhou · F. Wan
Power China Northwest Engineering Corporation Limited, Xi'an, China
e-mail: shaotian@nwh.cn

T. Shao · F. Wan
Shaanxi Union Research Center of University and Enterprise for River and Lake Ecosystems Protection and Restoration, Xi'an, China

H. Xu (ed.), *Proceedings of the 8th International Symposium on Water Resource and Environmental Management*, Environmental Science and Engineering,
https://doi.org/10.1007/978-3-032-19088-8_13

Keywords Small and medium rivers · Seasonal flow · Environmental flows · Ecological water requirements · Environmental flow management

1 Introduction

Since the 18th National Congress of the Communist Party, under the guidance of Xi Jinping's Thought on Ecological Civilization, the safeguarding of river and lake ecological flow, a key element in building an ecological civilization, is assuming growing strategic significance. To implement the overarching design, China has successively established ecological flow targets for key control sections in major river basins, including the Yangtze, Yellow, Huaihe, Pearl Rivers, and Taihu Lake, through basin-wide plans and specific regulations at the national level. Under this guidance, provincial-level water resources authorities nationwide have also undertaken systematic efforts in safeguarding and evaluating ecological flows for key local rivers (Li et al. 2025).

In the determination of ecological flows, water authorities are often driven to set unduly low assessment targets in pursuit of higher water utilization efficiency and compliance with the 90% assurance rate requirement. This practice is particularly prevalent in seasonally water-deficient rivers of northern China, where the minimum flow recorded in extremely dry years is frequently adopted as the ecological flow standard. This established value, representing a very small fraction of the multi-year average river runoff, proves insufficient to meet in-stream ecological water demands and consequently exacerbates the ecological degradation of these seasonal waterways (Chen and Li 2025; Wang et al. 2025).

Taking the Qingjian River, a primary tributary of the Yellow River, as a case study, this research calculates instream ecological flow through multiple methods based on the needs of different water users and the basin's natural runoff characteristics. The findings provide a theoretical foundation for ecological flow management in the Qingjian River Basin, while the methodology offers valuable insights for determining ecological flows in other seasonally water-deficient rivers across northern China.

2 Study Area

2.1 Overview of the Study Area

The Qingjian River is a first-order tributary on the right bank of the middle reaches of the Yellow River. It spans two cities (Yan'an and Yulin) and six counties (districts), with a total main stream length of 175 km and a basin area of 4078 km^2. The river channel has an average gradient of 4.5‰, and the mean annual runoff is 150 million m^3. The basin is predominantly characterized by hilly and gully regions of the Loess

Plateau and river valley terraces. The multi-year average temperature is 10.6 °C, with extreme minimum and maximum temperatures recorded at − 22.5 and 41.5 °C. The multi-year average precipitation is 490.4 mm, while the multi-year average evaporation is 1228.6 mm. The river typically begins to freeze in late November and thaws in late February of the following year, with an average freezing duration of 63 days.

The river runoff in the basin is primarily recharged by precipitation. There are two hydrological stations along the main stream—Zichang and Yanchuan (II)—both established in the 1950s. Their controlled drainage areas account for 22.4 and 85.1% of the total basin area. Runoff distribution throughout the year is uneven. At the Yanchuan (II) Station, the runoff during the flood season (July to October) constitutes 65% of the annual total, with the peak month, August, alone contributing 22% of the yearly runoff volume.

The Zhongshanchuan Reservoir is located at the headwaters of the main stream. It became operational in 1977 and serves as a medium-sized reservoir with key functions of urban water supply and also with comprehensive benefits such as flood control. The controlled catchment area above the dam site is 143 km^2, with a multi-year average runoff of 5.7 million m^3. The reservoir has an active capacity of 31.23 million m^3, a flood control capacity of 12.24 million m^3, and a dead storage of 0.83 million m^3.

The Shaanxi Qingjian River Wetland, a designated important wetland in Shaanxi Province, encompasses the section from Wangjaya Village in Zhejiaping Town to the confluence of the Qingjian River with the Yellow River at Suyahe Village in Tugang Township, Yanchuan County. This area includes the river course itself, its floodplains, the riparian zone, and artificial wetlands within 1 km on both banks. An investigation at the confluence of the Qingjian River and the Yellow River documented the presence of key protected fish species in Shaanxi Province, including the Lanzhou catfish, northern snakehead, and common carp (in the Yellow River basin). Additionally, two small spawning grounds were identified in the middle and lower reaches of the main stream: the Lijiawan Village spawning ground and the spawning ground at the estuary into the Yellow River (Fig. 1 and Table 1).

2.2 Implementation Status of Ecological Flow in the Basin

In 2025, the Shaanxi Provincial Department of Water Resources issued the Implementation Plan for Ensuring Ecological Flow in the Yan River, Kuye River, Yiluo River, Xingzi River, Zhou River, Qingjian River, Huaining River, Qishui River, Weishui River, Daxi River, Ju River, Feng River, and Qing River (Trial). Two ecological flow control sections have been established in the main stream of the Qingjian River: the Majiaban Section and the Baijiajian Section, both of which are boundary sections between Yulin City and Yan'an City. Additionally, the Zhongshanchuan Reservoir, as a regulating reservoir in the main stream, has specified ecological flow requirements during the environmental impact assessment for its reinforcement. The

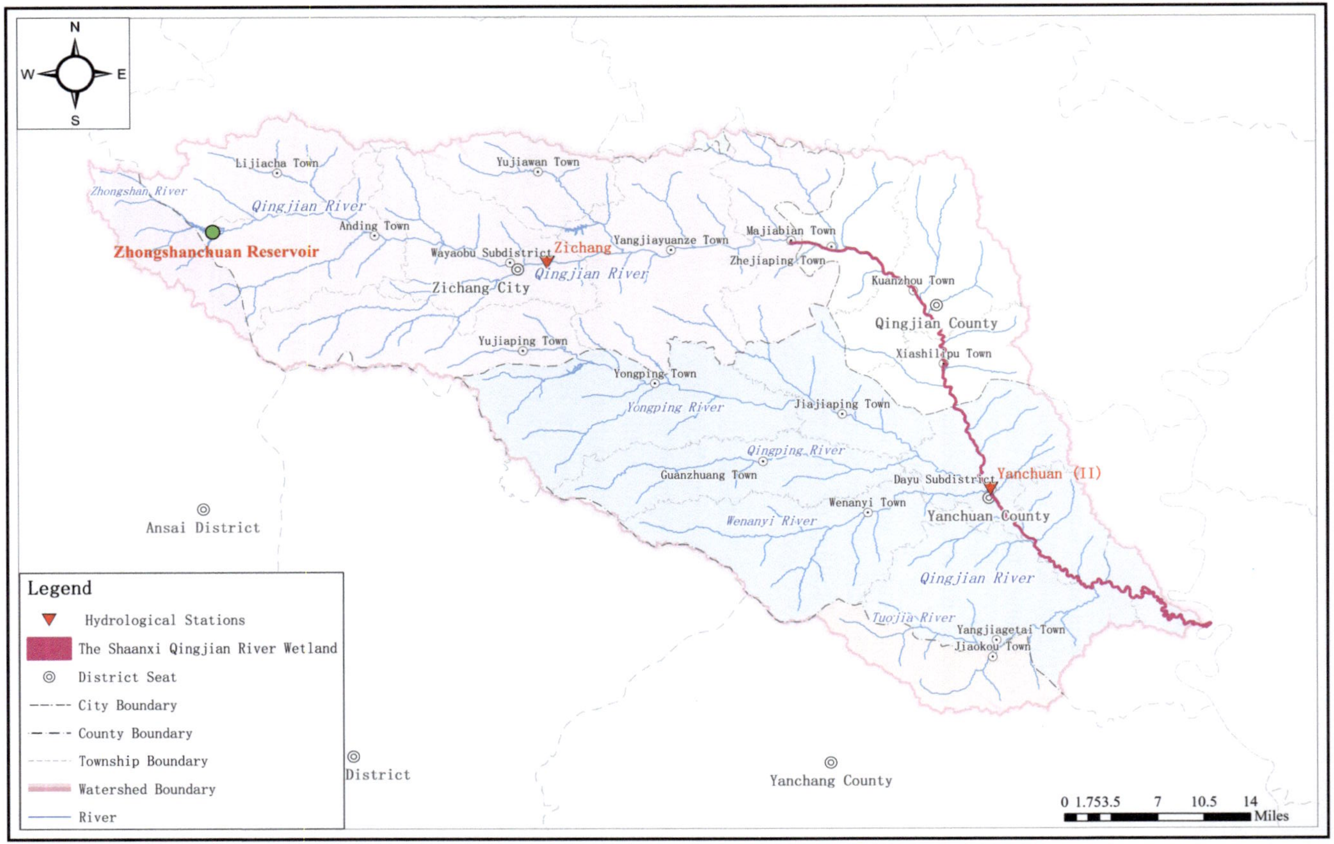

Fig. 1 Map of the River system and hydrometric station distribution in the Qingjian River Basin

Table 1 Hydrological stations in the Qingjian River Basin

Station	Location	Catchment area (km^2)	Mean annual runoff (km^2)	Observation data time period
Zichang	E109°42′N37°09′	913	3553	1959–2023
Yanchuan (II)	E110°11′N36°53′	3468	12912	1956–2023

ecological flow control indicators for the major control sections in the Qingjian River Basin are detailed in Table 2.

Currently, the main issues in the establishment and implementation of ecological flow at the three control sections in the basin are as follows:

(1) The ecological flow for the Zhongshanchuan Reservoir Dam Site Section, an engineering control section among the three control sections, was established by the environmental protection department, while the Majiaban and Baijiajian Sections were designated by the water resources department. The ecological flow requirements for these three sections vary significantly, with the values for the Majiaban and Baijiajian Sections even falling below 10% of the multi-year average flow at these sections.
(2) Different departments responsible for assessment and management

The ecological flow at the Zhongshanchuan Reservoir Dam Site Section is assessed and managed by the environmental protection department, whereas the Majiaban and Baijiajian Sections are overseen by the water resources department.

(3) Early reservoir projects on tributaries did not account for downstream ecological water needs

The Weijiacha Reservoir and Hanshashi Reservoir, both constructed in the 1970s, were built without considering the ecological water requirements of downstream river sections.

Table 2 Table of ecological flow control indicators for major control sections in the Qingjian River Basin

Section name	Section type	Ecological flow (m^3/s)	Percentage of flow at the corresponding cross-section (%)
Zhongshanchuan Reservoir Dam Site	Engineering control section	0.018	10
Majiaban Section	Municipal boundary section	0.090	5.5
Baijiajian Section		0.135	4.6

Table 3 Table of minimum ecological flow calculations (m^3/s)

Section name	Montana method	Q90	Q10	The minimum ecological flow
Zhongshanchuan Reservoir Dam Site	0.018	0.011	0.010	0.018
Majiaban Section	0.164	0.104	0.090	0.164
Baijiajian Section	0.295	0.154	0.135	0.295

3 Methodology

Due to significant variation in water availability between wet and dry periods in the basin, ecological flow studies were conducted considering different time phases. First, the Montana Method, Q90, and Q10 were applied to determine the minimum ecological flow requirements for the basin. Subsequently, ecological flow requirements for specific critical periods were defined based on the water demands of sensitive ecological targets (Yufei et al. 2023; Guo et al. 2025; Tennan't 1976; Ruiz et al. 2025; Wang et al. 2025; Gan and Jiang 2023; Chen et al. 2025; Fu et al. 2025).

According to field surveys of the surrounding environment, the water demands of sensitive ecological targets in the basin include the requirements of the Shaanxi Qingjian River Wetland and the water needs of fish species.

3.1 *Study on Minimum Ecological Flow*

The minimum ecological flow was determined using the Montana Method, Q90, and Q10 methods, respectively, and the calculation results are presented in Table 3. The maximum value obtained from these methods was adopted, with the minimum ecological flow for each control section set at 10% of the multi-year average runoff of the respective section.

3.2 *Water Demand of the Shaanxi Qingjian River Wetland*

The ecological water demand of the Shaanxi Qingjian River Wetland primarily considers the water requirements for wetland evapotranspiration, soil moisture, vegetation, biological habitats, and groundwater.

According to the data from the third national land survey in the basin, the total area of the Shaanxi Qingjian River Wetland is 3508.83 hectares, of which the river surface area accounts for 3381.40 hectares, and the inland beach area accounts for 127.43 hectares.

(1) Water Demand for Wetland Evapotranspiration

The primary consideration is the evapotranspiration volume in non-vegetated areas, calculated as follows (Xu et al. 2024; LYS 2024; Qiu et al. 2023):

$$dW_{P1} = A(t)ET_m \tag{1}$$

Among these, dW_{P1} represents the water demand of non-vegetated areas, A(t) denotes the area of non-vegetated wetland regions, ET_m is the evapotranspiration amount, and 't' stands for time. ETm is set as 1228.6 mm.

(2) Water Demand of the Soil

The soil water requirement is determined based on the field water capacity, calculated as follows:

$$Q_t = \alpha H_t A_t \tag{2}$$

Among these, Q_t represents the soil water requirement, α is the field water-holding capacity, H_t denotes the soil thickness, and A_t refers to the wetland soil area. α is set at 50%, and H_t is set at 50 cm.

(3) Water Demand of the Vegetation

The water demand of vegetation primarily considers the evapotranspiration volume in vegetated areas, calculated as follows:

$$dW_{P2}/d_t = A(t)ET_m \tag{3}$$

Among these, dW_{P2} represents the water demand of the vegetated area, A_t denotes the area of the vegetated wetland region, ET_m is the evapotranspiration amount, and 't' stands for time. ET_m is set as 1228.6 mm.

(4) Water Demand of the Biological Habitat

The Shaanxi Qingjian River Wetland serves as a critical habitat for birds and fish. The ecological water demand for biological habitats can be determined using the keystone species indicator method, calculated as follows:

$$dW_q/d_t = A(t)\%\ H_t \tag{4}$$

Among these, dW_q represents the water demand of the biological habitat, A_t denotes the area of the natural vegetated wetland zone, % indicates the percentage of water surface area, H_t is the water depth, and 't' refers to the water replacement cycle. During the main growing season of wetland vegetation, % is set at 80%, and in the non-growing season, it is set at 30%. H_t is set at 40 cm during the main growing season and 30 cm during the non-growing season.

(5) Water Demand of the Groundwater

Assuming the area recharging groundwater is the open water area of the wetland, the formula is as follows:

$$W_b = kIAT \tag{5}$$

Among these, W_b represents the groundwater recharge volume, 'k' is the permeability coefficient, 'I' denotes the hydraulic gradient, 'A' refers to the natural wetland area, and 't' stands for time. 'k' is set as 0.2 m/d, and I is 0.0048.

In northern Shaanxi, the growing season for wetland vegetation primarily spans from April to September. Therefore, the ecological water demand of the wetland is accounted for in two distinct periods, with the calculation results presented in Table 4.

Table 4 Calculation table of ecological water demand for the Shaanxi Qingjian River Wetland (10^4 m^3)

Water demand	Main growing season of wetland vegetation (April–September)	Other periods (October to March of the following year)
Water demand for wetland evapotranspiration	24.63	24.63
Water demand of the soil	54.80	54.80
Water demand of the vegetation	26.76	26.76
Water demand of the biological habitat	305.83	68.81
Water demand of the groundwater	22.39	22.26
Total	409.78	172.63

Based on the above calculations, during the main growing season of wetland vegetation (April–September), the water demand of the wetland is 4.0978 million m^3, equivalent to an ecological flow of 0.26 m^3/s, accounting for 15.9% of the multi-year average flow at the starting section of the Shaanxi Qingjian River Wetland, i.e., the Majiaban Section. During other periods (October–March of the following year), the wetland water demand is 1.497 million m^3, equivalent to an ecological flow of 0.11 m^3/s, accounting for 6.7% of the multi-year average flow at the Majiaban Section.

3.3 Water Demand of Fish

The R2-CROSS method and the wetted perimeter method were used to determine the ecological water demand for fish during sensitive periods.

(1) The R2-CROSS method

Based on field surveys and literature review, the ecological habits of fish in the study river reach were identified, including the species, body length, feeding habits, and reproductive behavior of protected fish, the distribution of "three critical habitats" (spawning, feeding, and overwintering grounds), and the flow regimes preferred by fish in the study reach. Subsequently, by investigating the current habitat conditions of fish in the study reach under natural conditions, the key hydraulic parameters influencing fish habitats were determined. Finally, the suitable hydraulic parameter standards for fish habitats in the study reach were established.

The R2-CROSS method assesses fish habitat conditions using hydraulic habitat parameters such as the average river depth, flow velocity, and wetted perimeter ratio. It can be used to calculate and analyze the ecological base flow of small rivers. The hydraulic habitat parameters are calculated as follows:

$$Q = AC\sqrt{RJ} \tag{6}$$

$$C = \frac{1}{n}R^{1/6} \tag{7}$$

Among these:

Q Flow rate (m^3/s);
A Cross-sectional area of flow (m^2);
R Hydraulic radius (m);
J Hydraulic gradient (m/m);
C Chezy coefficient ($m^{1/2}/s$);
n Roughness coefficient.

Calculations were conducted for three key sections: the Majiaban Section at the starting point of the Shaanxi Qingjian River Wetland, the Lijiawan Village Spawning Ground, and the Qingjian River Estuary Spawning Ground.

Calculation parameters: The downstream channel roughness coefficient (n) is set as 0.03, and the average slope (J) is 0.0018. The results are shown in Table 5. Based on this method, the ecological flow is calculated to be 9.7%.

(2) Wetted perimeter method

By establishing a flow–wetted perimeter relationship curve, the inflection point discharge is determined as the ecological flow for fish during the reproductive period. The sections selected remain the three ecologically significant sections mentioned earlier: the Majiaban Section, the Lijiawan Village Spawning Ground, and the

Table 5 Computed results of ecological flow in the Qingjian River Basin using the R2-CROSS method (flow rate: m^3/s)

Calculation parameters	Majiaban Section	Lijiawan village Spawning ground	Qingjian River Estuary Spawning ground
River width	5.8	6.7	11
Water depth	0.06	0.09	0.09
Cross-sectional area	0.359	0.627	1.014
Wetted perimeter	6.18	7.27	11.57
Hydraulic radius	0.06	0.09	0.09
Roughness coefficient	0.03	0.03	0.03
River gradient	0.0048	0.0048	0.0048
Chezy coefficient	20.74	22.16	22.22
Calculated minimum ecological flow value	0.12	0.28	0.46
Percentage of flow at the corresponding cross-section (%)	7.60	9.70	9.60

Qingjian River Estuary Spawning Ground. Flow rates were set at 5, 10, 15, 20, 25, 30, 35, 40, 45, and 50% of the multi-year average flow at each section (Chen et al. 2023; Shui et al. 2017), and the fitted points are illustrated in Fig. 2.

Based on the fitted curves, the inflection points were identified using the slope-of-1 method. The corresponding flow rates for the Majiaban Section, Lijiawan Village Spawning Ground, and Qingjian River Estuary Spawning Ground are 16.7%, 17.0%, and 16.3%, respectively. Adopting the highest value among the three sections, the recommended ecological flow during the reproductive period is set at 17%.

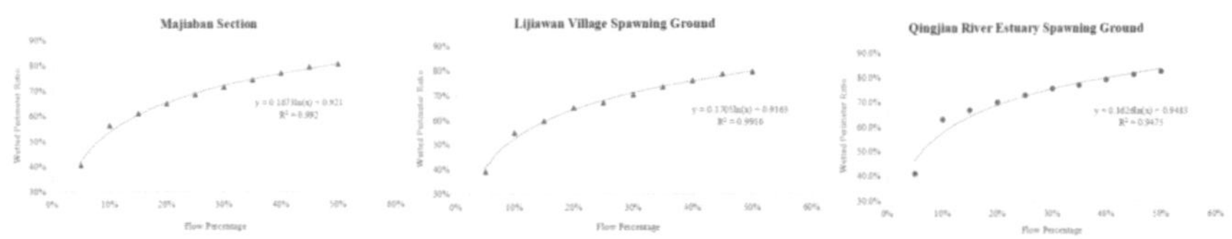

Fig. 2 Relationship curve diagram of wetted perimeter corresponding to different flow rates in the Qingjian River Basin

4 Results and Discussion

Based on the comprehensive analysis above, and taking into account the ecological water requirements of wetlands, the ecological water needs of fish during their reproductive period, and seasonal water availability characteristics, the recommended ecological flow strategy for the Qingjian River Basin is as follows:

During the high-flow period (April–September), which coincides with the fish reproductive period and the main growing season of wetland vegetation, ecological flow should be maintained at 10% of the multi-year average flow. During the low-flow period (October–March of the following year), ecological flow should be maintained at 17% of the multi-year average flow.

Additionally, as the Zhongshanchuan Reservoir serves water supply functions, when natural inflows are insufficient to meet the ecological flow requirements, the reservoir should release the available natural inflow without impounding water.

Currently, the monitoring sections assessed by the water resources department and the environmental protection department in the basin are not fully aligned. It is recommended to unify the assessment sections. The Zhongshanchuan Reservoir is both a key control project on the main stream and a national water quality monitoring section. Therefore, it is advisable to include it as an ecological flow assessment section, in addition to the existing Majiaban and Baijiajian assessment sections.

Currently, the water resource utilization rate in the Qingjian River Basin is only 13.55%. According to the Comprehensive Plan for the Qingjian River Basin, the surface water resource development and utilization rate is projected to increase to 16.55% by 2035. Given the relatively low intrinsic water resource development intensity in the basin, ensuring ecological flow should prioritize key regulatory structures such as the Zhongshanchuan Reservoir. Online flow monitoring facilities should be installed at both the inflow and outflow points of the reservoir to guarantee the release of adequate ecological flow.

In the middle and lower reaches of the river, where the catchment area increases and water resource development remains low, the issue of ecological flow being compromised by water resource development is less pronounced. However, emphasis should be placed on maintaining ecological flow during dry periods. In particular, the unregulated construction of check dams on tributaries may lead to reduced flow in the main stream, especially during dry seasons when ecological flow may not be guaranteed. Therefore, it is essential to rationally plan the placement of check dams, ensure the implementation of drainage facilities for these structures, and prioritize the use of permeable dam designs to prevent abrupt reductions in downstream river flow.

5 Conclusion

As a typical seasonal water-scarce river in northern China, the Qingjian River exhibits significant disparities between wet and dry season flows, making water resources difficult to utilize and resulting in a relatively low baseline level of water resource development. To meet the assessment requirement of ensuring ecological flow at a 90% assurance rate, water authorities have set relatively low ecological flow standards, which are insufficient to support the basic survival needs of fish and have exacerbated ecological degradation in river sections.

This study calculated the ecological water demand of the Qingjian River based on the water requirements of key wetlands, essential hydraulic characteristics for aquatic organisms, and the water needs of critical "three habitats" (spawning, feeding, and overwintering grounds). Considering the pronounced seasonal variability in river flow, a time-based ecological flow strategy was proposed. This approach aligns with the natural flow patterns of the basin while safeguarding the basic requirements of sensitive aquatic ecosystems, thereby promoting the healthy recovery of aquatic ecology in the watershed.

The findings of this study provide a theoretical foundation for river basin management authorities and offer valuable insights for establishing ecological flow standards in seasonal water-scarce rivers.

Based on the findings of this study, the following recommendations are proposed for the establishment and assessment of ecological flow in seasonal rivers in northern China:

(1) In the management of basin ecological flow, water resources and environmental protection departments are often involved. Due to differing mandates of the responsible authorities, discrepancies frequently arise in ecological flow assessments, such as inconsistent ecological flow standards and uncoordinated monitoring sections, leading to implementation challenges. It is recommended that relevant departments collaborate during the formulation of basin ecological flow requirements to jointly determine ecological assessment criteria.

(2) To meet the 90% assurance rate target for ecological flow, water resources departments often set relatively low ecological flow standards, failing to adequately consider the basic water requirements of river aquatic ecosystems. The original intent of the ecological flow assurance rate was to prevent other water uses from encroaching on ecological water needs. In basins like the Qingjian River, where significant encroachment on ecological water use is not a pressing issue, arbitrarily pursuing assurance rates by setting low ecological flow standards exacerbates ecological degradation in the basin, which runs counter to the original purpose of establishing ecological flow. Therefore, it is advised that the responsible authorities, when formulating ecological flow standards, take into account the characteristics of different basins, balance ecological water needs with basin development requirements, and establish reasonable ecological flow standards.

References

Chen, S.J., Li, Q.: Study on the ecological flow guarantee system for medium and small rivers in the yellow River Basin—a case study of the Zhouhe River Basin in Shaanxi Province. Groundwater **47**(04), 123–125 (2025). https://doi.org/10.19807/j.cnki.DXS.2025-04-038

Chen, T., Yu, W., Xingtao, F.: Research on the improvement of wetted perimeter method and application in seasonal rivers-a case study of Fenhe River, China. Environ. Sci. Pollut. Res. Int. **30**(37), 88009–88021 (2023). https://doi.org/10.1007/S11356-023-28331-0

Chen, X., Zhang, L., Tang, L.: A practical method for ecological flow calculation to support integrated ecological functions of the lower yellow river, China. Water **17**(15), 2326–2326 (2025). https://doi.org/10.3390/W17152326

Fu, Y., Liu, Z., Zhang, Z., et al.: Ecological flow study of the gorge section of Yongding River based on hydrological methods. River Res. Appl. **41**(8), 1658–1672 (2025). https://doi.org/10.1002/RRA.70013

Gan, Y.Y., Jiang, Y.M.: Assessment of ecological flow guarantee and impact attribution in the Huangpu River. J. Yangtze River Sci. Res. Inst. 1–9 (2023)

Guo, W., Wang, G., Hong, F., et al.: Evaluation of river ecological flow based on baseflow separation in Xiangjiang River, China. J. Water Clim. Change **16**(4), 1529–1550 (2025)

LYS: Critical Ecological Water Requirement Patterns and Utilizable Water Resources in Arid Regions. Northwest A&F University (2024). https://doi.org/10.27409/d.cnki.gxbnu.2024.001287

Li, Y., Yang, Z.K., Z.Q., et al.: Determination of Ecological Flow in Rivers and Lakes and Implementation Pathways for Its Guarantee, pp. 1–12. Yangtze River (2025)

Qiu, M.Q., Han, M., Jiao, C.T., et al.: Estimation of ecological water requirement in the Yellow River Delta wetland. Acta Ecol. Sin. **43**(21), 9096–9105 (2023). https://doi.org/10.20103/j.stxb.202210082850

Ruiz, M., et al.: Correction to "Diversity and eco-epidemiology of ectoparasites and *Rickettsia* spp. associated with the opossums Didelphis albiventris Lund in livestock farms from Argentinian Pampas region". Med. Veterin. Entomol. **3** (2025). https://doi.org/10.1111/MVE.70023

Shui, Y., Liu, J., Lou, Y.: Calculation of ecological flow in key sections of major rivers in the Huaihe River Basin using the wetted perimeter method. Huai River Manage. **11**, 35–36 (2017)

Tennan't, D.L.: Instream flow regimens for fish, wildlife, recreation and related environmental resources. Fisheries **1**(4), 6–10 (1976)

Wang, Z., Zhao, Y., Shang, J., et al.: Improved monthly frequency method based on copula functions for studying ecological flow in the Hailang River Basin, Northeast China. Atmosphere **16**(9), 1110–1110 (2025). https://doi.org/10.3390/ATMOS16091110

Wang, D.Y., Tang, X.Q., Wu, X.M., et al.: Analysis of ecological flow compliance at key control sections in the Yangtze River Basin and optimization recommendations. J. Yangtze River Sci. Res. Inst. 1–10 (2025)

Xu, H., Zhang, D., Wu, Y., et al.: Suitable ecological water demand for wetlands restored to different historical periods in a latitude area and their response to changing environments. Water Resour. Manage **38**(14), 5683–5700 (2024). https://doi.org/10.1007/S11269-024-03929-2

Yufei, J., Jia, L., Chuanzhe, L., et al.: Refined calculation of multi-objective ecological flow in rivers, North China. Water **15**(5), 1003–1003 (2023). https://doi.org/10.3390/W15051003

Research on Ecological Response and Carbon Budget Balance Under Microclimate Regulation of Photovoltaic Gobi Base in Hainan Prefecture, Qinghai Province

Yaning Liu, Quan Quan, Deli Ye, Xiaomei Kou, and Qi Li

Abstract To clarify the impact of photovoltaic construction in arid areas on the fragile Gobi ecosystem, taking the photovoltaic Gobi base in Hainan Prefecture, Qinghai Province as the research object, based on continuous observation data from February 2023 to February 2024, combined with random forest models and other methods, this study explores the microclimate regulation and the response of carbon and water fluxes as well as the driving mechanism of the carbon cycle. The results show that photovoltaic panels regulate soil temperature in a "peak shaving and valley filling" manner, increasing the moisture content of shallow soil during the growing season. The ecosystem has a stable carbon sink function. NEE has the strongest carbon absorption in August, Re shows a single-peak pattern, and GPP has a "U" shaped pattern. PAR and Rn are the core driving factors of carbon and Carbon–water flux. The multiple linear regression model can explain 64.37% of the NEE variation, providing support for the ecological sustainable management of photovoltaic power stations in arid areas.

Keywords Photovoltaic Gobi base · Microclimate regulation · Random forest model · Carbon and water flux · Carbon sink function · Carbon cycle driving mechanism

---

Y. Liu · Q. Quan (✉)
State Key Laboratory of Water Engineering Ecology and Environment in Arid Area, Xi'an University of Technology, Xi'an, China
e-mail: qq@xaut.edu.cn

D. Ye
Huanghe Hydropower Development Co, Ltd., Xining, China

X. Kou · Q. Li
Northwest Engineering Corporation Limited, Xi'an, China

H. Xu (ed.), *Proceedings of the 8th International Symposium on Water Resource and Environmental Management*, Environmental Science and Engineering,
https://doi.org/10.1007/978-3-032-19088-8_14

1 Introduction

Against the backdrop of global climate change, deteriorating ecological environment, increasing energy demand and shortage of fossil energy, the development of clean and pollution-free new energy sources has become an important way for sustainable development (Armstrong et al. 2020). In 2020, the net increment of renewable energy power generation surpassed that of fossil fuels (Al-Shetwi 2022). Wind and solar power generation are low-cost, easily accessible and have huge development potential. Nowadays, they have become an important choice for the world to move away from fossil fuels and towards clean energy. The utilization of wind and solar resources, as the core content of energy transition and an important way to address climate change, will continue to maintain a high proportion of development in the future (Zhu et al. 2018). Over the past two decades, the global wind and solar power generation industry has developed rapidly, with installed capacity continuously increasing. Wind power has risen from 7.5 GW in 1997 to 1021 GW in 2023, while photovoltaic power has grown from 6.6 GW in 2008 to 1411 GW in 2023 (International Renewable Energy Agency 2024). At present, wind and solar power have become an important means for both domestic and international parties to build a green and low-carbon clean energy system and achieve the "dual carbon" goals (carbon peak and carbon neutrality) (Huang and Xie 2021; Yang et al., 2023). The construction of photovoltaic power stations can modify the local surface albedo, temperature and humidity, and thereby have an impact on the local climate. The construction of photovoltaic power stations can modify the local surface albedo, temperature and humidity, and thereby have an impact on the local climate. Previous studies have found that the increase in the albedo of photovoltaic panels leads to changes in radiative forcing. After replacing fossil fuels, the radiative forcing is reduced by approximately 30 times (Nemet 2009). This discovery indicates that photovoltaic power generation has a significant climate regulation effect compared with traditional energy power generation. Zhang et al. (2023) studied the impact of solar photovoltaic power stations on broader ecosystem processes, such as changes in carbon flux balance and soil moisture retention capacity in farmland and grassland ecosystems, which led to changes in ecosystem services across spatial scales. The construction of photovoltaic power stations not only affects the local ecosystem functions but also may have an impact on biodiversity. The deployment of photovoltaic arrays alters the physical properties of the land and may affect the habitat of plants and animals. Photovoltaic panels can affect air humidity and soil moisture content by regulating the amount of radiation received by plants during photosynthesis (Wang et al. 2024), while significantly reducing wind speed and turbulence (Armstrong et al. 2016). These effects further influence plant communities, altering plant biomass, species richness and vegetation coverage (Zhai et al. 2020). Professor Zhao Pengjun's team, through geospatial data analysis, evaluated the potential conflicts between the development of photovoltaic power and the global conservation of endangered birds, and revealed the mechanism by which the development of renewable energy affects global biodiversity (Zhang et al. 2024). At present, research both at home and abroad on the impact of photovoltaic power stations on the microclimate environment has not yet reached a consistent conclusion. Foreign research mainly focuses on the impact of photovoltaic power stations on climate, ecosystems and biodiversity, while domestic research mostly concentrates on the ecological effects of photovoltaic power stations in regions such as Gobi (Deng et al. 2015), deserts (Gao et al. 2016; Hua et al. 2022). Studies on high-altitude and special ecological areas are still relatively scarce. In the future, research in this area should be strengthened, especially in ecologically sensitive regions such as the Qinghai-Xizang Plateau In the affected areas, further assess the long-term impact of photovoltaic power stations on the ecological environment.

This study takes the typical photovoltaic Gobi base in Hainan Prefecture, Qinghai Province as the research object. Based on the continuous observation data of one year (February 2023–February 2024), it aims to: reveal the regulatory effects and spatio-temporal differentiation characteristics of photovoltaic panels on microclimate elements such as soil temperature and soil moisture content at different levels in the Gobi area. Clarify the dynamic laws of carbon and water fluxes such as evapotranspiration (ET) and net ecosystem carbon exchange (NEE) under the influence of photovoltaics. The driving contributions and action paths of key environmental factors to ET and NEE were quantified through methods such as random forest model, path analysis and multiple linear regression. The research results are expected to deepen the understanding of the ecological and environmental effects of photovoltaic power stations, providing theoretical basis and data support for the sustainable development and ecological adaptability management of the photovoltaic industry in arid areas.

2 Overview of the Study Area and Research Methods

2.1 Overview of the Study Area

The annual average temperature is 3.2 °C, and the annual precipitation is 310 mm. The precipitation is concentrated from June to September. The underlying surface is mainly composed of Gobi gravel soil, with a soil bulk density of 1.6 to 1.8 g cm^{-3} and a field water holding capacity of 18–22%. The vegetation is mainly composed of camel hair grass and needle grass, with a coverage rate of less than 3% (Fig. 1).

2.2 Research Methods

2.2.1 Correlation Analysis

The linear relationship between NEE and environmental factors was quantified using the Pearson correlation coefficient (Kim and Lee 2025), and the formula is as follows:

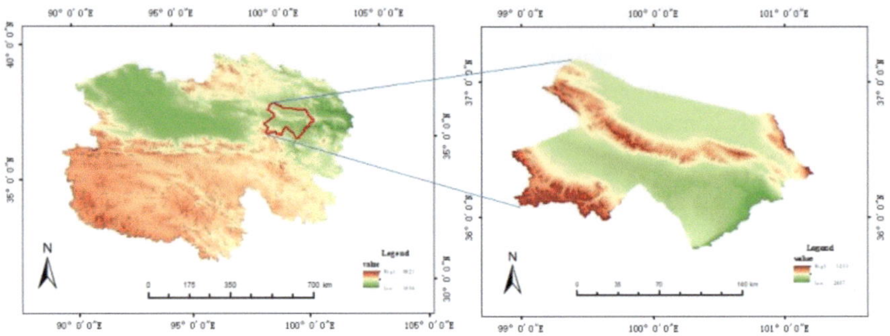

Fig. 1 Overview map of the study area

$$r = \frac{\sum_{i=1}^{n}(x_i - \overline{x})(y_i - \overline{y})}{\sqrt{\sum_{i=1}^{n}(x_i - \overline{x})^2 \sum_{i=1}^{n}(y_i - \overline{y})^2}} \tag{1}$$

In the formula: x_i is the i-th observed value of the environmental factor; y_i is the i-th observed value of NEE. $\overline{x}$ and $\overline{y}$ are the mean values of the environmental factor and NEE, respectively. n is the sample size (n = 17,236 in this study).

The significance levels were set as: $p < 0.001$, $p < 0.01$, $p < 0.05$, and ns (not significant).

2.2.2 Path Analysis

Based on the multiple linear regression model, the direct and indirect effects of environmental factors on NEE are decomposed. The core formulas are as follows:

Direct path coefficient (P), Standardized partial regression coefficient, reflecting the intensity of the direct influence of the factor on NEE. The formula is:

$$P_{x \to y} = \frac{\beta_x \sigma_x}{\sigma_y} \tag{2}$$

In the formula: β_x is the partial regression coefficient of the environmental factor x; σ_x and σ_y are the standard deviations of x and NEE, respectively.

Indirect path coefficient (I),The indirect effect of factor x on NEE through factor z, with the formula:

$$I_{x \to z \to y} = r_{x,z} P_{z \to y} \tag{3}$$

Total effect (T),The sum of the direct effect and all indirect effects:

$$T_x = P_{x \to y} + \sum I_{x \to z \to y} \tag{4}$$

The reliability of the results is verified by "total effect ≈ correlation coefficient" (an error less than 0.01 is considered reasonable).

2.2.3 Random Forest Model

Construct a random forest regression model for ET and NEE, identify the dominant driving factors and evaluate the prediction accuracy.

Model parameters: Number of decision trees n_estimators = 100, random seed random_state = 42, parallel computing n_jobs = − 1.

Feature importance: Calculated using the Gini coefficient, the formula is:

$$\mathrm{Gini}(S) = 1 - \sum_{i=1}^{c} p_i^2 \tag{5}$$

In the formula: S represents the node sample set; c represents the number of categories; p_i represents the proportion of samples of type i. The feature importance is the average of the reduction in node impurity of this feature in all decision trees.

Model accuracy evaluation: Three indicators are adopted.

Coefficient of determination (R^2):

$$R^2 = 1 - \frac{\sum_{i=1}^{n}(y_i - \hat{y}_i)^2}{\sum_{i=1}^{n}(y_i - \overline{y})^2} \tag{6}$$

Root mean square error (RMSE):

$$RMSE = \sqrt{\frac{1}{n}\sum_{i=1}^{n}(y_i - \hat{y}_i)^2} \tag{7}$$

Lin's harmonic coefficient (LCCC):

$$LCCC = \frac{2r\sigma_y\sigma_{\hat{y}}}{\sigma_y^2 + \sigma_{\hat{y}}^2 + (\overline{y} - \hat{y})^2} \tag{8}$$

In the formula: y_i represents the predicted value; y and σ_y are the mean and standard deviation of the predicted values, respectively. r is the correlation coefficient between the observed value and the predicted value.

Data processing is conducted using Pycharm, and the dependent libraries include pandas, numpy, scipy (correlation/path analysis), scikit-learn (Random Forest), and matplotlib (visualization).

3 Microclimate and Energy Flux Characteristics

3.1 Microclimate Regulation Effects

It can be directly observed from the figure that the installation of solar panels has changed the thermal environment of the soil beneath them, making the temperature change curve more gentle and presenting a stable effect of "peak shaving and valley filling". During the warm season (April to October), the soil temperature at depths of 10 and 40 cm under the solar panels is lower than that of the same layer in the no-panel area, demonstrating a significant shading and cooling effect in summer. During the cold season (from November to February of the following year), the soil temperature curve under the slab tends to be close to or even slightly higher than that in the no-slab area, indicating that it plays a buffering and maintaining role in the low temperature of the soil.

From the above phenomena, it can be inferred that the installation of solar panels has a significant regulatory effect on the surface microclimate, creating a milder soil thermal environment. This change has a significant impact on the local ecological environment, especially the carbon budget. Soil temperature is a key factor in controlling the activity of soil microorganisms and the rate of organic matter decomposition. A decrease in temperature is likely to directly inhibit soil respiration, reduce carbon dioxide emissions, and thereby have a positive carbon sequestration effect on the local carbon budget. Therefore, the pattern revealed in this diagram indicates that wind and solar power generation facilities not only provide clean energy but may also become a potential "carbon sink" by altering the physical environment of the underlying surface. However, such temperature changes may also have a chain reaction on the habitat of soil animals, the germination of plant seeds and the growth of root systems. Therefore, when assessing its comprehensive ecological effects, a more comprehensive study needs to be conducted in combination with data such as soil moisture, vegetation and carbon fluxes (Fig. 2).

From the trend chart of soil moisture content at different depths from February 25, 2023 to February 25, 2024, it can be seen that the soil moisture content at the corresponding depth of the photovoltaic panel coverage area is generally higher than that of the unobstructed area under the tower. Among them, the 10 cm shallow soil layer is particularly affected by the photovoltaic panels. For instance, after May 2023, when the growing season began, the moisture content of the 10 cm soil layer beneath the photovoltaic panels increased much more than that of the 10 cm soil layer beneath the tower, and the overall fluctuation was also more obvious. The moisture content of the 40 cm deep soil layer not only fluctuates more gently overall than that of the shallow soil layer, but also the difference in moisture content between the soil layer beneath the photovoltaic panel and the tower is smaller than that of the shallow soil layer. The core reason for this phenomenon lies in the fact that the shading of photovoltaic panels reduces the evaporation of water from the surface soil, making the shallow soil layer more capable of retaining water. As a result, the increase in moisture content is more significant and the fluctuation with environmental changes is more

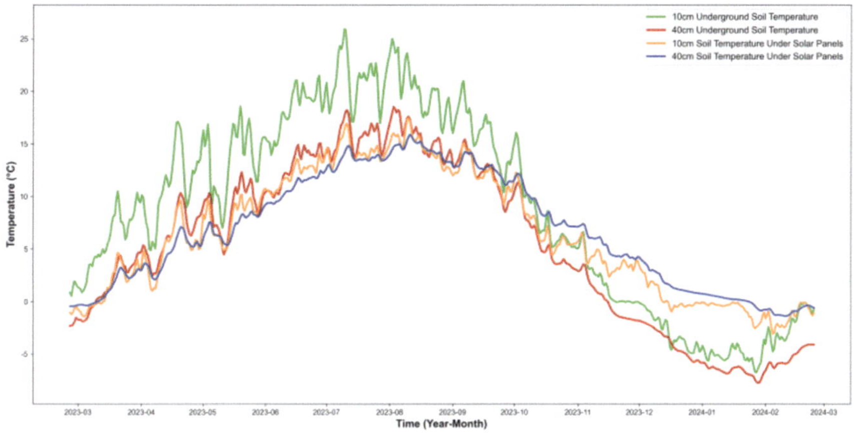

Fig. 2 The fluctuation trend of temperature over time at different positions

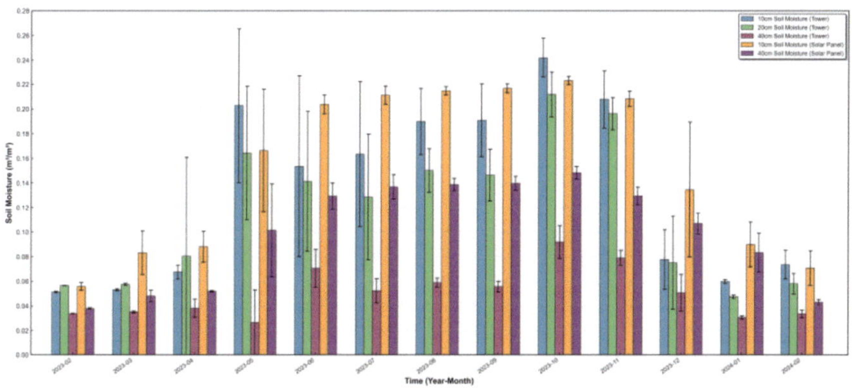

Fig. 3 Water content variations at different depths and locations

prominent. However, the 40 cm deep soil layer is less directly affected by surface evaporation, and its moisture state is relatively stable. Therefore, the regulatory effect of photovoltaic panels on its moisture content is even weaker (Fig. 3).

3.2 *Dynamic Characteristics of Energy Flux*

The soil heat flux and energy trend charts from February 25, 2023 to February 25, 2024 clearly show that the soil heat flux (including 5 and 10 cm underground heat fluxes and the surface soil heat storage capacity) and the surface soil energy flux exhibit a highly synchronized seasonal rhythm. The soil thermal process and energy dynamics are all in the high value range in summer, while they significantly drop to the trough value in winter, confirming that the soil thermal process and energy dynamics are co-driven by temperature and solar radiation at the seasonal scale. Meanwhile, there are obvious vertical stratified response differences in underground heat flux. The fluctuation range of heat flux in the 10 cm layer is slightly greater than that in the 5 cm layer, and the occurrence times of peak and valley values in both are basically the same. It indicates that the vertical transmission of soil thermal signals is synchronous, while the thermal response intensity of deep soil is relatively higher. The heat storage capacity of the surface soil demonstrates extremely strong stability, with fluctuations throughout the year remaining within a narrow range. This forms a sharp contrast with the intense seasonal fluctuations in underground heat flux and surface energy flux, highlighting its regulatory function in buffering seasonal heat fluctuations and maintaining its own thermal state stability in soil-atmosphere heat exchange. It is worth noting that the energy flux of the surface soil is the most sensitive element among all the indicators, and its fluctuation range throughout the year is much greater than that of other indicators, indicating that the surface soil is the core interface for heat-energy exchange between the soil system and the atmosphere, and it responds most directly to changes in environmental thermal factors. These results not only clearly clarify the seasonal coupling laws of soil heat and energy dynamics, but also define the functional differentiation characteristics of different soil layers and different heat indicators (Fig. 4).

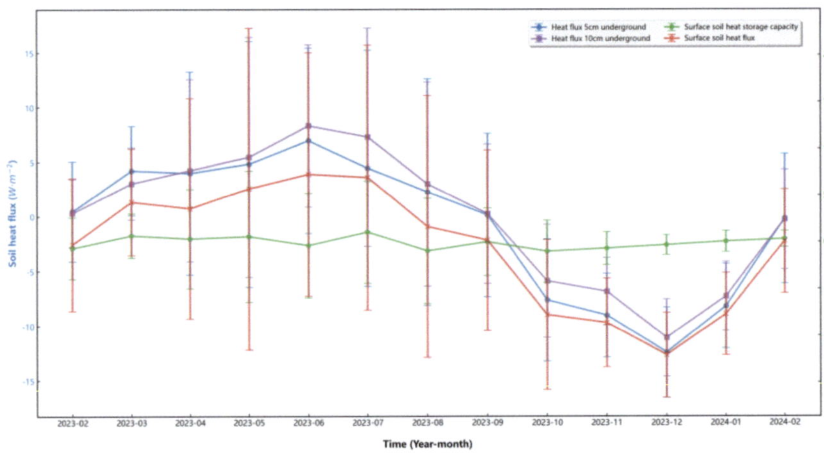

Fig. 4 Heat flux (at different depths) and energy variation trends

4 Carbon Flux Dynamics and Driving Mechanisms

4.1 Spatio-Temporal Dynamics of Carbon Fluxes

Based on continuous observations from February 2023 to February 2024, the carbon flux of this ecosystem shows significant seasonal differentiation and intrinsic coupling patterns. Ecosystem respiration (Re) exhibits a typical temperature-driven unimodal pattern, reaching a peak (0.11 mg CO_2 m^{-2} s^{-1}) in July of summer and a trough (0.01 mg CO_2 m^{-2} s^{-1}) in winter. Its dynamics are highly synchronous with temperature changes. The total primary productivity (GPP) shows a "U" -shaped seasonal pattern, with the first peak (approximately − 0.145 mg CO_2 m^{-2} s^{-1}) formed in May of spring. However, in July and August of summer, due to potential high temperatures or water stress, significant photosynthetic inhibition occurs, dropping to the lowest point of the year (approximately − 0.15 CO_2 m^{-2} s^{-1}). Autumn saw a significant recovery as environmental stress eased. Net ecosystem carbon exchange (NEE) remained negative in the vast majority of months, especially reaching the strongest carbon absorption intensity (approximately − 0.06 mg CO_2 m^{-2} s^{-1}) in August, indicating that despite the suppression of photosynthesis and the enhancement of respiration in summer, the ecosystem still maintained a stable net carbon sink function. In winter, all fluxes approach zero, and the system enters a state of carbon budget balance. Overall, this ecosystem demonstrated a continuous carbon sink during the observation period, with a clear seasonal rhythm of the carbon cycle, jointly regulated by temperature, moisture and phenology, reflecting the dynamic game and seasonal adaptation strategy of photosynthesis and respiration processes (Fig. 5).

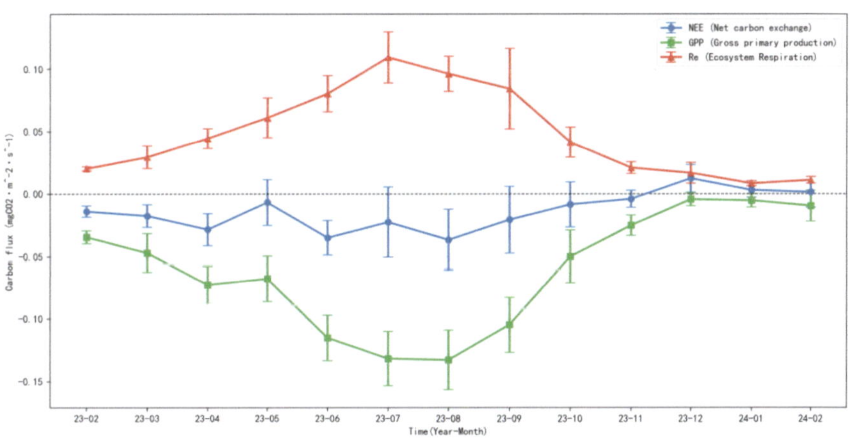

Fig. 5 Monthly seasonal variations of carbon fluxes

4.2 Establishment of Multiple Linear Regression Model

To quantify the comprehensive impact of multiple environmental factors and their interactions on NEE, this study constructed a multiple linear regression model. As can be seen from the residual distribution graph of the model, the residual distribution is close to the normal distribution, which is in line with one of the core assumptions of the multiple linear regression model, indicating that the statistical fitting basis of this model is robust. The normality characteristics of the residuals directly support the reliability of the model, meaning that the statistical inference results of the NEE (Net Ecosystem Carbon Exchange) impact analysis based on this model have strong rationality and credibility (Fig. 6).

From the relative influence intensity graph of variables, it can be seen that there are significant differences in the regulatory effects of different wind-solar ecological variables on NEE: photosynthetically active radiation is the dominant factor affecting NEE, and its standardized regression coefficient is negative, indicating that this factor will significantly promote carbon absorption in the ecosystem. Variables such as soil water under the photovoltaic panels and air temperature at 15 m also have varying degrees of influence. Some interaction terms (such as light × soil water under the photovoltaic panels) are also involved in the carbon absorption or release process of NEE, clearly presenting the coordinated regulation pattern of multiple variables on NEE. The basis of the analysis can be further supplemented by combining descriptive statistical tables: the values of each variable cover a relatively wide range (for example, the maximum value of photosynthetically active radiation reaches 2020.17), and the fluctuation range of some variables is relatively large (for

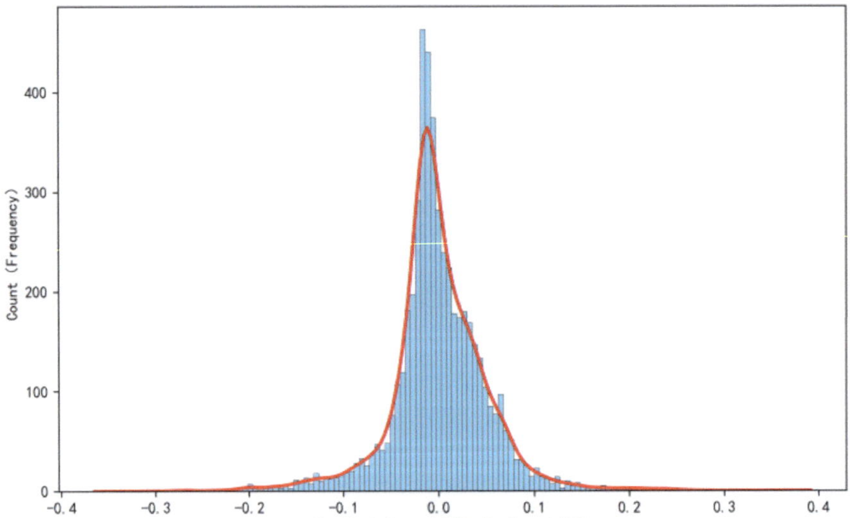

Fig. 6 Model residual distribution

example, the standard deviation of air temperature at 15 m is 9.36). This data feature not only reflects the coverage of the sample to the actual environment, but also explains the potential reasons for the differences in the intensity of variable influence (such as the high fluctuation of photosynthetically active radiation making its regulatory role more prominent), further solidifying the practical representativeness of the analysis results of this time (Table 1; Fig. 7).

The specific equation is:

$$\begin{aligned} NEE = -0.01 - 0.0658\,Par - 0.0034\,Ws - 0.0049\,VWC + 0.0085\,Rh \\ - 0.0005\,Pr - 0.0052\,Ts - 0.0195\,PV - 0.0030\,PW \end{aligned} \tag{9}$$

Among them, *Par* represents photosynthetically active radiation, *Ws* represents wind speed at 2 m, *VWC* represents the volumetric water content of the soil 10 cm

Table 1 Basic statistical characteristics of carbon flux (NEE) and micro-meteorological indicators

Element	Mean	Standard deviation	Min	Max
NEE	– 0.015	0.087	– 0.339	0.31
PAR	389.209	545.111	0	2020.17
2 m Wind speed	2.185	1.788	0	34.321
VWC_panel_10 cm	0.159	0.064	0	0.231
15 m Air temperature	4.804	9.36	– 20.97	27.88
Precipitation	0.006	0.049	0	0.7
Ts_tower_10 cm	8.73	9.553	– 9.359	31.664

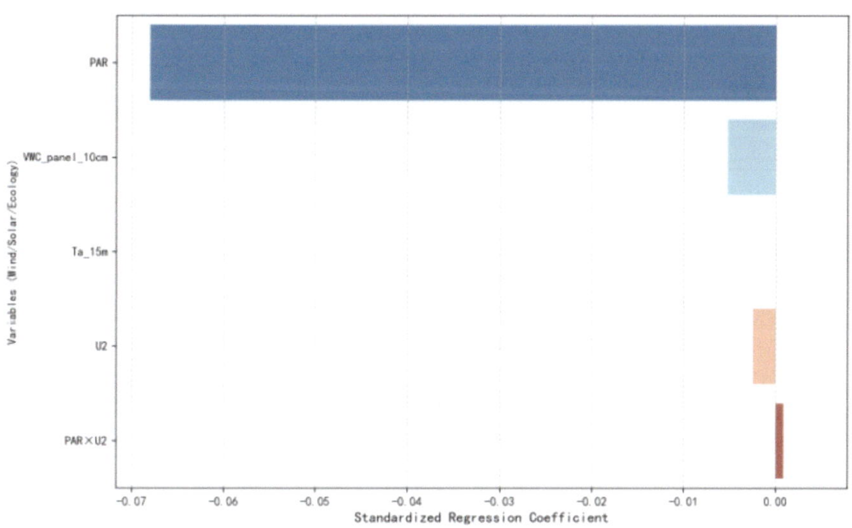

Fig. 7 The relative influence of each wind and solar variable on the carbon budget

underground under the solar panel, *Rh* represents air temperature at 15 m, *Pr* represents precipitation, *Ts* represents soil temperature 10 cm underground under the tower. *PV* represents the interaction term between photosynthetically active radiation and the volumetric water content of the soil 10 cm underground under the solar panel, and *PW* represents the interaction term between photosynthetically active radiation and 2 m wind speed. In this equation, a positive coefficient represents the promotion of carbon release, and a negative coefficient represents the promotion of carbon absorption. The model is constructed based on 17,236 samples, with an adjusted R^2 of 0.6437, an F-statistic of 3892.83, and a *p*-value of 0. This indicates that the model as a whole is extremely significant, capable of explaining 64.37% of the NEE variation, and has a good fitting effect (Tables 2 and 3).

Table 2 Parameters and confidence interval results of the NEE-related factor regression model

	Coefficient	Standard error	t value	*p* value	95% Lower limit of confidence interval	95% Upper limit of confidence interval
Intercept	– 0.0136	0.0004	– 33.5535	0	– 0.0144	– 0.0128
PAR	– 0.0658	0.0005	– 134.5477	0	– 0.0668	– 0.0648
2 m Wind speed	– 0.0034	0.0004	– 8.0705	0	– 0.0042	– 0.0025
VWC_panel_10 cm	– 0.0049	0.0005	– 9.8189	0	– 0.0059	– 0.0039
15 m Air temperature	0.0085	0.0013	6.6874	0	0.006	0.011
Precipitation	– 0.0005	0.0004	– 1.2434	0.2137	– 0.0013	0.0003
Ts_tower_10 cm	– 0.0052	0.0012	– 4.4466	0	– 0.0076	– 0.0029
PAR × VWC_panel_10 cm	– 0.0195	0.0004	– 45.9556	0	– 0.0203	– 0.0186
PAR × 2 m Wind speed	– 0.003	0.0005	– 6.503	0	– 0.0039	– 0.0021

Table 3 Goodness-of-fit of the NEE model and key parameters for statistical testing

Indicator	Value
R^2	0.643847
Adjustment R^2	0.643682
F Statistics	3892.832755
F The statistic *p* value	0
Sample size	17,236

4.3 Carbon Flux-Driven Mechanism

4.3.1 Correlation Between NEE and Environmental Factors

Based on the correlation analysis of NEE (Net Ecosystem Carbon Exchange) and environmental factors from the 2023 observation data, we have found the following results.

First, it has been clarified that net radiation, photosynthetically active radiation, air temperature and humidity, and soil water and heat conditions are the core direct factors driving the changes in carbon flux in the Gobi photovoltaic area. The analysis results show that NEE is associated with net radiation ($r = -0.713, p < 0.001$), photosynthetically active radiation ($r = -0.7105$, $p < 0.001$), air humidity at 15 m ($r = 0.327, p < 0.001$), and air temperature at 15 m ($r = -0.2876, p < 0.001$), soil temperature 10 cm below the tower ($r = -0.1672$, $p < 0.001$), and soil moisture content 10 cm below the tower ($r = 0.1216$, $p < 0.001$) were all significantly correlated, but the correlation with wind speed 2 m was not significant ($r = 0.0102$, $p > 0.05$). This discovery reveals that in the Gobi ecosystem with extremely low vegetation coverage, the carbon–water coupling process is not limited to the subsurface. The radiation, temperature and humidity conditions of the near-surface atmosphere also participate in regulating the carbon sink or carbon source function of the ecosystem. It is particularly worth noting that soil moisture content is significantly positively correlated with NEE, which is contrary to the general rule in humid areas that "water promotes photosynthesis and enhances carbon sinks". This indicates that in the Gobi region, an increase in water may preferentially stimulate the respiration of soil microorganisms, thereby making the ecosystem tend towards a carbon source in the short term (Table 4).

Second, it reveals a strong collinearity among key environmental factors, among which soil temperature serves as a crucial hub connecting atmospheric and underground processes. Further heat map analysis revealed that there was a very strong positive correlation between the soil temperature 10 cm below the tower and the air temperature 15 m below ($r = 0.94$), and at the same time, the net radiation was

Table 4 The correlation and significance results between environmental factors and NEE

Serial number	Environmental factors	Correlation coefficient (r)	*p* value	Significance
1	Net Radiation	− 0.713	0	***
2	PAR	− 0.7105	0	***
3	15 m Air Humidity	0.327	0	***
4	Ta_15 m	− 0.2876	0	***
5	Ts_tower_10 cm	− 0.1672	0	***
6	VWC_tower_10 cm	0.1216	0.000106	***
7	2 m wind speed	0.0102	0.745042	ns

***Show significant differences

highly synchronized with the photosynthetically active radiation (r = 0.97). This indicates that under the open Gobi Desert, the surface soil temperature changes almost synchronously with the near-surface air temperature, and the soil temperature can be regarded as a direct manifestation of the change in air temperature. Therefore, although air temperature itself has a direct and significant correlation with NEE, the strong collinearity between the two also means that air temperature can exert a closer indirect regulation on carbon flux by driving changes in soil temperature (Fig. 8).

Path analysis shows that the NEE regulation of photovoltaic power stations in arid areas is dominated by temperature factors, and the soil moisture content 10 cm below the tower has a weak effect. The direct diameter coefficient of soil temperature at 10 cm below the tower is 0.9107, which is the strongest direct driving factor of NEE. The air temperature at 15 m directly and strongly inhibited NEE (− 1.1014), but produced a positive indirect effect through soil temperature, and the total effect was inhibitory (− 0.2876). The direct effect of soil moisture content 10 cm below the tower is only 0.1121, and the indirect effects cancel each other out. The interaction of temperature factors constitutes the core of NEE regulation, and the analysis results are completely consistent with the correlation coefficient, providing key theoretical

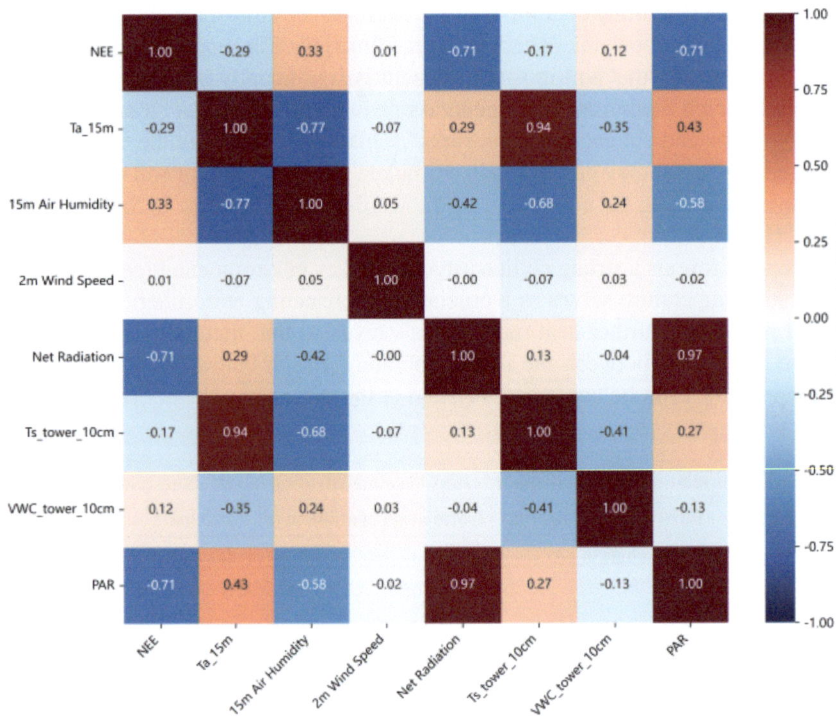

Fig. 8 The thermal correlation matrix of the correlation coefficient between NEE and environmental factors

Table 5 Path analysis of NEE-related variables: Direct coefficient and indirect effect statistics

Serial number	Independent variable	Direct diameter coefficient	Indirect path coefficient (by VWC_ tower_10 cm)	Indirect path coefficient (by Ts_ tower_10 cm)	Indirect path coefficient (by Ta_15 m)	Total indirect effect	Overall effect
1	VWC_ tower_10 cm	0.1121	–	– 0.376	0.3856	0.0096	0.1216
2	Ts_tower_ 10 cm	0.9107	– 0.0463	–	– 1.0317	– 1.0779	– 0.1672
3	Ta_15 m	– 1.1014	– 0.0392	0.8531	–	0.8138	– 0.2876

support for accurately predicting the carbon source/sink effect of photovoltaic power stations and ecological optimization design (Table 5).

4.3.2 Identification of ET and NEE Driver Factors and Model Evaluation Based on Random Forest Model

To deeply reveal the driving mechanisms of evapotranspiration (ET) and net ecosystem carbon exchange (NEE), this study constructed a random forest regression model, systematically evaluated the model performance, quantified the contributions of key environmental factors, and analyzed the differences in sensitivity of ET and NEE to different factors.

The random forest model demonstrates excellent fitting and predictive capabilities for both ET and NEE. The coefficient of determination (R^2) of the test set of the ET model is 0.734, and that of the NEE model is even higher, reaching 0.783. The Linn's consistency correlation coefficients (LCCC) of the two were 0.841 and 0.875 respectively, indicating that the predicted values of the model were highly consistent with the measured values. The lower root mean square error (RMSE, ET: 0.3916 mm/d; NEE: 0.0162 mg CO_2 m^{-2} s^{-1}) further confirms that the model has reliable accuracy. In the time series prediction throughout the year, the model can accurately capture the seasonal fluctuations and short-term change characteristics of ET and NEE. The predicted curve is highly consistent with the measured curve, which proves the effectiveness of the model in time series dynamic simulation (Table 6; Fig. 9).

Based on the variable importance graph, it can be clearly seen that although the core driving factors of ET and NEE both involve energy variables, there are detailed differences in the control logic. The ET process is dominated by energy availability. In the ET model, Photosynthetically active radiation (PAR, with an importance score of 0.225) and Rn (0.224) are the top two variables, highlighting that energy supply is the primary constraint of actual evapotranspiration and directly determines the potential capacity of evapotranspiration. The NEE process is the result of the coordinated

Table 6 Summary of fitting evaluation indicators for ET and NEE random Forest models

Model	R^2 (coefficient of determination)	RMSE	LCCC (Lin's Concordance correlation coefficient)	Unit	Number of test samples
ET (evaporation) random forest model	0.734	0.3916	0.841	mm/d	1984
NEE (net ecosystem exchange) random Forest model	0.783	0.0162	0.875	mg CO_2 m^{-2} s^{-1}	1984

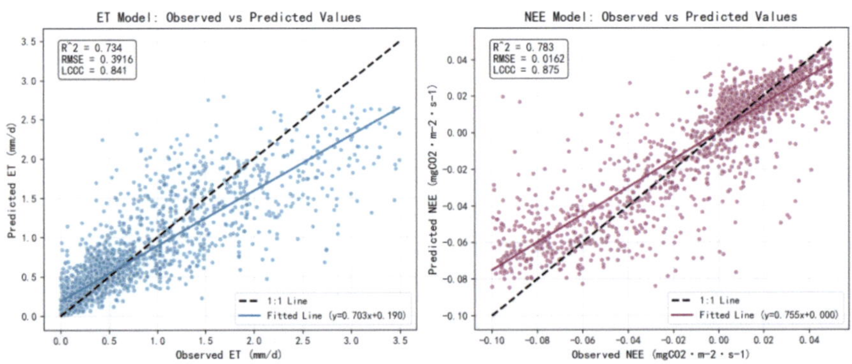

Fig. 9 Verification and comparison of observed and predicted values of ET and NEE random forest models

regulation of energy supply and environmental temperature. In the NEE model, Rn (0.316) and PAR (0.272) are the most critical variables. Meanwhile, the air temperature at 15 m (0.145) also has a certain driving effect, which not only reflects the core position of energy as a photosynthetic substrate, It also reflects the regulatory effect of temperature on the activity of carbon exchange, while soil factors (10 cm temperature and humidity) and low-level meteorological factors (2 m wind speed, etc.) have relatively weak influences on both, further distinguishing the core control mechanisms of ET and NEE. The full-year time series plot verified the dynamic simulation capability of the model: the fluctuation trends of the predicted values (red curves) and measured values (blue curves) of ET and NEE were highly synchronized, precisely matching the seasonal variation characteristics of ET being high in spring and summer and low in autumn and winter, as well as the fluctuations of carbon sinks/carbon sources of NEE at different times. Even short-term small dynamic deviations are consistent with the measured trend, indicating that this random forest model not only has reliable static fitting accuracy but also has strong adaptability in capturing the dynamic changes of ecological processes in the time dimension. The time series prediction effect is highly consistent with the actual observation (Figs. 10 and 11).

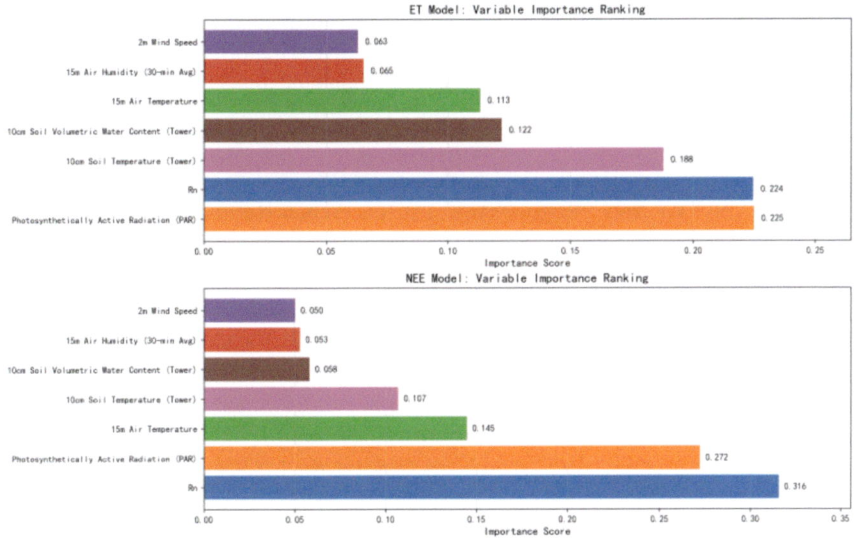

Fig. 10 Comparison of feature importance ranking between ET and NEE random forest models

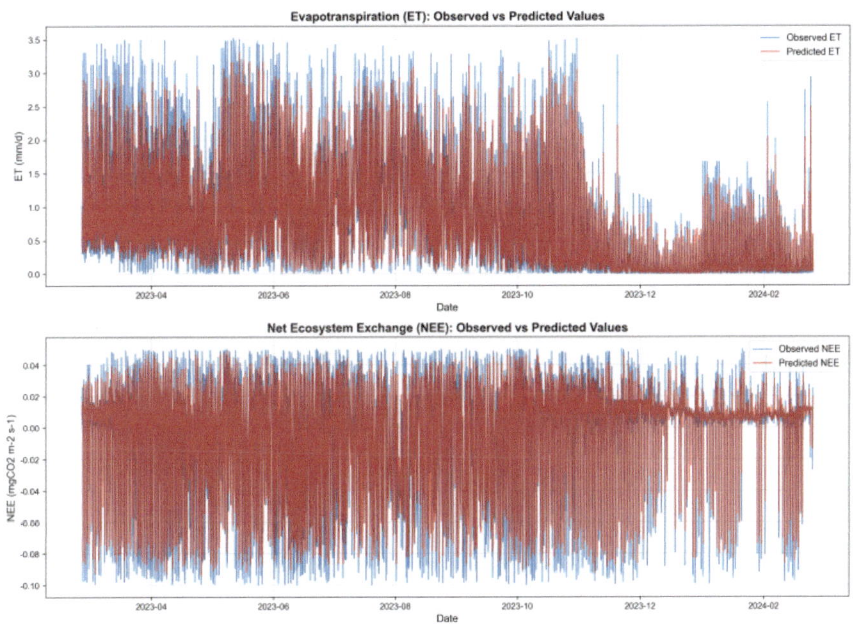

Fig. 11 Dynamic comparison of ET and NEE observed and predicted value time series

The results of model-based sensitivity analysis show that the response intensity and direction of ET and NEE to different environmental disturbances are completely different, clarifying the differentiated dominant mechanism of the carbon and water cycle processes. Response to energy input: The disturbance of photosynthetically active radiation (PAR) and net radiation (Rn) has a much stronger impact on NEE than on ET. For example, a 20% reduction in PAR leads to a change in NEE of − 11.719%, while ET only changes by − 1.928%. When Rn decreases by 20%, the sensitivity coefficient of NEE (29.947) is approximately ten times that of ET (2.834). This confirms that in the Gobi ecosystem, carbon exchange processes (especially photosynthesis) are far more dependent on energy availability than evapotranspiration processes. Response to turbulent exchange: The impact of 2 m wind speed disturbances on ET is generally weaker than that on NEE, which is consistent with the physical mechanism by which wind speed directly dominates water transport by enhancing turbulence. Response to temperature gradient: Temperature changes have a reverse regulatory effect on ET and NEE. Cooling (− 1 °C) significantly promotes ET (sensitivity coefficient + 13.347) but inhibits NEE (i.e., enhances carbon sink, coefficient − 18.326); A temperature increase of + 2 °C slightly inhibited ET (+ 2.268). This reveals the complex coupling relationship that temperature reduction may promote evaporation by increasing the vapor pressure difference between soil and air, while enhancing net carbon absorption by inhibiting microbial respiration (Fig. 12).

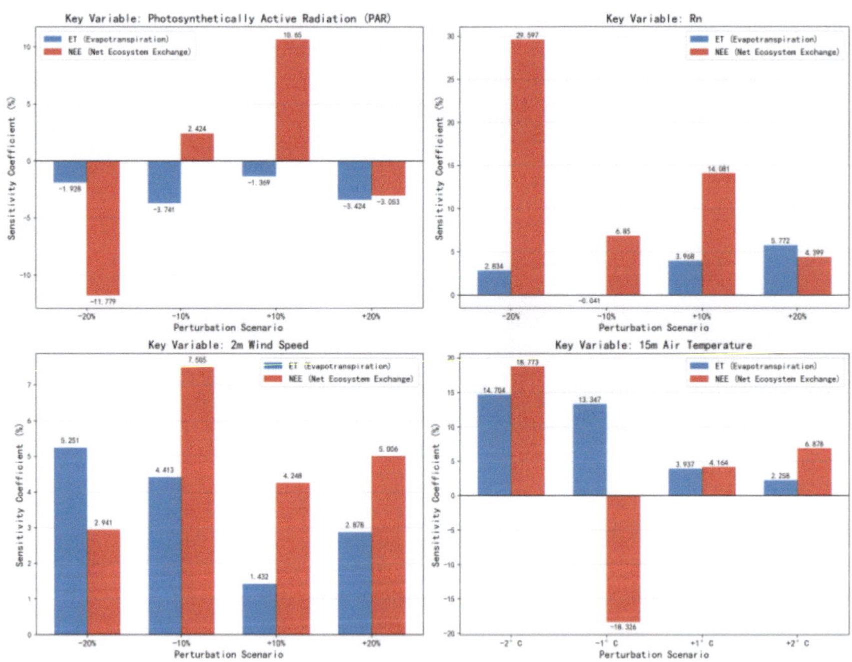

Fig. 12 Sensitivity analysis of a single disturbance of key environmental factors to ET and NEE

5 Discussion and Conclusion

Based on the continuous one-year positioning observation data of the photovoltaic Gobi base in Hainan Prefecture, Qinghai Province, this study, combined with correlation analysis, path analysis and random forest models, systematically revealed the microclimate regulation effect and the response mechanism of carbon and water fluxes of photovoltaic power stations. Research shows that photovoltaic panels have significantly altered the soil thermal environment through "peak shaving and valley filling" regulation. During the warm season, they have reduced the average soil temperature at 10 and 40 cm by 5.8 °C. In the cold season, they have maintained the soil's low-temperature stability, significantly increased the shallow soil moisture content during the growing season, and simultaneously weakened the near-surface wind speed. This is highly consistent with the difference characteristics between the area under the slab and the area under the tower in the dynamic chart of soil temperature and humidity. In terms of carbon and water fluxes, this ecosystem exhibits a stable carbon sink throughout the year. NEE shows significant seasonal differentiation, with the strongest carbon absorption in August and a tendency towards carbon budget balance in winter. Represents a temperature-driven single-peak pattern, while GPP shows a "U"-shaped seasonal pattern, which is consistent with the trend of the spatio-temporal dynamic graph of carbon fluxes. In terms of the driving mechanism, the random forest model shows that PAR and Rn are the core driving factors of ET (with importance scores of 0.225 and 0.224 respectively) and NEE (with importance scores of 0.272 and 0.316 respectively). Path analysis confirms that the 10 cm soil temperature (direct path coefficient 0.9107) is the strongest direct driving factor of NEE. After integrating key factors and interaction terms, the multiple linear regression model can explain 64.37% of the NEE variation. Moreover, it has discovered a special rule that soil moisture content in the Gobi region is positively correlated with NEE, revealing the unique carbon cycle feature of arid regions where increased moisture preferentially stimulates soil respiration. In summary, this study systematically clarified the "peak shaving and valley filling" regulatory effect of photovoltaic panels on microclimate and the seasonal differentiation law of carbon and water fluxes in the photovoltaic Gobi base in Hainan Prefecture, Qinghai Province. It also revealed the driving mechanism centered on photosynthetically active radiation and soil temperature, as well as the unique water-carbon cycle coupling characteristics of arid areas. It provides key theoretical support and practical guidance for the coordinated advancement of the large-scale development of the photovoltaic industry in arid areas and the protection and restoration of fragile Gobi ecosystems under the "dual carbon" vision.

Acknowledgements This project is supported by the science and technology projects of State Power Investment Corporation Limited Huanghe Hydropower Development Co, Ltd. (KY-C-2024-GF04, KY-C-2025-HB04)

References

Al-Shetwi, A.Q.: Sustainable development of renewable energy integrated power sector: trends, environmental impacts, and recent challenges. Sci. Total. Environ. **822**, 153645 (2022)

Armstrong, A., Ostle, N.J., Whitaker, J.: Solar park microclimate and vegetation management effects on grassland carbon cycling. Environ. Res. Lett. **11**(7), 074016 (2016). https://doi.org/10.1088/1748-9326/11/7/074016

Armstrong, A., Alona, A., Trevor, P., et al.: Integrating environmental understanding into freshwater floatovoltaic deployment using an effects hierarchy and decision trees. Environ. Res. Lett. **15**(11), 114055 (2020).https://doi.org/10.1088/1748-9326/abbf7b

Deng, X., Shi, Q., Zhang, Q., et al.: Impacts of land use and land cover changes on surface energy and water balance in the Heihe River Basin of China, 2000–2010. Phys. Chem. Earth Parts a/b/c **79**, 2–10 (2015)

Gao, X.Q., Yang, L.W., Lv, F., et al.: Observational study on the impact of the large solar farm on air temperature and humidity in desert areas of Golmud. Acta Energiae Solaris Sin **11**, 2909–2915 (2016)

Hua, Y., Chai, J., Chen, L., et al.: The influences of the desert photovoltaic power station on local climate and environment: a case study in Dunhuang photovoltaic industrial park, Dunhuang City, China in 2019. Atmosphere **13**(8), 1235 (2022)

Huang, Z., Xie, X.: Energy revolution under vision of carbon neutrality. Bull. Chin. Acad. Sci. **36**(9), 1010–1018 (2021)

International Renewable Energy Agency: Renewable capacity statistics 2024. International Renewable Energy Agency, Abu Dhabi (2024)

Kim, S.H., Lee, W.J.: AI-based prediction of SPPB scores using questionnaires of abilities: findings from the national health and aging trends study. BMC Med. Inf. Decis. Making **25**(1), 434 (2025). https://doi.org/10.1186/S12911-025-03263-6

Nemet, G.F.: Net radiative forcing from widespread deployment of photovoltaics. Environ. Sci. Technol. **43**(6), 2173–2178 (2009)

Wang, X., Zhou, Q., Zhang, Y., et al.: Diurnal asymmetry effects of photovoltaic power plants on land surface temperature in Gobi deserts. Remote Sensing **16**(10), 1711 (2024)

Yang, W., Zeng, L., Li, X.: Advances in research of carbon sinks and their influencing factors evaluation. Adv. Earth Sci. **38**(2), 151–167 (2023)

Zhai, B., Dang, X.H., Chen, X., et al.: Difference regularity of precipitation redistribution and soil water evapotranspiration in photovoltaic panels in typical steppe areas of Inner Mongolia. J. Chin. Agric. Univ **25**, 144–155 (2020)

Zhang, Y., Tian, Z., Liu, B., et al.: Effects of photovoltaic power station construction on terrestrial ecosystems: a meta-analysis. Front. Ecol. Evol. **11**(11), 1151182 (2023). https://doi.org/10.3389/fevo.2023.1151182

Zhang, H., Zhao, P., Zhang, W., et al.: Promoting sustainable solar-energy development in harmony with global threatened bird ranges. Nexus **1**(2) (2024)

Zhu, R., Shi, W., Wang, Y., et al.: Research suggestions for ecological and climatic environmental effects of wind power development in China. Strat. Study Chin. Acad. Eng. **20**(3), 39–43 (2018)

Study on the Threshold of Water Supply–Demand Imbalance in the Yellow River Basin

Hong Lv, Yuting Zhao, Siyu Yang, Xiaokang Zheng, and Wenxiu Shang

Abstract The Yellow River Basin, a critical water-bearing region in China, is facing increasingly prominent conflicts between water supply and demand. Based on Maslow's hierarchy of needs theory, this study categorizes the socio-economic water use in the Yellow River Basin into three levels: rigid water use, semi-flexible water use, and flexible water use. Furthermore, the concept of bottom-line water use is proposed to ensure basic water security under extreme drought conditions. By constructing a water resources allocation model and using the actual water usage data from 2022 as the baseline, this study simulates the supply–demand responses under different water inflow scenarios and identifies the critical thresholds for water supply–demand imbalance in the Yellow River Basin. The results indicate that when the minimum natural water inflow of the Yellow River is 30.8 billion m^3, the system is in a state of severe imbalance, with an associated economic loss of 259.27 billion CNY. An inflow of 44.6 billion m^3 corresponds to a moderate-severe imbalance, 52.6 billion m^3 to a moderate imbalance, and 56.2 billion m^3 to a mild imbalance. The findings of this study can provide a scientific basis and decision-making support for precise regulation, risk warning, and sustainable utilization of water resources in the Yellow River Basin.

Keywords Yellow River Basin · Water supply–demand imbalance · Threshold · Bottom-line water use · Economic loss

H. Lv (✉) · Y. Zhao · X. Zheng · W. Shang
Yellow River Engineering Consulting Co., Ltd., Zhengzhou, China
e-mail: 13253610702@163.com

S. Yang
School of Water Conservancy and Transportation, Zhengzhou University, Zhengzhou, China

H. Xu (ed.), *Proceedings of the 8th International Symposium on Water Resource and Environmental Management*, Environmental Science and Engineering,
https://doi.org/10.1007/978-3-032-19088-8_15

1 Introduction

Water resources are strategic and fundamental assets that underpin socio-economic development and ecological environmental protection (Chen et al. 2024; Ding et al. 2010). Factors such as population growth, economic development-driven supply–demand imbalances, upgraded water use in upstream regions, water quality deterioration, inefficient water management practices, and rising temperatures due to climate change collectively exacerbate water stress and the trend of supply–demand imbalance (Gao et al. 2021). Concurrently, China's water resources situation aligns with global trends, characterized by uneven spatiotemporal distribution and insufficient per capita water availability. The impacts of drought are intensifying, particularly in major river basins where hydrological imbalances and human activities further compound associated risks (Huang et al. 2021). With rapid socio-economic development in recent years, the conflict between water supply and demand has become one of China's most severe water resource challenges.

As a crucial ecological barrier and the lifeline for Northwest and North China, the water resources of the Yellow River Basin are a fundamental factor influencing population growth, economic development, and even national security (Huang et al. 2009). Currently, under the combined pressures of climate change and intensive human activities, the Yellow River Basin faces severe challenges: inherent resource scarcity, over-exploitation (with a utilization rate reaching 80% in 2020, far exceeding the international warning line of 40%), and the destabilization of hydrological regimes. These factors lead to sharply intensified supply–demand conflicts during exceptionally dry years. The water security of the basin directly affects the well-being of 230 million people and regional ecological security.

Research on the thresholds of water resources imbalance in river basins, both domestically and internationally, primarily focuses on two aspects: Firstly, quantitative studies based on water balance analysis, utilizing methods such as system dynamics models (Yu 2025) and water resources carrying capacity models (Wang 2019) to identify critical points of regional water scarcity. Secondly, criticality studies based on economic loss and risk assessment, emphasizing the nonlinear impacts of water scarcity on economic systems (Liu 2025). However, these studies often concentrate on aggregate or structural supply–demand analysis, paying less attention to the differentiation of societal demand hierarchies and their impact mechanisms on water security. Traditional methods struggle to reflect the shifting priorities of water allocation for different societal functions under extreme drought conditions.

To address this gap, this paper introduces Maslow's hierarchy of needs theory (Zhong 2018). From the perspective of human needs, it categorizes socio-economic water use in the Yellow River Basin into three levels: rigid, semi-flexible, and flexible water use. Furthermore, it proposes the concept of "bottom-line water use" to define the minimum safe water supply under extreme drought scenarios. Based on actual water use data from the Yellow River Basin in 2022, a water resources allocation model constrained by bottom-line water use is constructed. By simulating supply–demand responses under different natural inflow scenarios, this study quantitatively identifies the critical thresholds of water supply–demand imbalance in the basin and assesses the corresponding economic losses.

2 Study Area and Data Sources

2.1 *Overview of the Study Area*

The Yellow River, the second-longest river in China, flows through nine provinces and autonomous regions. The specific provinces it flows through are indicated in Fig.1. It serves not only as a vital "Mother River," ecological barrier, energy base, and key water source (Miao 2019), but also undertakes the task of supplying water to 15% of China's farmland and 12% of its population with only 2% of the nation's total runoff (Wang et al. 2025). Furthermore, while supplying water to areas outside the basin, it faces a severe water resources crisis (Wang and Jia 2025). However, the per capita water resources possession in the Yellow River Basin is only about 20% of the national average, classifying it as a region experiencing resource-driven water scarcity. Precipitation within the basin is highly unevenly distributed throughout the year, primarily concentrated between June and September.

Based on data from the Third National Water Resources Survey and Assessment, the average annual natural runoff of the Yellow River Basin for the period 1956–2016 was 49.0 billion cubic meters. This represents a decrease of 4.5 billion cubic meters, or 8.4%, compared to the baseline period of 1956–2000 (Zhou et al. 2021).

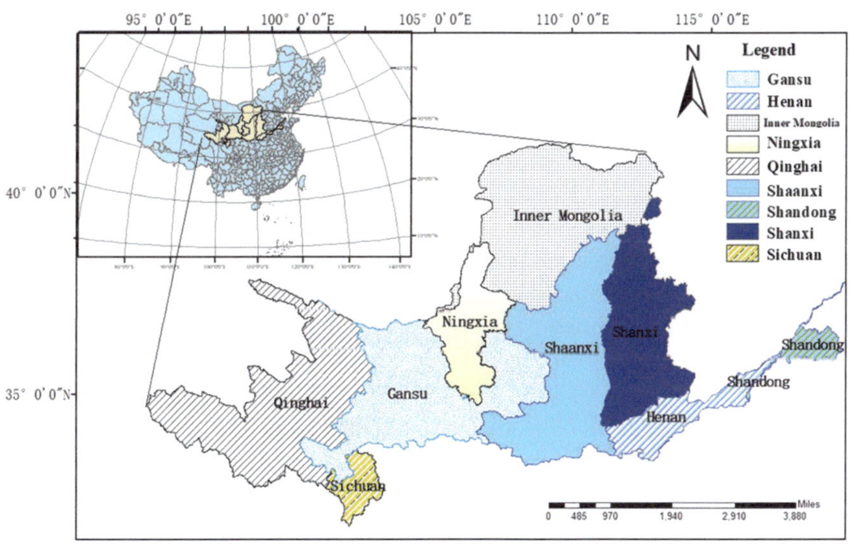

Fig. 1 Study area location

Time series analysis indicates that the basin experienced two significant continuous dry periods during the study period: 1969–1974 and 1990–2002. Extreme drought events were concentrated in 1997 and 2000–2002, during which the annual average runoff deviation ranged from -2% to -40%. Over the past 60 years, the natural runoff of the Yellow River has significantly decreased (Si et al. 2019). The average annual natural runoff for 2001–2016 was 4.18 billion m^3 less than that for 1956–2000 (Jiang et al. 2023). Accurately identifying the supply–demand imbalance thresholds for different demand levels and accordingly formulating graded response strategies for extreme droughts to effectively enhance the water security guarantee level in the Yellow River Basin is a critical issue urgently needing resolution in current water resources management.

2.2 Data Sources

Water supply and usage data for the Yellow River Basin in this paper were obtained from the Yellow River Water Resources Bulletin (1990–2002). Socio-economic statistical data for the provinces (autonomous regions) within the basin were sourced from the China Statistical Yearbook. Data on urban and rural populations, industrial value-added, and Gross Domestic Product (GDP) were selected from the Statistical Bulletin of National Economic and Social Development. Water use quotas for various sectors were determined according to the Yellow River Protection Law of the People's Republic of China.

3 Methodology

3.1 Hierarchical Water Demand Theoretical Framework

The water supply–demand relationship in the Yellow River Basin is a complex issue involving a coupled socio-economic-natural system. Traditional water resources allocation studies primarily focus on the supply side, calculating water supply capacity under different inflow scenarios using water balance equations or system dynamics models. However, they often pay less attention to the differentiation in societal demand hierarchies and their practical impacts on basin security. To more scientifically describe the necessity and priority of different types of water use for societal functioning, this paper introduces Maslow's hierarchy of needs theory to construct a hierarchical model for societal water demand.

According to Maslow's theory, human needs can be categorized into levels such as basic physiological, safety, social, esteem, and self-actualization needs. The satisfaction of lower-level needs is a prerequisite for fulfilling higher-level ones. Drawing on this concept, and considering the characteristics and requirements of water security under extreme drought conditions, this paper categorizes the socio-economic water use in the basin into three levels: First, Rigid Water Use: the water volume required to meet residents' basic survival needs, industrial enterprises' operational production, grain safety production, and the basic health of rivers and lakes. Second, Semi-flexible Water Use: the water volume needed to improve residents' quality of life, expand industrial production scale, increase grain yield, and shape a suitable ecological environment. Third, Flexible Water Use: the water volume required to sustain residents' luxury consumption, high water-consuming industries, ensure grain exports, and create high water-consuming landscapes (Wang et al. 2023).

3.2 Definition of Bottom-Line Water Use

Within rigid water use, to further identify the guaranteed water quantity that must be unconditionally secured and not compromised under extreme drought conditions, this study proposes the concept of bottom-line water use quantity. The bottom-line water use quantity includes domestic, industrial, and ecological bottom-line water use for maintaining residents' basic quality of life, socially stable production scale, and minimum ecological functions in river channels, with the critical water shortage value leading to accelerated reduction in agricultural water use efficiency serving as the agricultural bottom-line water use.

Domestic bottom-line water use is calculated using the per capita daily minimum water use standard method, combined with urban and rural population data from each province, to account for the most basic domestic water use maintaining social stability. Industrial bottom-line water use adopts the key industry minimum operational load method, targeting core industrial sectors crucial to national economy and people's livelihood, to calculate the minimum water use required for maintaining basic operational rates. Agricultural bottom-line water use employs the critical water shortage value method, aiming to avoid a cliff-like drop in grain yield and ensure absolute staple food security, to determine the minimum irrigation quota for core irrigation areas. Ecological bottom-line water use utilizes the Tennant method or the minimum monthly average flow method to verify the in-stream minimum ecological water requirement preventing irreversible degradation of river ecosystem functions. Based on the aforementioned methods, the calculated total bottom-line water use quantity for the nine provinces in the Yellow River Basin is 252.72 billion m^3, with specific bottom-line water use quantities shown in Table 1.

3.3 Water Resources Allocation Model Construction

A water resources allocation model for the Yellow River Basin was constructed with the objective of meeting bottom-line water use. The current water usage situation

Table 1 Bottom-line water use in the nine provinces of the Yellow River Basin (unit: 100 million m^3)

Province	Domestic	Industrial	Agricultural	Ecological	Total
Qinghai	2.3	0.52	5.99	1.24	10.05
Sichuan	0.14	0.01	0.13	0.00	0.28
Gansu	6.77	1.60	15.98	3.45	27.80
Ningxia	3.7	2.26	26.44	4.53	36.93
Inner Mongolia	5.45	6.34	18.63	9.77	40.19
Shaanxi	16.49	6.48	24.67	5.98	53.62
Shanxi	10.14	4.50	22.26	3.35	40.25
Henan	8.13	4.31	14.46	4.14	31.04
Shandong	2.9	1.79	6.57	1.30	12.56
Yellow River Basin	42.15	27.80	135.14	33.76	252.72

in 2022 was taken as the water demand. For water supply sources, groundwater, reclaimed water, and transferred water adopted the actual water usage scenario from 2022, while the natural Yellow River inflow was used as the variable, and the bottom-line water shortage was set as the objective function, as shown in the following formula.

$$\min \varepsilon = \min(F(r) - \sum_{n=1}^{9} w_n) \tag{1}$$

In the formula, ε is the difference between the allocated water volume and the bottom-line water use threshold; F is the water resources allocation model for the Yellow River Basin; r is the set natural Yellow River inflow; w is the bottom-line water use of the 9 provinces; n is the province code.

This model is a mathematical programming model. Its core is to minimize the bottom-line water shortage across the entire basin by optimizing the spatial allocation of water resources under various physical and policy constraints. In the model construction, the bottom-line water use volume w for each province uses the calculated results from Table 1 as fixed input parameters. Regarding water supply sources, groundwater, reclaimed water, and transferred water use the actual water usage scenario from 2022 as the baseline supply, while the natural Yellow River inflow r is set as the key variable, simulating different drought scenarios by systematically adjusting the value of r.

The model solution and threshold identification are achieved by systematically adjusting the natural Yellow River inflow r. An error limit of less than 0.001 is set as the convergence criterion. The value of r is gradually changed, and for each inflow scenario, the allocation model is run to calculate the corresponding total system water shortage. When the total system water shortage first approaches zero, meaning all bottom-line water demands are met within the error limit, the corresponding critical inflow *r* is defined as the severe imbalance threshold. By sequentially replacing the water demand input parameters in the model with rigid water use, semi-flexible water use, and total water use, and repeating the aforementioned solution process, the complete supply–demand imbalance threshold system for the Yellow River Basin can be systematically identified.

4 Results and Discussion

4.1 Identification of Critical Thresholds for Water Supply–Demand Imbalance in the Yellow River Basin

By running the water resources allocation model and systematically adjusting the variable of natural Yellow River inflow (r), this study accurately identified the critical inflow volumes required to satisfy different levels of water demand, i.e., the series of thresholds for water supply–demand imbalance in the Yellow River Basin, as shown in Table 2.

Aiming for the minimum degree of satisfaction of bottom-line water use and continuously changing the natural inflow of the Yellow River Basin, the minimum natural inflow of the Yellow River Basin was sought, with an error limit set to less than 0.001. The results show that when the minimum natural inflow of the Yellow River is 30.8 billion m^3, it can meet the current bottom-line water demand of 25.3 billion m^3, defining the severe imbalance threshold for water supply and use in the Yellow River Basin. When the minimum natural inflow is 44.6 billion m^3, it can meet the current rigid water demand of 31.1 billion m^3, defining the moderate-severe imbalance threshold. When the minimum natural inflow is 52.6 billion m^3, it can meet the current semi-flexible water demand of 35.8 billion m^3, defining the moderate imbalance threshold. When the minimum natural inflow is 56.1 billion m^3, it can meet the current total water demand (including flexible use) of 40.9 billion m^3, defining the mild imbalance threshold.

Table 2 Critical thresholds for water supply–demand imbalance in the Yellow River Basin

Imbalance level	Minimum natural inflow (100 million m^3)	Satisfied demand level	Guaranteed water demand (100 million m^3)
Severe imbalance	308	Bottom-line water use	252.7
Moderate-severe imbalance	446	Rigid water use	311.3
Moderate imbalance	526	Semi-flexible water use	357.9
Mild imbalance	562	Total water use	409.0

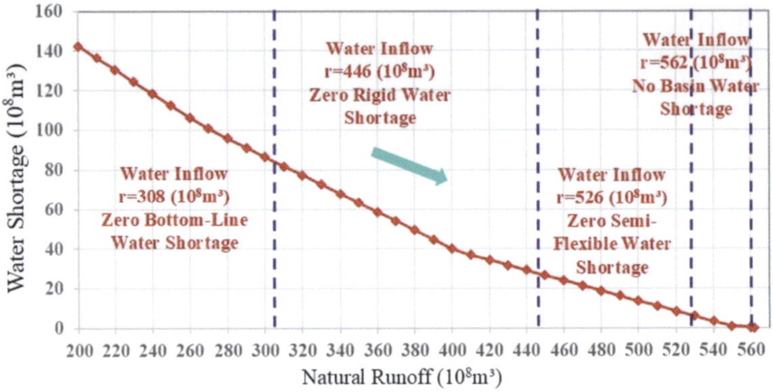

Fig. 2 Critical water inflow under different critical water demands

The identification of these thresholds provides clear early-warning water levels for water resources management in the Yellow River Basin. The critical water inflows under different critical water demands are shown in Fig. 2. A significant non-linear growth relationship exists between the critical water demand and the critical water inflow. This means that satisfying higher levels of water demand requires progressively larger unit marginal increases in inflow. This highlights that, against the backdrop of water scarcity, improving water use efficiency and optimizing the demand structure are more urgent and effective than merely seeking additional water sources.

4.2 *Economic Loss Assessment of Water Supply–Demand Imbalance*

To translate the physical risk of water resources supply–demand imbalance into tangible economic impacts, thereby providing a quantitative basis for watershed water resources risk management and regulatory decision-making, this study directly converts water shortages under different imbalance states into significant economic losses (Qian et al. 2022; Jiao et al. 2020; Han and Ruan 2007). Based on the model output of water shortage volumes and the economic loss coefficient per unit of water, this study quantified the economic losses borne by the Yellow River Basin when reaching each critical threshold under the 2022 baseline scenario. The results are shown in Table 3.

The economic losses under different minimum natural inflow scenarios in the Yellow River Basin in 2022 under the bottom-line water use scenario demonstrate the varying impacts of water supply–demand imbalance on the regional economy. Across the different scenarios, the economic losses in different provinces show relatively clear spatial differences and gradient changes.

Table 3 Critical economic losses from water supply–demand imbalance in the Yellow River Basin under the 2022 bottom-line water use scenario

Province	Minimum Natural Inflow (100 million m^3)			
	308	446	526	562
	Economic loss (100 million CNY)	Economic loss (100 million CNY)	Economic loss (100 million CNY)	Economic loss (100 million CNY)
Qinghai	32.3	13.61	6.66	2.22
Sichuan	17.14	10.23	4.83	1.64
Gansu	47.68	14.25	5.61	2.32
Ningxia	18.55	9.52	4.09	1.79
Inner Mongolia	15.16	9.1	3.98	1.36
Shaanxi	30.47	9.73	4.51	1.99
Shanxi	46.27	13.54	6.63	1.83
Henan	18.41	10.09	4.65	1.38
Shandong	33.29	12.34	6.77	2.84
Total	259.27	102.41	47.73	17.37

According to the data in Table 3, under the scenario of a minimum natural inflow of 308 billion m^3, the total economic loss in the Yellow River Basin was 259.27 billion CNY. Gansu Province suffered the highest economic loss, reaching 47.68 billion CNY, while the losses in Qinghai and Shaanxi were relatively lower, at 32.3 billion CNY and 30.47 billion CNY respectively, reflecting differences in water source allocation among these regions. Under the 446 billion m^3 scenario, the economic loss decreased to 102.41 billion CNY, with losses in Gansu and Shanxi still being particularly prominent at 14.25 billion CNY and 13.54 billion CNY respectively. Under the 526 billion m^3 and 562 billion m^3 scenarios, economic losses further decreased to 47.73 billion CNY and 17.37 billion CNY, where losses in upstream areas such as Gansu, Qinghai, and Shaanxi dropped to 5.61 billion CNY, 2.22 billion CNY, and 1.99 billion CNY, respectively.

In summary, as the natural inflow increases, the economic losses in the Yellow River Basin show a significant decreasing trend, especially in the upstream and midstream regions. However, even under higher water inflow scenarios, some regions still face relatively high economic losses, indicating that the risk of water resources supply–demand imbalance persists in these areas. Overall, the quantitative analysis results show that the risk is higher in the upstream and midstream regions, and the regulatory capacity and water supply security in these areas under drought scenarios still need further enhancement.

Using the water supply–demand imbalance risk assessment method and the loss quantification method, the critical thresholds for water supply–demand imbalance in the Yellow River Basin under the 2022 water use scenario were determined. When the minimum natural Yellow River inflow is 308 billion m^3, the quantified loss from water supply–demand imbalance under the bottom-line water use scenario is 259.27

billion CNY; when the minimum natural inflow is 446 billion m^3, the quantified loss is 102.41 billion CNY; when the minimum natural inflow is 526 billion m^3, the quantified loss is 47.73 billion CNY; when the minimum natural inflow is 562 billion m^3, the quantified loss is 17.37 billion CNY.

5 Conclusion

Focusing on the core issue of the water resources supply–demand imbalance threshold in the Yellow River Basin, this study quantitatively identified the critical points of supply–demand balance under different natural inflow scenarios and assessed their economic impacts based on Maslow's hierarchy of needs theory and a water resources allocation model. The main conclusions are as follows:

(1) A hierarchical water demand framework based on Maslow's hierarchy of needs theory was constructed, clarifying the three levels of water use: rigid, semi-flexible, and flexible, and quantifying the water use structure at each level in the Yellow River Basin in 2022 and the total bottom-line water use volume of 252.72 billion m^3.

(2) Key thresholds for supply–demand imbalance in the Yellow River Basin were identified. By simulating different natural inflow scenarios, four critical thresholds for water resources supply–demand imbalance in the Yellow River Basin were determined: when the natural inflows are 308 billion m^3, 446 billion m^3, 526 billion m^3, and 562 billion m^3, the system state manifests as severe imbalance, moderate-severe imbalance, moderate imbalance, and mild imbalance, respectively. These results clearly delineate the critical process of the basin's transition from high risk to stable supply and demand, providing a quantitative basis for graded regulation and risk early warning of water resources in the basin.

(3) The pattern of economic losses under imbalanced states was quantified. The research shows that as the natural inflow increases from 308 billion m^3 to 562 billion m^3, the total economic loss in the Yellow River Basin decreases significantly from 259.27 billion CNY to 17.37 billion CNY, showing an exponential deceleration trend. The spatial distribution of economic losses has distinct regional characteristics, with Gansu Province and Shanxi Province suffering the most severe losses under extreme drought conditions, making them highly sensitive areas and key management and control regions for water resources security in the basin.

Acknowledgements This research is supported by the National Key Research and Development Program of China (No. 2022YFC3202303), National Natural Science Foundation of China (52509038) and Henan Province Youth Talent Support Project (2025HYTP030). Special thanks are owed to editors and anonymous reviewers whose comments helped to improve the paper.

Declaration of Competing Interest The authors declare that they have no known competing financialinterests or personal relationships that could have appeared to influencethe work reported in this paper.

References

Chen, J., Wang, L., Lu, J.: Habitat health assessment of the inland section of the Pinglu canal based on the CRITIC method. China Rural Water and Hydropower (9), 21–28 (2024)

Ding, H., Li, Z., Wang, J.: Comprehensive index assessment of water safety based on entropy weight coefficient method. J. Chengdu Univ. Inf. Technol. (2010)

Gao, J., Liu, Y., Zheng, H.: Land use multifunctional diagnosis based on entropy weight and matter-element model. J. China Agric. Resour. Regional Plan. **42**(9), 52–61 (2021)

Han, Y., Ruan, B.Q.: Economic loss assessment of shortage risk of water resources. **10**, 1253–1257 (2007)

Huang, Y., et al.: Study on assessment methods of evaluate indexes for complex system: taking urban water safety as case. Hydrology **29**(2), 45–49 (2009)

Huang, X., et al.: Evaluation of water resource security based on combination weight cloud model. J. Econ. Water Resour. (39), 60–65 (2021)

Jiang, S., et al.: Coordinated development of water, food, and energy in upper and middle reaches of the Yellow River. Strat. Study Chin. Acad. Engineering. **25**(4), 180–190 (2023)

Jiao, S., et al.: Study on economic loss and its driving factors caused by the risk of water resources imbalance in Anyang City. Yellow River **42**, 57–61 (2020)

Liu, C.: Construction of climate security evaluation indicator system, risk assessment and adaptation strategies in China. Yuejiang Acad. J. **4**, 43–61 (2025)

Miao, B.Z.Y.X.W.G.G.T.J.J.L., The trend in streamflow of the Yellow River Basin during 1956–2016. J. Water Resour. Water Eng. **30**, 52–57 (2019)

Qian, L., et al.: Model for water shortage risk econimic losses based on M-Copula and its application **30**, 907–917 (2022)

Si, Y., et al.: Revealing the water-energy-food nexus in the upper Yellow River basin through multi-objective optimization for reservoir system. Sci. Total. Environ. **682**, 1–18 (2019)

Wang, L., Jia, D.: Spatiotemporal analysis of habitat quality and driving factors in the middle reaches of the Yellow River Basin. Environ. Monit. Assess. **197**(4) (2025)

Wang, Y., et al.: Water allocation of the first phase of South-to-North water diversion western route project based on balanced provisioning of water resources in the Yellow River basin. **34**(3), 336–348 (2023)

Wang, Y., et al.: Research on resilience regulation of water resource systems in the Yellow River basin under during extreme drought years. **36**(3), 359–371 (2025)

Wang, H.R., Qian, L.X., Zhao, Z.Y., et al.: Water resources risk analysis theory and assessment methods. J. Hydraul. Eng. **50**(8), 980–989 (2019)

Yu, L.C.Y.: Evaluation of groundwater resource carrying capacity based on game theory-matter-element extension model. Water Resour. Plan. Des. **12**, 66–72 (2025)

Zhong, H.: Water resources management based on Maslow's hierarchy of needs. China Water Resour. **1**, 35–37 (2018)

Zhou, Z., et al.: Attribution analysis of the natural runoff evolution in the Yellow River basin. **33**(1), 27–37

Analysis of Spatiotemporal Vegetation Changes and Future Trends in the Shiyang River Basin from 2000 to 2024

Hong Lv, Shu Li, Huiliang Wang, Ziyu Wang, Xiaokang Zheng, and Wenxiu Shang

Abstract Based on MODIS NDVI data from 2000 to 2024, this study employs linear trend analysis, Mann–Kendall significance testing (MK), and Hurst index analysis to investigate the spatiotemporal characteristics of vegetation cover changes and future trends in the Shiyang River Basin over the past 25 years. Results indicate: (1) Overall NDVI in the Shiyang River Basin showed a significant upward trend from 2000 to 2024, reflecting an overall improvement in vegetation cover; (2) Spatially, vegetation improvement was concentrated in the oases and mountainous areas of the middle and upper reaches, while degradation occurred in the arid desert areas downstream and some peripheries of irrigated zones; (3) The Mann–Kendall -test indicates improvement trends in most areas of the basin; (4) Hurst index results demonstrate strong persistence in the basin's NDVI time series, suggesting continued vegetation improvement overall, though downstream regions face certain degradation risks. These findings reveal the phased achievements and spatial variations in ecological restoration within the Shiyang River Basin, providing scientific basis for regional ecological construction and water resource management.

Keywords Shiyang River Basin · NDVI · Mann–Kendall test · Hurst exponent · Vegetation change · Spatiotemporal analysis

H. Lv (✉) · S. Li · X. Zheng · W. Shang
Yellow River Engineering Consulting Co., Ltd., Zhengzhou, China
e-mail: 13253610702@163.com

H. Wang · Z. Wang
School of Water Conservancy and Transportation, Zhengzhou University, Zhengzhou, China

H. Xu (ed.), *Proceedings of the 8th International Symposium on Water Resource and Environmental Management*, Environmental Science and Engineering,
https://doi.org/10.1007/978-3-032-19088-8_16

1 Introduction

The Shiyang River Basin, located in the arid region of northwestern China, is a typical inland river basin and an important ecological and economic zone within the Hexi Corridor. Due to its extremely arid climate, limited precipitation, and increasing intensity of human activities, the ecological environment of the basin is highly fragile, and vegetation dynamics are particularly sensitive to variations in climate and water resources (Miao et al. 2023). In recent years, under the dual influence of global climate change and regional water resource reallocation (Jin et al. 2023), the ecological environment of the Shiyang River Basin has undergone noticeable changes. Consequently, vegetation dynamics have become a key indicator for assessing ecosystem health and sustainable development in this region.

The Normalized Difference Vegetation Index (NDVI), as an important remote sensing parameter reflecting surface vegetation cover (Wang et al. 2020), provides a quantitative means to characterize the spatiotemporal patterns of vegetation growth, ecological restoration, and responses to climate change (Lan et al. 2024). In recent decades, the integration of remote sensing technology and time-series NDVI data has become a powerful tool for studying vegetation changes in arid and semi-arid regions (Yu et al. 2023). Numerous studies have investigated NDVI variations in river basins such as the Heihe River (Zhu et al. 2020; Han et al. 2023), Yellow River (Lu et al. 2021; Liiliu and Hongmei 2007), providing valuable insights into regional ecological evolution and environmental management. However, long-term investigations on the spatiotemporal dynamics and persistence of vegetation changes in the Shiyang River Basin (2000–2024) remain limited, particularly those focusing on quantitative prediction of future vegetation trends.

Therefore, based on MODIS NDVI data from 2000 to 2024, this study systematically analyzes the spatiotemporal evolution and persistence of vegetation cover in the Shiyang River Basin by integrating linear regression analysis, the Mann–Kendall (MK) trend test, and Hurst exponent analysis. The objectives are to reveal the spatiotemporal patterns of vegetation change over the past twenty-five years, evaluate the effectiveness of ecological restoration efforts, and predict future vegetation evolution trends. The findings aim to provide a scientific basis for regional ecological protection and sustainable water resource management in arid inland basins.

2 Study Area

The Shiyang River Basin, located in central Gansu Province, northwestern China (101° 30′–104° 30′ E, 37° 30′–39° 30′ N), covers a total area of approximately 41,600 km^2. It is primarily composed of eight rivers, including the Dajing, Gulang, Huangyang, Zamu, Jinta, Xiying, Dongda, and Xida Rivers. The basin's water resources are mainly derived from precipitation and snowmelt in the Qilian Mountains, with an annual runoff of approximately 1.591×10^9 m^3. The terrain of the basin slopes from south to north, with the upstream region consisting of the high-altitude, cold Qilian Mountains, the midstream region characterized by the Wuwei–Liangzhou oasis agricultural area, and the downstream region forming the arid Minqin desert basin. The regional climate is a temperate continental arid type. Vegetation coverage is relatively high in the upstream area, whereas the midstream region experiences intensive human activities and significant land-use changes. The downstream ecosystem is fragile and has long been affected by groundwater overexploitation and desertification. The geographical location and topography of the study area are shown in Fig. 1. In recent years, ecological restoration projects such as the conversion of farmland to forest and ecological water replenishment have markedly improved ecological conditions in the middle and upper reaches of the basin; however, the downstream area remains at risk of degradation. Overall, the ecological environment of the Shiyang River Basin exhibits pronounced spatial heterogeneity and vulnerability, making it a representative region for studies on ecological restoration and sustainable development in arid areas (Fig. 1).

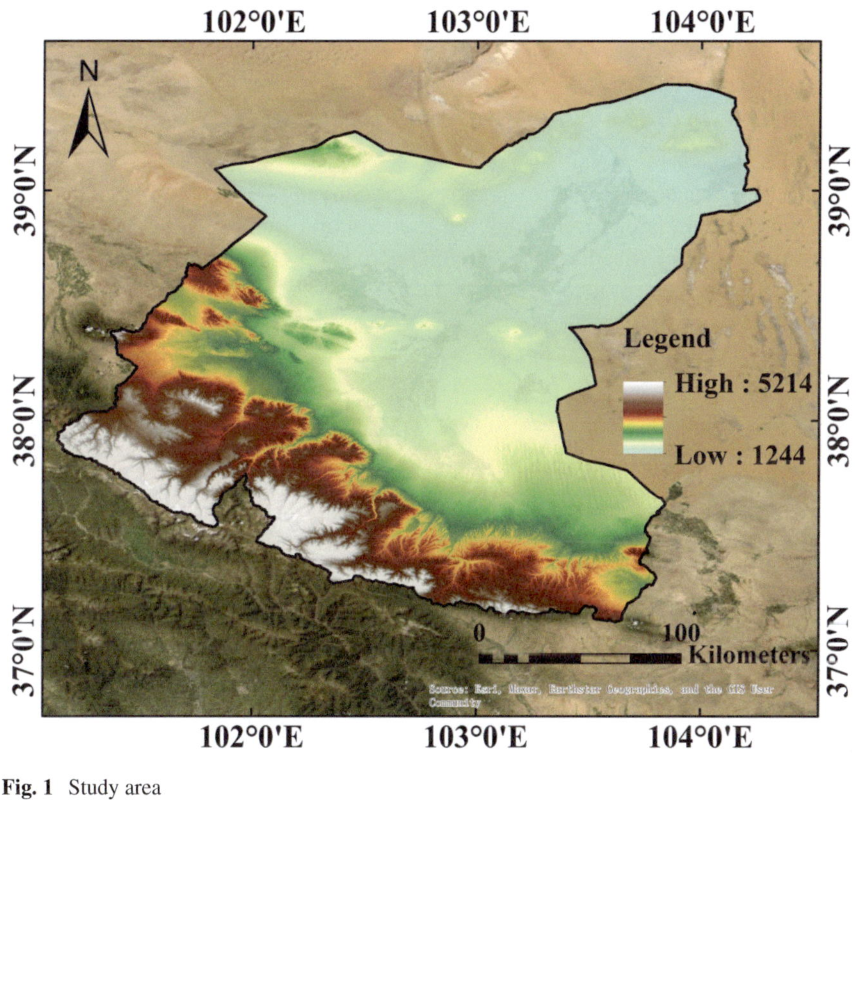

Fig. 1 Study area

3 Data Collection

In this study, MODIS NDVI data (MOD13Q1),spatial resolution 250 m, temporal resolution 16 days) for the period 2000–2024 were downloaded via the Google Earth Engine (GEE) platform. Annual maximum NDVI was generated using the Maximum Value Composite (MVC) method (Chen and Li 2023). During data processing, the NDVI images were clipped to match the boundary of the Shiyang River Basin, and all datasets were projected to a unified coordinate system (WGS_1984_UTM_Zone_47N). Ultimately, a total of 25 annual maximum NDVI datasets for 2000–2024 were obtained. The digital elevation model (DEM) was derived from the SRTMGL1_003 dataset provided by the U.S. Geological Survey (USGS), with a spatial resolution of 30 m. Basin boundary data were obtained from the Institute of Tibetan Plateau Research, Chinese Academy of Sciences, and were used for clippg and masking operations.

4 Methodology

4.1 Analysis of the Timing Variation Trend of NDVI

To reveal the vegetation cover dynamics in the Shiyang River Basin from 2000 to 2024, this study quantitatively analyzed the temporal trends of annual maximum NDVI using a pixel-wise linear regression approach. Let the NDVI value in year t be y_t, and the sequence of years be $t = 1, 2, \ldots, n$, The linear regression model is expressed as:

$$y_t = a + bt + \varepsilon_t \tag{1}$$

Here, b is the regression slope, representing the rate of NDVI change. A positive $b(b > 0)$ indicates an improving vegetation trend, a negative b ($b < 0$) indicates vegetation degradation, and $b = 0$ denotes a relatively stable condition. In addition, the multi-year mean NDVI was calculated to characterize the overall vegetation status.

4.2 Mann–Kendall Trend Significance Test

To assess the statistical significance of NDVI trends, the Mann–Kendall (MK) non-parametric test was employed (Feng et al. 2025; Guo et al. 2025). This method does not require any assumptions regarding the distribution of the time series and is effective for detecting long-term trends. The MK test statistic is defined as:

Table 1 Trend classification table

$b > 0$	$p < 0.05$	Significant improvement
$b > 0$	$p \geq 0.05$	Slight improvement
$b = 0$		Basically unchanged
$b < 0$	$p < 0.05$	Significant degradation
$b < 0$	$p \geq 0.05$	Slight degradation

$$S = \sum_{i=1}^{n-1} \sum_{j=i+1}^{n} \operatorname{sgn}\left(x_j - x_i\right) \tag{2}$$

The standardized test statistic Z was then calculated to determine the significance of the trend:

$$Z = \begin{cases} \frac{S-1}{\sqrt{Var(S)}} & S > 0 \\ 0 & S = 0 \\ \frac{S+1}{\sqrt{Var(S)}} & S < 0 \end{cases} \tag{3}$$

A trend is considered significant when $|Z| > Z_{1-\alpha/2}$, (in this study, the significance level was set at $\alpha = 0.05$ (in this study, the significance level was set at Table 1).

4.3 Hurst Index

To assess the persistence of NDVI trends and their potential future direction, the Hurst exponent (H) was employed (Qi et al. 2024; Lai et al. 2023). Originating from the rescaled range (R/S) analysis, the Hurst exponent characterizes the long-term memory of a time series and is defined as:

$$H = \lim_{n \to \infty} \frac{\log\left(\frac{R_{(n)}}{S_{(n)}}\right)}{\log(n)} \tag{4}$$

n is the length of the time series (or segment length), $R_{(n)}$ is the range of the time series, and $S_{(n)}$ is the standard deviation of the time series.

Combining the results of the linear trend slope estimation with the Hurst exponent, future NDVI change trends were classified into four types, as shown in Table 2.

Table 2 Hurst trend grade classification

	Classification of changing trends
Hurst < 0.5, slopes < 0	Continuous degradation
Hurst < 0.5, slopes > 0	Future degradation
Hurst > 0.5, slopes < 0	Future improvement
Hurst > 0.5, slopes > 0	Continuous improvement

5 Results

5.1 *Spatio-Temporal Variation Characteristics of NDVI*

5.1.1 Temporal Variation Characteristics of NDVI

Figure 2 illustrates the temporal variation of the annual mean NDVI in the Shiyang River Basin from 2000 to 2024. Overall, NDVI exhibited a fluctuating upward trend during the study period, and the linear regression results indicate a general improvement in regional vegetation cover. Between 2000 and 2008, NDVI values were relatively low and highly variable, reflecting limited early-stage vegetation restoration. From 2009 onwards, NDVI showed a steady increasing trend, reaching a peak in 2019 (maximum value of 0.204), indicating a significant improvement in the ecological environment. However, NDVI declined notably during 2022–2024, which may be associated with abnormal precipitation, drought events, or intensified human activities. Overall, vegetation conditions in the Shiyang River Basin have generally improved over the past 25 years, but interannual fluctuations and short-term degradation risks remain.

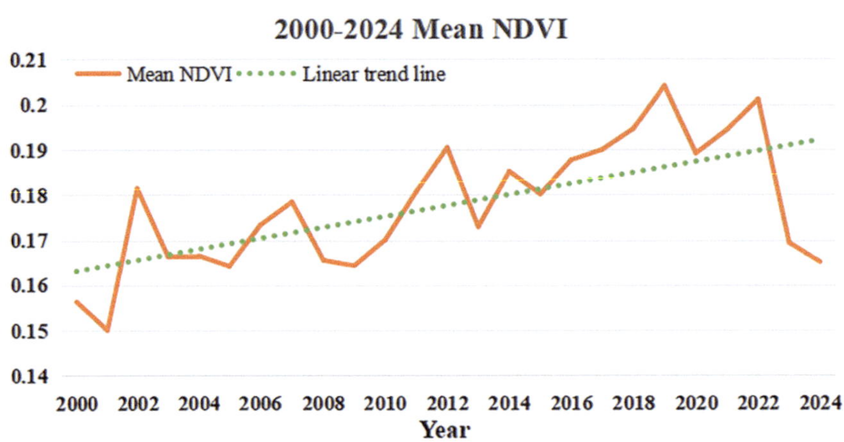

Fig. 2 Time change chart from 2000 to 2024

Figure 3 illustrates the spatial distribution of mean normalized difference vegetation index (NDVI) in the Shiyang River Basin from 2000 to 2024. Overall, NDVI values range from 0.005 to 0.857, indicating significant spatial variation in vegetation cover across the basin. The figure shows that the southern and southwestern mountainous areas, mainly the Qilian Mountains, exhibit higher NDVI values (Wu et al. 2023), with dense and continuous vegetation cover. In contrast, the central and northern parts of the basin, particularly the downstream plains and arid regions, show markedly lower NDVI values (< 0.2), reflecting sparse vegetation and pronounced arid-desert characteristics. In general, NDVI exhibits a south-to-north decreasing gradient, which is closely related to topographic relief, precipitation distribution, and the intensity of human activities.

Figure 4 presents the spatial distribution of NDVI slopes in the Shiyang River Basin from 2000 to 2024. Overall, positive NDVI slopes dominate the study area, indicating a general improvement in vegetation over the past 25 years. The regions showing improvement are mainly located in the midstream and parts of the upstream areas, where higher NDVI slope values reflect significant enhancement in vegetation growth. These improvements are closely associated with ecological restoration measures, such as the conversion of farmland to forest and grassland, water resource regulation, and oasis ecological construction.

In contrast, areas with negative NDVI slopes are primarily concentrated in the central and southern Hexi Corridor and parts of the downstream regions, particularly

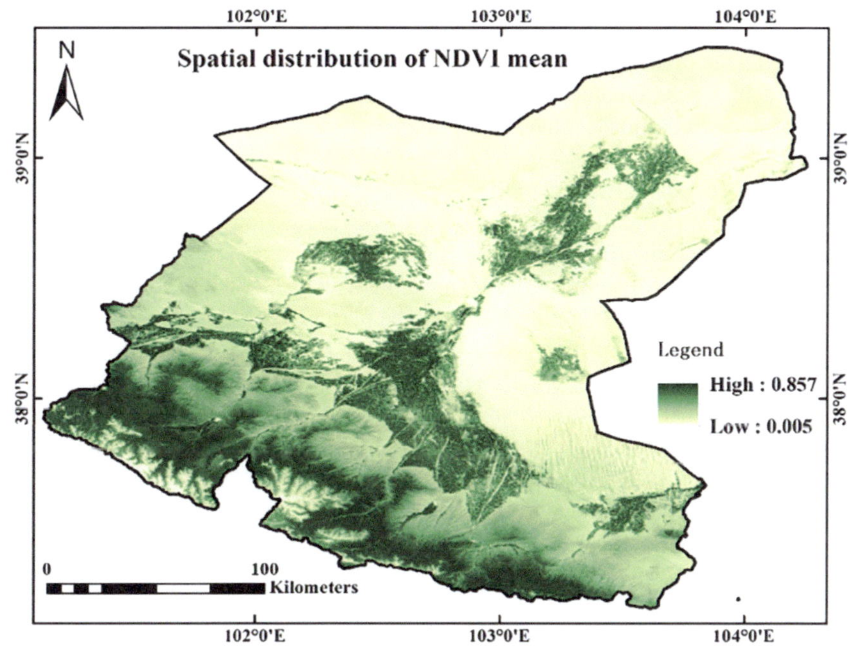

Fig. 3 Spatial distribution map of NDVI mean

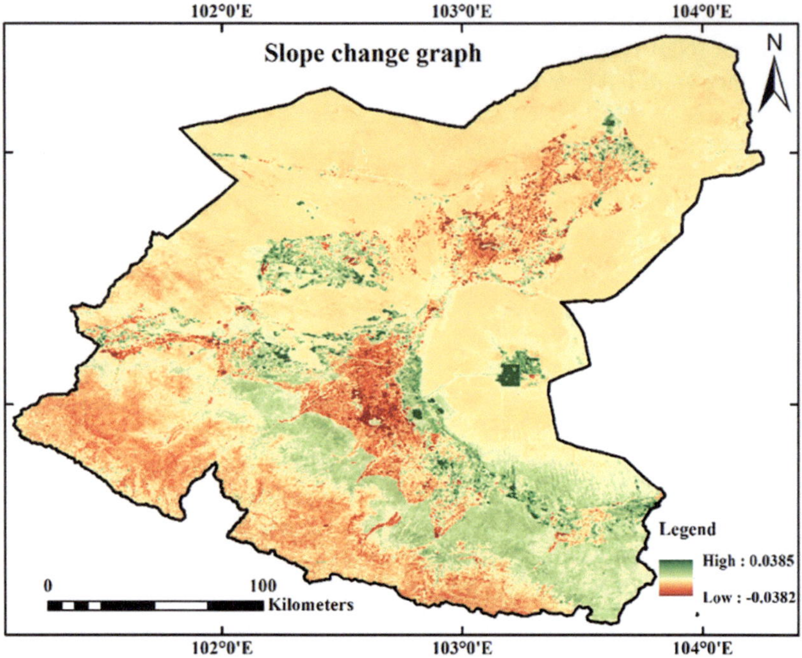

Fig. 4 Spatial distribution map of NDVI slope from 2000 to 2024

along the edges of arid desert zones and the periphery of irrigation areas. The relatively low slope values indicate vegetation degradation, likely resulting from a combination of increased aridity, intensified human activities, and declining groundwater levels.

Based on the NDVI trend classification map from 2000 to 2024 (Fig. 5) derived using the Mann–Kendall (MK) test and linear regression, areas of significant improvement are primarily concentrated in the midstream oasis and upstream mountainous regions. Slightly improved areas are more widely distributed, mainly covering the transitional zones of the midstream and the peripheries of local irrigation areas. Relatively stable areas are located in the central Hexi Corridor and parts of the arid desert regions, indicating that these regions maintain stable ecological conditions with minimal vegetation change. In contrast, areas of significant and slight degradation are mainly distributed in the downstream regions, including Minqin, Gulang, and parts of the desert margins. These areas are characterized by scarce water resources, arid climate, and high intensity of human activities, resulting in pronounced vegetation degradation trends.

The Hurst exponent reflects the persistence characteristics of a time series, where $H > 0.5$ indicates a persistent trend, and $H < 0.5$ indicates a tendency for reversal. As shown in Fig. 6a, the Hurst exponent in the Shiyang River Basin is generally high, suggesting that future vegetation changes are likely to follow a persistent trend. In particular, the upstream mountainous areas and midstream oasis regions exhibit

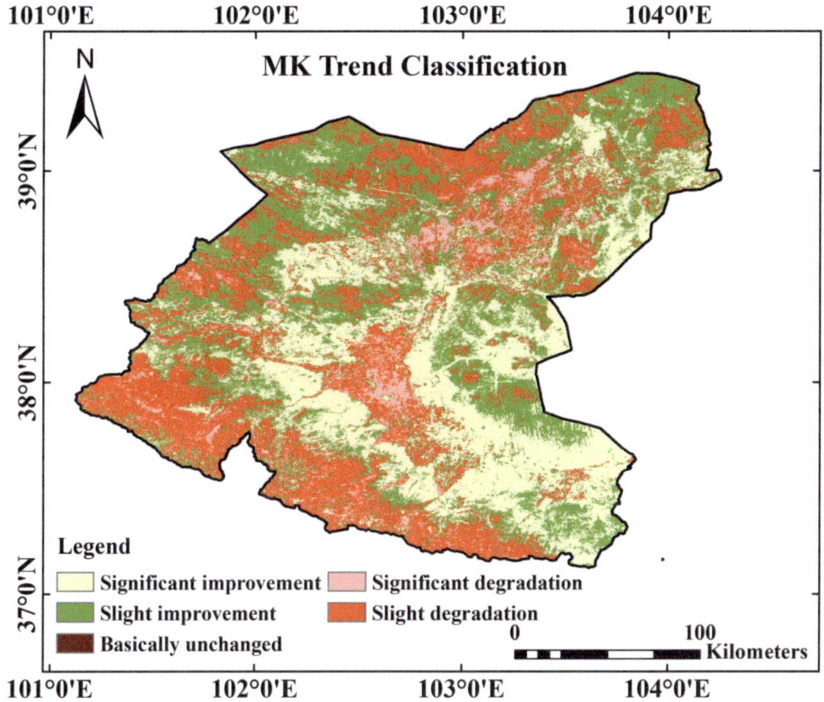

Fig. 5 MK trend classification

relatively high Hurst values, indicating that the upward NDVI trends in these regions are expected to remain stable. In contrast, the downstream desert areas have lower Hurst values, implying greater uncertainty in vegetation dynamics.

Based on a cross-analysis of the Hurst exponent and linear trends (Fig. 6b), future vegetation changes are predominantly expected to show sustained improvement,

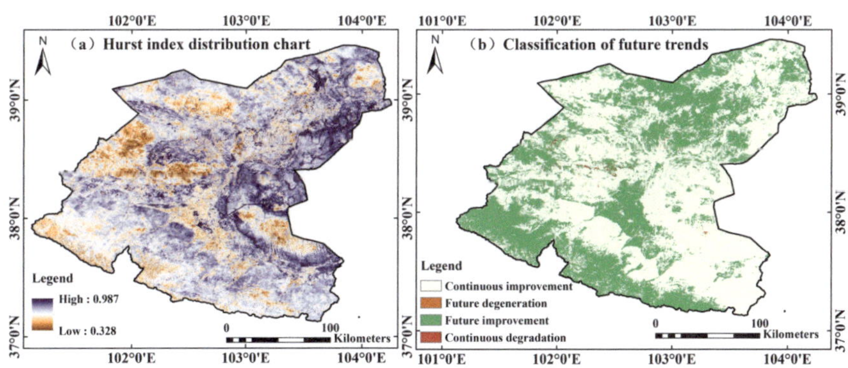

Fig. 6 Hurst index and future trend classification

covering 60.57% of the basin area, mainly in the upstream Qilian Mountains (Li et al 2025) and midstream oasis farmland region. Areas of future improvement are primarily located along oasis margins and transitional zones, where vegetation may continue to improve but with potential fluctuations. Future degradation and persistent degradation areas are concentrated in the downstream Minqin Basin and the edges of arid desert zones. These regions are constrained by limited water resources and fragile ecosystems, indicating that vegetation is likely to continue declining in the future.

6 Discussion

6.1 Spatiotemporal Patterns of Vegetation Change and Driving Mechanisms

This study reveals a significant improvement in vegetation cover in the Shiyang River Basin from 2000 to 2024, with NDVI increasing from approximately 0.12 to 0.21. This upward trend is consistent with observations in other arid inland river basins in northwestern China (Sun et al. 2015; Wang et al. 2025), such as the Heihe (Han et al. 2023; Yuan et al. 2019) and Tarim River Basins (Li et al. 2021), where ecological restoration projects have similarly enhanced regional vegetation. The observed improvements can be attributed to the combined effects of climatic factors and human interventions. The southern mountainous areas of the Qilian Mountains exhibit persistently high NDVI values (> 0.6), benefiting from relatively abundant precipitation and snowmelt (Li et al. 2016). In contrast, the downstream Minqin Basin maintains low NDVI values (< 0.2), reflecting severe water scarcity and desertification pressures. This north–south gradient is a typical feature of arid inland river basins, where water resource distribution fundamentally controls vegetation patterns.

6.2 Vegetation Degradation in Downstream Areas: Causes and Impacts

Mann–Kendall test results indicate sustained vegetation degradation in the downstream regions, particularly around the Minqin Basin, concentrated along desert margins and the peripheries of irrigation areas. This pattern contrasts sharply with the improvements observed in upstream and midstream areas, suggesting that current water resource allocation strategies may inadvertently exacerbate spatial inequalities in vegetation cover. Downstream degradation is likely driven by multiple mechanisms. First, agricultural water use in the midstream reduces the availability of water for downstream ecosystems. Second, increased human activities, including urbanization, place additional pressure on already limited water resources. Furthermore, vegetation loss may generate positive feedback loops, as reduced vegetation cover diminishes soil water retention capacity, further limiting the potential for vegetation recovery (Pan et al. 2018). These cascading effects underscore the urgency of implementing integrated watershed management strategies to balance socio-economic development with ecological protection (Li et al. 2025).

6.3 Future Vegetation Trends: Opportunities and Challenges

Hurst exponent analysis indicates that 60.57% of the basin exhibits sustained improvement trends, suggesting that ongoing ecological restoration efforts have established positive trajectories in most areas, particularly in upstream and midstream oasis regions (Lu et al. 2025). The high Hurst values in the improved areas indicate strong persistence, demonstrating that ecological restoration investments have yielded sustainable outcomes. However, the lower Hurst values and negative slopes in downstream areas highlight high uncertainty and potential reversibility of vegetation conditions, reflecting the marginal stability of desert ecosystems. The spatial heterogeneity of persistence emphasizes the need for differentiated management strategies tailored to regional characteristics.

6.4 Limitations and Future Directions

Although the study period is relatively long, it may not fully capture long-term climatic cycles or the lagged effects of ecological restoration measures. Extending time-series analyses by integrating historical records and future projections would provide a more comprehensive perspective on vegetation dynamics (Wang et al. 2023). While pixel-based analyses offer high spatial resolution, they do not account for landscape connectivity or ecosystem function. Integrating landscape ecology approaches and ecosystem service assessments could provide a more holistic evaluation of ecological restoration outcomes.

Future research should prioritize the following directions: (1) conducting comprehensive assessments of ecological restoration effectiveness using multiple indicators beyond vegetation cover, including biodiversity, soil quality, and ecosystem services; (2) projecting vegetation dynamics under different water resource management and climate change scenarios; and (3) conducting comparative studies across multiple inland river basins to identify generalizable patterns and management strategies. These efforts will advance both theoretical understanding and practical management of arid land ecosystems in dryland regions.

7 Conclusions

(1) Between 2000 and 2024, NDVI in the Shiyang River Basin exhibited a fluctuating upward trend, indicating a significant improvement in regional vegetation cover. The annual mean NDVI increased from approximately 0.156 in 2000 to 0.201 in 2022, reflecting the positive effects of ecological engineering and vegetation restoration measures.

(2) Spatially, the mean NDVI shows a pronounced south-to-north gradient. Vegetation cover in the upstream Qilian Mountains is generally high, with NDVI values exceeding 0.6. Significant vegetation improvements are observed in the midstream oasis areas, whereas NDVI values are relatively low in the downstream Minqin Basin and arid desert regions, indicating fragile ecological conditions. Overall, vegetation cover exhibits a clear gradient corresponding to topographic variation and water–energy conditions.

(3) NDVI trend analysis indicates that over the past 25 years, most of the basin has experienced vegetation improvement, with areas of significant improvement mainly distributed in the upstream mountainous regions and midstream oasis, accounting for approximately 65% of the total area. Degraded areas are primarily concentrated in the downstream arid regions, where vegetation decline is closely related to groundwater depletion and intensified human activities.

(4) Hurst exponent analysis suggests that NDVI changes in the Shiyang River Basin generally exhibit strong persistence. Areas of sustained improvement account for 60.57% of the basin and are mainly distributed in the upstream and oasis agricultural regions. Future degradation and persistent degradation areas are predominantly located in the downstream arid regions, where the uncertainty of future vegetation changes is relatively high.

Overall, the ecological environment of the Shiyang River Basin has improved, but significant north–south differences persist. Strengthening water resource management and ecological restoration in the downstream areas is essential to enhance vegetation stability and achieve coordinated and sustainable development of the basin's ecosystems.

Acknowledgements This research is supported by the National Key Research and Development Program of China (No. 2023YFC3206800), National Natural Science Foundation of China (52509038) and Henan Province Youth Talent Support Project (2025HYTP030). Special thanks are owed to editors and anonymous reviewers whose comments helped to improve the paper.

Declaration of Competing Interest The authors declare that they have no known competing financialinterests or personal relationships that could have appeared to influencethe work reported in this paper.

References

Chen, C., Li, G.: Temporal and spatial variation of grassland NDVI in Kunlun mountains, Altun mountains and its responses to temperature and precipitation from 1981 through 2020. Chin. J. Grassl. **45**(2), 13–25 (2023)

Feng, P., Yang, W., Li, J.-Z., Yang, N.-J., Duan, J.-C.: Vegetation cover changes and their responses to natural and anthropogenic factors in the Luanhe River Basin. [English Abstract]. Ying yong sheng tai xue bao = J. Appl. Ecol. **36**(4), 1222–1232 (2025). https://doi.org/10.13287/j.1001-9332.202504.022

Guo, J., Liao, W., Qimuge, H., Xu, Y., Wang, J.: Seasonal analysis of spatial and temporal variations in NDVI and its driving factors in inner Mongolia during the vegetation growing season (1999–2019). Front. For. Glob. Chang. **8** (2025). https://doi.org/10.3389/ffgc.2025.1555385

Han, Z., Meng, J., Zou, Y., Zhu, L.: Vegetation dynamics and their response to climate change and ecological protection projects in the Heihe River Basin from 1982 to 2017. J. Desert Res. **43**(3), 96–106 (2023)

Jin, Y.-S., Jin, K., Wang, F., Liu, C.-X., Qin, P., Zong, Q.-L., et al.: Impacts of climate change and human activities on NDVI change in eastern coastal areas of China. [English Abstract]. Huan jing ke xue= Huanjing kexue **44**(6), 3329–3342 (2023). https://doi.org/10.13227/j.hjkx.202207039

Lai, J.-L., Qi, S., Cui, R.-R., Liao, R.-E., Tang, Y., Li, P.: Analysis of vegetation change and influencing factors in Southwest Alpine Canyon area. [English Abstract]. Huan jing ke xue= Huanjing kexue **44**(12), 6833–6846 (2023). https://doi.org/10.13227/j.hjkx.202212190

Lan, M., Zhou, D., Guan, Y.: Spatiotemporal variation and driving factors of normalized difference vegetation index (NDVI) in Yan'an City. J. Zhejiang A&F Univ. **41**(6), 1293–1302 (2024)

Li, Y., Chen, Y., Ye, Z., Wang, F., Sun, F., Qin, J.: Ecological responses of ecological water conveyance in the lower reaches of Tarim River for 20 years. Arid. Land Geogr. **44**(3), 700–707 (2021)

Li, X., Cao, G., Chen, Z., Yuan, J., Sun, Z., Tang, J.: Spatiotemporal dynamics and future sustainability of vegetation NDVI on the southern slope of the Qilian mountains from 2001 to 2022. Res. Soil Water Conserv. **32**(3), 196–205 (2025)

Li, A., Yin, S., Li, N., Shi, C.: Comprehensive analysis of the driving forces behind NDVI variability in China under climate change conditions and future scenario projections. Atmosphere **16**(6), 738 (2025)

Li, H., Jiang, J., Chen, B., Li, Y., Xu, Y., Shen, W.: Pattern of NDVI-based vegetation greening along an altitudinal gradient in the eastern Himalayas and its response to global warming. Environ. Monit. Assess. **188**(3) (2016). https://doi.org/10.1007/s10661-016-5196-4

Liiliu, L.I.U., Hongmei, X.U.: Change of NDVI of main vegetations and their relationship with meteorological factors in Yellow River Basin. Chin. J. Agrometeorol. **28**(3), 334–337 (2007)

Lu, C., Hou, M., Liu, Z., Li, H., Lu, C.: Variation characteristic of NDVI and its response to climate change in the middle and upper reaches of Yellow River Basin, China. IEEE J. Sel. Top. Appl. Earth Obs. Remote. Sens. **14**, 8484–8496 (2021). https://doi.org/10.1109/jstars.2021.3105897

Lu, M., Sun, H., Yang, Y., Xue, J., Ling, H., Zhang, H., et al.: Assessing recovery time of ecosystems in China: insights into flash drought impacts on gross primary productivity. Hydrol. Earth Syst. Sci. **29**(3), 613–625 (2025). https://doi.org/10.5194/hess-29-613 2025

Miao, L., Zhang, Y., Chuai, X., Bao, G., He, Y., Zhu, J.: Effects of climatic factors and their time-lag on grassland NDVI in Asian drylands. Chin. J. Plant Ecol. **47**(10), 1375–1385 (2023)

Pan, F., Xie, J., Lin, J., Zhao, T., Ji, Y., Hu, Q., et al.: Evaluation of climate change impacts on wetland vegetation in the Dunhuang Yangguan National Nature Reserve in Northwest China using Landsat derived NDVI. Remote. Sens. **10**(5) (2018). https://doi.org/10.3390/rs10050735

Qi, R., Li, Q., Ren, J., Zou, M., Yang, H., Wei, Y., et al.: Study on the characteristics of changes in vegetation cover and its driving forces in the three-North Shelterbelt program regions: taking Ningxia as example. Arid Zone Res. **41**(10), 1740–1752 (2024)

Sun, W., Song, H., Yao, X., Ishidaira, H., Xu, Z.: Changes in remotely sensed vegetation growth trend in the Heihe Basin of Arid Northwestern China. Plos One **10**(8) (2015). https://doi.org/10.1371/journal.pone.0135376

Wang, Y., Chen, T., Wang, Q., Peng, L.: Time-lagged and cumulative effects of drought and anthropogenic activities on China's vegetation greening from 1990 to 2018. Int. J. Digit. Earth **16**(1), 2233–2258 (2023). https://doi.org/10.1080/17538947.2023.2224086

Wang, Z., Che, T., Dai, L., Jiang, Y.: Comparative analysis on the spatiotemporal changes of vegetation and drivers in Inland River Basins of the Hexi Corridor, China. IEEE J. Sel. Top. Appl. Earth Obs. Remote. Sens. **18**, 8112–8124 (2025). https://doi.org/10.1109/jstars.2025.3548982

Wang, X., Liu, Z., Jiao, K.: Spatiotemporal dynamics of normalized difference vegetation index (NDVI) and its drivers in forested region of Northeast China during 2000–2017. Chin. J. Ecol. **39**(9), 2878–2886 (2020)

Wu, J., Jiao, L., Zhang, H., Du, D., Zhu, X., Che, X.: Vegetation coverage variation in the Qilian mountains before and after ecological restoration. Acta Ecologica Sinica **43**(1), 408–418 (2023)

Yu, S., Du, W., Zhang, X., Hong, Y., Liu, Y., Hong, M., et al.: Spatiotemporal changes in NDVI and its driving factors in the Kherlen River Basin. Chin. Geogr. Sci. **33**(2), 377–392 (2023). https://doi.org/10.1007/s11769-023-1337-1

Yuan, L., Chen, X., Wang, X., Xiong, Z., Song, C.: Spatial associations between NDVI and environmental factors in the Heihe River Basin. J. Geogr. Sci. **29**(9), 1548–1564 (2019). https://doi.org/10.1007/s11442-019-1676-0

Zhu, L., Meng, J., Zhu, L.: Applying Geodetector to disentangle the contributions of natural and anthropogenic factors to NDVI variations in the middle reaches of the Heihe River Basin. Ecological Indicators **117** (2020). https://doi.org/10.1016/j.ecolind.2020.106545

A Study to Evaluate the Positive Impacts of River Management and Greening Projects on Public Water Recreation, Sports Environments, Health, and Well-being

Hsiao-Hsien Lin, Chien-Hung Wu, Chin-Hsien Hsu, and Yong-Zhan Zheng

Abstract The primary aim of this study was to examine the effects of river restoration and greening projects on individuals' environmental experiences and well-being during water-based recreational activities. The study involved surveys and interviews with both residents and academic experts along the Mulan River in China. Data were collected through random and purposive sampling. Data analysis was performed using questionnaires, interviews, and multivariate statistical methods. The findings indicated that the public perceived river restoration and greening projects as having a positive effect on their environmental experience and well-being related to water-based recreation. However, challenges such as the complex implementation procedures and technical requirements of these projects, combined with residents' economic constraints and livelihood priorities limited the effectiveness of river conservation efforts and hindered the public's positive perception of the recreational environment and well-being. To address these issues, the government should establish a river conservation incentive program that publicly recognizes individuals or enterprises successfully implementing environmental conservation measures, enhances advocacy efforts, and encourages active participation in conservation activities. These would help overcome current policy challenges and enhance the public's positive experience and well-being related to river-based recreational activities.

H.-H. Lin
School of Physical Education, Jiaying University, Meizhou, China

H.-H. Lin · C.-H. Hsu (✉)
Department of Healthcare Industry Technology Development and Management, National Chin-Yi University of Technology, Taichung, Taiwan
e-mail: hsu6292000@yahoo.com.tw

C.-H. Wu
Department of Marine Recreation, National Penghu University of Science and Technology, Magong, Taiwan

Y.-Z. Zheng (✉)
Department of Recreation and Sports Management, Tajen University, Yanpu, Pingtung, Taiwan
e-mail: m04370001@tajen.edu.tw

H. Xu (ed.), *Proceedings of the 8th International Symposium on Water Resource and Environmental Management*, Environmental Science and Engineering,
https://doi.org/10.1007/978-3-032-19088-8_17

Keywords Environmental conservation efforts · Water recreation · Environmental experience value · Well-being · Friendly business

1 Introduction

Rivers are generally classified into three types based on flow velocity, area, and water volume: main streams, tributaries, and distributaries. Although river lengths vary (Bernauer and Böhmelt 2020), they all share characteristics such as large drainage areas and stable water volumes (Peng et al. 2010). Rivers are vital water resources that humans rely on for survival and development. However, global climate change and meteorological anomalies have led to irregular rainfall patterns across many regions. As a result, river water volume and flow velocity fluctuate due to changing climatic conditions. When river flow and velocity remain stable, and water quality, cleanliness, and sanitation are maintained at a high standard, river water resources can meet the essential needs for human survival and urban development (Loucks and Van Beek 2017). Such stability contributes to a higher quality of life, greater security, and improved economic conditions (Aselina 2014; Babatunde 2019). Conversely, instability can lead to ecological and environmental damage in surrounding communities, posing risks to public safety and disrupting recreational activities. Therefore, improving river stability, ensuring safe environments, and promoting sustainable communities that support physical and mental well-being remain critical challenges.

The Mulan River, 105 km long, runs between Licheng and Chengxiang in China. It serves as a vital water source for the livelihoods and irrigation needs of the surrounding cities and communities (Zeng et al. 2019). However, early water conservancy engineering and technology in China were limited (Abdul Quadir et al. 2016). Factors such as climate change, irregular rainfall patterns (Mandal et al. 2021; Guo-Zhen 2017), and erosion caused by heavy rains have significantly altered the topography and soil structure around the river. As a result, the waterway can no longer accommodate the volume of water brought by intense rainfall (Zhi-Zhonglu 2017; Lv et al. 2020). In response, the local government has integrated the river's scenic value to develop riverside spaces that emphasize leisure, livability, tourism, historical culture, and green energy industries (Zhu and Chen 2020). They have implemented river management and greening water conservancy projects (Zeng et al. 2019) to address these challenges. These initiatives aim to resolve issues related to embankment breaches, water resource management, and ecological pollution, while creating a safe environment for community leisure and sports activities (Al-Azzawi 2019), ultimately enhancing the well-being of both the city and its residents (Calp 2020).

Differences in time, location, and demographics may lead to discrepancies between policy development goals, historical research findings, and actual outcomes. Consequently, scholars suggest that gathering opinions from on-site participants through various perspectives and methods can enhance understanding of decision-making effectiveness and yield the most insightful and direct answers (Lin et al. 2021). Additionally, by applying environmental experience theory and considering perceptions of logical, emotional, and practical value, it is possible to analyze the public's physiological and psychological evaluations of daily life and their surrounding environment (Dreyer et al. 2018). Personal subjective feelings can also serve as indicators of current emotional states and levels of satisfaction, helping determine whether material needs are being met and whether a positive interactive atmosphere exists between the living environment and the public (Averill and More 1993; Berridge 2018; Oliveira-Castro 2021). These criteria offer an authentic evaluation of local policies, building facilities, leisure and recreational environments, spatial improvements, and people's overall life experience (Dinç Uyaroğlu 2021). Therefore, this study aimed to analyze the impact of river improvement and greening projects on the happiness index of both the public and the city through case studies, in order to assess how these projects influence the urban water recreation environment and public well-being.

2 Methodology

Based on cognitive theory, this study employed quantitative research methods and drew upon relevant literature on environmental experience perception (Dreyer et al. 2018; Oliveira-Castro 2021; Dinç Uyaroğlu 2021; Chua et al. 2019; Alexander and Cano 2019; Kang 2020; Lin 2016) and happiness (Averill and More 1993; Berridge 2018; Diener et al. 1985; Hu 2021; Tech and Theophilos 2020) to develop a questionnaire. The questionnaire consisted of three sections. The first section collected demographic information, including gender (male or female) and age groups (under 20, 21–30, 31–40, 41–50, 51–60, and over 61). The second section focused on environmental experience value, containing six questions, while the third section addressed well-being with seven questions. The related items are presented in Table 1. According to scholars, a Kaiser–Meyer–Olkin (KMO) value greater than 0.6 and a Bartlett's test p-value less than 0.01 ($p < 0.01$) indicate that the scale is suitable for factor analysis (Ajzen 1991). Items with a Cronbach's alpha above 0.60 demonstrate good reliability and can be further analyzed (Bower et al. 2019).

The assessment of the water recreation environment experience comprised 6 questions. Analysis revealed a KMO value of 0.919, a Bartlett's test approximation χ^2 of 1525.31 with 15 degrees of freedom, and a significance level of $p < 0.001$, indicating that factor analysis was appropriate. The scale accounted for 72.68% of the total variance. Following factor analysis, reliable items were retained and categorized into three factors: Education (2 questions), Escapism (2 questions), and Aesthetics (2 questions). The Cronbach's alpha coefficients for these factors ranged from 0.924 to 0.935, with an overall alpha of 0.941 for the entire scale.

The happiness questionnaire contained 7 items. Analysis revealed a Kaiser–Meyer–Olkin (KMO) measure of 0.937 and a Bartlett's test of sphericity with a chi-square value of 1730.426, degrees of freedom (df) of 21, and a significance level of $p < 0.001$, indicating that the data were suitable for factor analysis. The total variance explained by the scale was 70.616%. After factor analysis, reliable items

Table 1 Analysis of differences in public perception of the environment and well-being related to water recreation

Facets (M)	Issue	Public		Gender				
		M	Rank	Male (M)	Rank	Female (M)	Rank	p
Perceived water recreation and sports environment	It helps promote environmental conservation awareness	3.82	4	3.75	5	3.90	2	0.116
	Provide and learn environmental protection knowledge	3.74	6	3.68	6	3.80	6	0.459
	The riverbanks around the river are beautiful	3.81	5	3.77	4	3.86	5	0.281
	Being here allows you to forget your troubles	3.83	3	3.79	3	3.87	4	0.050*
	I like the riverbank landscape design style	3.85	2	3.81	2	3.89	3	0.543
	The overall riverbank environmental space planning is complete	3.88	1	3.86	1	3.91	1	0.888
Well-being (3.92)	Trying new things	3.87	5	3.82	5	3.94	4	0.034*
	Having goals in life	3.90	3	3.87	4	3.94	4	0.494
	Being younger and more energetic	3.98	2	3.94	1	4.03	2	0.015*
	Being more optimistic	3.99	1	3.93	2	4.06	1	0.008*
	Developing different skills and abilities	3.89	4	3.90	3	3.87	6	0.023*
	Enjoying my life	3.89	4	3.82	5	3.97	3	0.023*
	Social interaction	3.90	3	3.87	4	3.93	5	0.499

*$P < 0.05$

were retained. The Cronbach's alpha coefficients ranged from 0.933 to 0.937, with an overall alpha of 0.944. Based on these results, the KMO, Bartlett's test, and degrees of freedom for perceived environmental value and happiness all met the required standards. Therefore, the authors conclude that this questionnaire demonstrates good reliability.

After finalizing the questionnaire, the researchers distributed 500 copies in July and August 2025, receiving 300 valid responses. They analyzed the data using SPSS 27.0, employing basic statistical tests, t-tests, ANOVA, and Pearson's product-moment correlation. Subsequently, interviews were conducted with experts in policy analysis, environmental management, leisure management, ecology, and hydraulic engineering, as well as with residents aged 20–50, to gather their personal insights. Background information and interview topics are provided in Appendix 1. Finally, a multivariate analysis method was applied to examine the data from multiple perspectives (Strauss and Corbin 1990; Janesick 2000). The research framework is illustrated in Fig. 1.

Hypothesis 1: It is hypothesized that following the river restoration and greening project, the public will consistently have a positive perception of their experience (H1-1) and an enhanced sense of well-being (H1-2) related to the river's recreational and aquatic environment.

Hypothesis 2: It is assumed that following the river restoration and greening project, people of different genders will consistently have a positive perception of their experience (H2-1) and a heightened sense of well-being (H2-1) related to the river's recreational and aquatic environment.

Hypothesis 3: It is assumed that experiencing recreational and aquatic environments significantly influences the perception of well-being.

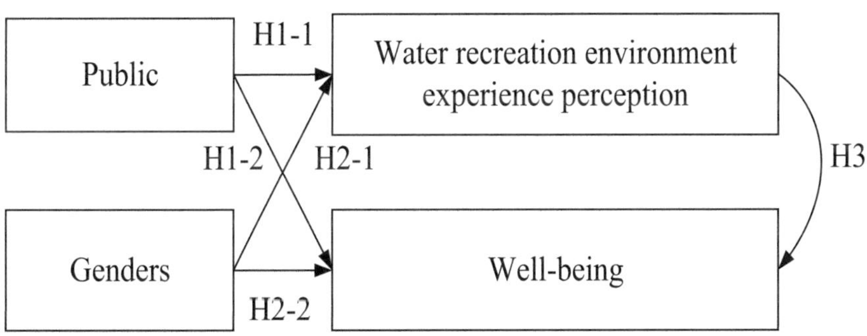

Fig. 1 Framework

3 Results

3.1 Analysis of the Public's Experience and Well-being in River Recreation and Sports Environments

First, basic statistical tests were conducted to analyze the public's perceived experience and well-being related to the water recreation environment, as presented in Table 1. The analysis of perceived experience indicated that the highest rating was for the "well-planned overall riverbank environment" (3.88), while the lowest rating was for "providing environmental protection knowledge" (3.74). Subsequently, the well-being analysis showed that the highest score was for "more optimistic about life" (3.99), and the lowest was for "trying new things" (3.87).

These results indicate that the public perceived the river improvement and greening project as having the greatest impact on the overall spatial planning of the riverbank environment and on enhancing their sense of optimism about life in relation to the water recreation environment. In contrast, the project has the least impact on "providing environmental protection knowledge." These findings did not support Hypothesis 1.

3.2 Analysis of the Experience and Well-being in River Recreation and Sports Environments by Gender

Secondly, basic statistical tests were conducted to analyze the perceptions of male and female respondents regarding their experiences and well-being related to river recreational activities, as presented in Table 1. For male respondents, the highest-rated aspect of their river recreational activities was the "well-planned overall riverbank environment" (3.86), while the lowest-rated was "providing opportunities to learn about environmental protection" (3.80). In terms of well-being, the highest score was for "feeling younger and more energetic" (3.94), and the lowest was for "trying new things and enjoying my life" (3.82). For female respondents, the highest-rated aspect of river recreational activities was also the "well-planned overall riverbank environment" (3.91), with the lowest being "providing opportunities to learn about environmental protection" (3.68). Regarding well-being, female respondents rated "being optimistic about life" highest (4.06), and "developing different skills and abilities" lowest (3.87).

It is clear that male respondents perceive river improvement and greening projects as having the greatest impact on their experience of river recreational activities and personal well-being, especially in terms of a "well-planned overall riverbank environment" and feeling more energetic." Conversely, they reported the least impact from factors such as learning about environmental protection, trying new activities, and simply enjoying life. Female respondents, on the other hand, identified improved overall riverbank environmental planning and a more optimistic outlook on life as having the greatest impact, while gaining knowledge about environmental protection and developing new skills and abilities were seen as having the least impact. These findings did not support Hypothesis 2.

3.3 Analysis of Differences in Experience and Perceived Happiness of River Recreational Activities Between Genders

A t-test was conducted to examine gender differences in perceived experiences and well-being related to river recreational sports environments, as presented in Table 1. The results indicated a significant gender difference in the ability to forget worries while in these environments ($p < 0.05$), with women reporting greater satisfaction. Regarding well-being, significant differences were observed in experiences such as trying new things, feeling younger and more energetic, having a more optimistic outlook, enjoying life, and various skills and abilities ($p < 0.05$). Women reported higher satisfaction across most of these areas, whereas men reported greater satisfaction with specific skills and abilities.

This finding suggests that perceptions of river-based experiences differ by gender. Women reported greater satisfaction with certain aspects, whereas men expressed higher satisfaction with different skills and abilities. These results did not support Hypothesis 3.

3.4 Correlation Analysis of the Water Recreation Sports Environment Experience and Well-being Perception

Following this, Pearson's product-moment correlation analysis was conducted to examine the influence of the water recreation environment experience on perceived happiness. The results showed that the public's perception of the water recreation environment significantly affected happiness ($p < 0.05$). The strongest perception was the ability to forget one's troubles while in the area (0.467), whereas the lowest was the perception of the overall well-planned riverbank environment (0.352).

This finding suggests that the public feels the river improvement and greening project has positively influenced their view of the local water recreation environment and boosted their overall well-being. Notably, the ability to forget one's troubles while in the area has the greatest effect on enhancing happiness, whereas the well-planned riverbank environment has the least impact (Table 2).

Table 2 Analysis of cognitive differences in environmental experience and well-being perception in water recreation sports

	Perceived water recreation and sports environment		
Well-being	0.451*		
	It helps promote environmental conservation awareness	Provide and learn environmental protection knowledge	The riverbanks around the river are beautiful
	0.426*	0.377*	0.369*
	Being here allows you to forget your troubles	I like the riverbank landscape design style	The overall riverbank environmental space planning is complete
	0.467*	0.388*	0.352*

*P < 0.05

4 Discussions

4.1 *Public Perception of Riverside Recreation, Sports Environment, and Well-being*

River management and greening projects employ ecological engineering techniques alongside principles such as flood control, landscape beautification, and ecological conservation to address river design flaws and reduce pollution (Zeng et al. 2019). These policies and engineering approaches help mitigate risks associated with extreme rainfall events caused by global climate change and meteorological anomalies, which can result in levee breaches, urban flooding, and damage to public life and recreational environments. They also provide a safer living and recreational environment for riverside residents while supporting local economic development. Moreover, when people's living conditions are secure, they can have positive recreational experiences, fostering feelings of safety and enjoyment. However, despite a general appreciation for the environment, protecting river water and promoting ecological sustainability remain shared responsibilities among citizens. In practice, river management and greening projects often involve lengthy construction periods, complex processes, and extensive project areas (Huang and Cai 2016). These challenges can leave the public feeling overwhelmed and stressed, reducing their willingness to take actions to protect river water and ecological sustainability. Consequently, public awareness and motivation to participate in environmental conservation efforts have gradually diminished, sometimes leading to reluctance to engage with or support relevant conservation measures. While river restoration and greening projects improve the overall riverside environment and enhance people's outlook on life, they do not necessarily increase public interest in environmental education or encourage the adoption of new conservation behaviors.

Therefore, this demonstrates that Hypothesis 1: the assumption that the public will consistently have a positive perception of the riverside recreational sports environment (H1-1) and happiness (H1-2) following river restoration and greening projects—is invalid. This finding confirms that implementing such projects does not necessarily result in a consistently positive experience or enhanced well-being related to water recreational sports.

4.2 Perceptions of Well-being and Recreational Experiences in Riverside Recreation and Sports Settings across Different Genders

The policy objectives of river management and greening projects are to enhance river safety, address pollution issues, conserve freshwater resources, and promote the sustainable development of natural ecosystems. These efforts positively impact the stabilization of river water quality and flow, as well as the preservation of the natural environment (Liu et al. 2020). Additionally, they support public and government promotion of river-based recreational activities (Ou 2020), thereby increasing people's willingness to engage in such activities along rivers. These conditions further encourage individuals who enjoy recreational sports to participate more actively. Consequently, both men and women agree that river management and greening projects improve the environmental quality of riverbanks.

Furthermore, as the public's willingness to participate in recreational sports grows, the frequency of such activities tends to increase accordingly. This phenomenon encourages individuals to reap the health benefits of recreational sports, including stronger bones and muscles, enhanced cellular activity, healthy development, and a positive, healthy outlook on life. Consequently, male respondents believe that river management and greening projects not only boost personal vitality but also help women cultivate positive values.

Stable river flow, clean water quality, and the potential for diverse ecosystems create a favorable environment for safe recreational sports and encourage public participation in river activities. However, improving river water quality and conserving ecosystems require significant time and effort, and river management and greening projects are complex undertakings. Additionally, residents often face low incomes and poor living standards, necessitating considerable effort to sustain their personal income and meet basic quality-of-life needs. As a result, they may lack the motivation to invest time in learning about environmental protection, engaging in conservation efforts, or maintaining their well-being. Consequently, male respondents believe that river management and greening projects do not effectively increase the willingness to learn about environmental protection, try new activities, improve personal quality of life, or enhance women's environmental awareness and behavior.

Therefore, this demonstrates that research hypothesis 2, which posits that following river restoration and greening projects, individuals of different genders will share a consistently positive perception of the river recreation environment (H2-1) and their sense of well-being (H2-1), is invalid. This phenomenon confirms that implementing such projects does not necessarily result in a uniform positive experience of the river recreation environment or an enhanced sense of well-being across different genders.

4.3 The Impact of Water Recreation and Sports Environment Experiences on Well-being

River management and greening projects play a crucial role in improving water quality, maintaining flow, and restoring diverse ecosystems. These efforts help create abundant recreational resources and environments, providing comfortable and safe spaces for leisure activities. When governments effectively oversee river water and ecological development, they foster safe recreational environments that attract a wider range of people to engage in river-based activities. This issue not only diversifies the types of recreational activities available but also broadens the demographics of participants. Additionally, stable, ecologically rich surrounding areas offer comfortable, secure spaces for recreation, benefiting both water sports enthusiasts and those less confident in swimming. Experiencing the health benefits of river and water-based recreation also raises public awareness and understanding of water quality, ecology, and environmental conservation. These advantages help individuals reduce stress, maintain health, and enjoy their time by the river. Consequently, analysis shows that participation in water-based recreational sports significantly enhances well-being. Moreover, it confirms that such activities promote greater environmental awareness, motivate learning, increase acceptance of current river environment planning and management, and improve overall enjoyment.

Therefore, this confirms that hypothesis 3—that the experience of water-based recreational sports significantly impacts perceived well-being—is valid. It also demonstrates that river improvement and greening projects positively contribute to well-being to some extent.

5 Conclusions

This study confirms that river restoration and greening projects positively influence public environmental experience and subjective well-being. However, perspectives on improvements in vitality, positive values, quality of life, environmental awareness, and practical conservation actions varied between genders. Because such projects often involve long durations and complex construction, residents tend to prioritize stable employment and income to maintain their quality of life rather than participate in river conservation volunteer work. These factors may limit the effectiveness of such projects in enhancing the environmental experience and well-being associated with water recreation. Based on these findings, the researchers recommend that the government develop tailored conservation action plans, establish incentive mechanisms for river conservation, organize parent-child river environmental volunteer activities to encourage public engagement, create river ecological conservation organizations to promote environmental awareness and behaviors, and publicly recognize environmentally responsible businesses and individuals for their efforts. Implementing these measures will help raise public environmental awareness, improve the effectiveness of river management and greening initiatives, and enhance the positive environmental experiences and well-being of those involved in water recreation activities.

Acknowledgements We sincerely thank all the colleagues and respondents who participated in this research, and we are especially grateful to the Department of Research at Chiayi College for supporting the College Development Project (2022WRC11).

References

Abdul Quadir, D., Van Scheltinga, C.T., Ludwig, F.: Recent changes in temperature and rainfall trends and variability over Bangladesh. In: The Gobeshona Conference for Research on Climate Change in Bangladesh, pp. 1–17 (2016). https://research.wur.nl/en/publications/recent-changes-in-temperature-and-rainfall-trends-and-variability

Ajzen, I.: The theory of planned behavior. Organ. Behav. Human Decis. Process. **50**, 179–211 (1991). https://doi.org/10.1016/0749-5978(91)90020-T

Al-Azzawi, A.: Dubai happiness agenda: engineering the happiest city on earth. In: Samad, W., Azar, E. (eds.) Smart Cities in the Gulf. Palgrave Macmillan, Singapore (2019). https://doi.org/10.1007/978-981-13-2011-8_11

Alexander, B., Cano, M.B.: Futurising the physical store in the omnichannel retail environment. In: Piotrowicz, W., Cuthbertson, R. (eds.) Exploring Omnichannel Retailing. Springer, Cham (2019). https://doi.org/10.1007/978-3-319-98273-1_9

Aselina, M.: Levels of industrial pollutants and their effects on water resources and livelihoods along Msimbazi sub catchment Dar es Salaam, Tanzania. Master thesis, Kenyatta University, Nairobi city, Kenya (2014). http://www.taccire.sua.ac.tz/handle/123456789/537

Averill, J.R., More, T.A.: Happiness. In: Lewis, M., Haviland, J.M. (eds.) Handbook of Emotions, pp. 617–629. The Guilford Press (1993)

Babatunde, A.O.: Oil pollution and water conflicts in the riverine communities in Nigeria's Niger delta region: challenges for and elements of problem-solving strategies. J. Contemp. Afr. Stud. **32**(2), 274–293 (2019). https://doi.org/10.1080/02589001.2020.1730310

Bernauer, T., Böhmelt, T.: International conflict and cooperation over freshwater resources. Nat Sustain. **3**, 350–356 (2020). https://doi.org/10.1038/s41893-020-0479-8

Berridge, K.C.: Evolving concepts of emotion and motivation. Front. Psychol. **9**, 1647 (2018). https://doi.org/10.3389/fpsyg.2018.01647

Bower, I., Tucker, R., Enticott, P.G.: Impact of built environment design on emotion measured via neurophysiological correlates and subjective indicators: a systematic review. J. Environ. Psychol. **66**, 101344 (2019). https://doi.org/10.1016/j.jenvp.2019.101344

Calp, Ş.: Peaceful and happy schools: how to build positive learning environments. Int. Electron. J. Elem.Tary Educ. **12**(4), 311–320 (2020). Retrieved from https://iejee.com/index.php/IEJEE/article/view/1073

Chua, B.L., Kim, H.C., Lee, S., Han, H.: The role of brand personality, self-congruity, and sensory experience in elucidating sky lounge users' behavior. J. Travel Tour. Mark. **36**(1), 29–42 (2019). https://doi.org/10.1080/10548408.2018.1488650
Diener, E., Emmons, R.A., Larsen, R.J., Griffin, S.: The satisfaction with life scale. J. Pers. Assess **49**(1), 71–75 (1985). https://doi.org/10.1207/s15327752jpa4901_13
Dinç Uyaroğlu, İ.: A performance evaluation tool for inclusiveness in university campus outdoor spaces. Megaron **16**(2), 197–211 (2021). https://doi.org/10.14744/megaron.2021.35336
Dreyer, B.C., Coulombe, S., Whitney, S., Riemer, M., Labbé, D.: Beyond exposure to outdoor nature: exploration of the benefits of a green building's indoor environment on well-being. Front. Psychol. **9**, 1583 (2018). https://doi.org/10.3389/fpsyg.2018.01583
Xu, G.-Z.: Comprehensive control measures for soil erosion in Mulanxi Basin, Putian City **33**, 164–165 (2017). https://www.sk163.com/qikan/12c8cf55e09a85e1fef1189fcac2a4dd.html
Hu, H.: Harmonious relationship-management based on human relationships. In: Principles of Chinese Management. Management for Professionals. Springer, Singapore (2021). https://doi.org/10.1007/978-981-33-6522-3_5
Huang, R., Cai, Z.: Research on the establishment of an evaluation index system for the sustainable construction of river engineering. J. Chin. Civ. Eng. Hydraul. Eng. **28**(2), 105–115 (2016)
Janesick, V.J.: The choreography of qualitative research design: minuets, improvisations, and crystallization. In: Denzin, N.K., Lincoln, Y.S. (eds.) Handbook of Qualitative Research, pp. 379–399. Sage, Thousand Oak, CA (2000)
Kang, J.: Identifying antecedents and consequences of well-being: the case of cruise passengers. Tour. Manag. Perspect. **33**, 100609 (2020). https://doi.org/10.1016/j.tmp.2019.100609
Lin, H.-H., Ling, Y., Lin, J.-C., Liang, Z.-F.: Research on the development of religious tourism and the sustainable development of rural environment and health. Int. J. Environ. Res. Public Health **18**(5), 2731 (2021). https://doi.org/10.3390/ijerph18052731
Lin, M.-C.: A study of the relationship between the environmental experience, attitude and behavior intention of elementary school children—a case of participants in Jia-Jian-Tan environmental ecological park. Master thesis, TOKO University, Chiayi, Taiwan (2016)
Liu, Z., Su, H., Huang, S., Chen, Y., Zhang, Y., Zhang, J.: Research on the effects of sewage system construction on river water quality improvement and greenhouse gas reduction. J. Taiwan Agric. Eng. **2020**, 66(2) (2020)
Loucks, D.P., Van Beek, E.: Water Resource Systems Planning and Management: An Introduction to Methods, Models, and Applications. Springer, Berlin (2017). https://doi.org/10.1007/978-3-319-44234-1
Lv, H., Yang, L., ZhouabXianZhang, J., Wenjing, W., Li, Y., Jiang, D.: Water resource synergy management in response to climate change in China: from the perspective of urban metabolism. Resour. Conserv. Recycl. **163**, 105095 (2020). https://doi.org/10.1016/j.resconrec.2020.105095
Mandal, T., Das, J., Rahman, A.T.M.S., Saha, P.: Rainfall insight in Bangladesh and India: climate change and environmental perspective. In: Rukhsana, H.A., Alam, A., Satpati, L. (eds.) Habitat, Ecology and Ekistics. Advances in Asian Human-Environmental Research. Springer, Cham (2021). https://doi.org/10.1007/978-3-030-49115-4_3
Oliveira-Castro, J.M.: Behavior analysis and psychological concepts: commentary on Foxall's intentional Behaviorism. In: Zilio, D., Carrara, K. (eds.) Contemporary Behaviorisms in Debate. Springer, Cham (2021). https://doi.org/10.1007/978-3-030-77395-3_14
Ou, J.W.C.: A political ecological analysis of watershed management: a case study of Xinhai wetland in the lower reaches of the Dahan River. Master's thesis, National Taiwan Normal University, Taiwan (2020)
Peng, J., Chen, S., Dong, P.: Temporal variation of sediment load in the Yellow River basin, China, and its impacts on the lower reaches and the river delta. Catena **83**(2–3), 135–147 (2010). https://doi.org/10.1016/j.catena.2010.08.006
Strauss, A., Corbin, J.: Basics of Qualitative Research: Grounded Theory Procedures and Techniques. Sage, Newbury Park, CA (1990)

Tech, R.R.M., Theophilos, C.: Preventing Bullying: A Manual for Teachers in Promoting Global Educational Harmony. Balboa Press, Liberty Dr, Bloomington, U.S (2020)

Zeng, J.-H., Guo, X.-D., Fan, S.-Q.: Putian City continues to manage Mulan Creek: a blueprint drawn to the end. J. Contemp. Jiangxi 19–21 (2019). https://www.sk163.com/qikan/50dc8adf0b2abe18450369f81ab08bdc.html

Zhi-Zhonglu: Case design of rural geography teaching based on regional cognitive literacy—taking the Mulan River Basin in Putian as an example (2017). http://cnki.sris.com.tw/kcms/detail/detail.aspx?recid=&FileName=ZUKZ202103028&DbName=CJFDLASN2021&DbCode=CJFD. Accessed 1 Sept 2021)

Zhu, Y., Chen, J.-Q.: The Key Elements and Practical Logic of Ecological Governance Modernization—Taking the Governance of the Mulanxi Watershed in Fujian as an Example, vol. 6, pp. 17–23. Southeast Academic (2020). http://yuxiqbs.cqvip.com//Qikan/Article/Detail?id=7103209784&from=Qikan_Search_Index

How Can Retail Marketing Mitigate the Impact of Blockchain-Based Water and Carbon Footprint Disclosures on the Consumption of Sustainable Luxury Goods in China?

Wenyi Tang, Tengye Chen, Shuqi Xuan, Yuan Yuan, and Weilun Huang

Abstract While blockchain technology is widely touted as a solution for enhancing sustainability transparency in the luxury sector, its ability to influence consumer behavior hinges not on water-and-carbon data immutability alone, but on how that data is interpreted and delivered through retail channels. Drawing on survey data from 397 high-income luxury consumers in China and analyzed via Partial Least Squares Structural Equation Modeling (PLS-SEM), this study reveals that nearly half (48.67%) of blockchain's effect on green consumption intentions operates indirectly through three mediators: retail marketing adaptation, perceived water-and-carbon transparency, and green trust (i.e., resistance to greenwashing). Once these mechanisms are taken into account, the direct impact of blockchain water and carbon disclosure is small and statistically fragile. These findings challenge the assumption of technological optimism and emphasise the key role of the retail interface (physical and digital), as water and carbon technology transparency becomes a place for meaningful action. We have proposed a comprehensive framework for luxury brands to embed the water and carbon environmental narrative supported by blockchain into specific channels and people-oriented experiences, so as to build credibility, reduce doubt and promote sustainable purchasing. Theoretically, the study bridges information systems and consumer behavior literatures by positioning retail marketing as the pivotal moderator of blockchain's water-and-carbon behavioral efficacy; practically, it offers actionable strategies for aligning technological capability with market reality in one of the world's most influential luxury markets.

Keywords Blockchain transparency · Sustainable luxury consumption · Retail marketing adaptation · Water and carbon footprint · Green trust · Greenwashing skepticism

W. Tang · T. Chen · S. Xuan · Y. Yuan · W. Huang (✉)
School of Management, Wenzhou Business College, Zhejiang, China
e-mail: huangwl@wzbc.edu.cn

H. Xu (ed.), *Proceedings of the 8th International Symposium on Water Resource and Environmental Management*, Environmental Science and Engineering,
https://doi.org/10.1007/978-3-032-19088-8_18

1 Preface

The global luxury industry, which has long been known for its craftsmanship, tradition and desirable charm, now finds itself in an era defined not only by aesthetic innovation, but also by escalating ecological responsibility—especially in the heavy use of water and carbon emissions in complex global supply chains. Historically, the industry has been free from censorship because of its unique aura, but the new generation of conscious consumers' requirements for elegance are getting higher and higher. They seek verifiable evidence of responsibility. Given that luxury fashion is one of the industries that consumes the most water and carbon emissions, this change is particularly urgent: a handbag may contain thousands of litres of water and produce more carbon dioxide than a transatlantic flight. While blockchain technology has emerged as a promising tool to authenticate sustainability claims through immutable, end-to-end data recording—from raw material sourcing to final delivery—the mere existence of such data does little to shift behavior if it remains inaccessible or unintelligible at the moment of purchase (Alotaibi 2024). According to the research conducted by Viet (2023), a growing chasm exists between the technical capacity to disclose environmental information and the retail mechanisms capable of translating that data into meaningful consumer experiences. Brands may proudly announce blockchain-verified water footprint metrics in corporate reports, yet fail to integrate them into the tactile, visual, or digital environments where purchasing decisions are actually made: a Gucci boutique in Milan might house a blockchain terminal few customers notice, a Farfetch product page may tuck sustainability credentials beneath layers of styling imagery, and a secondhand platform like Vestiaire Collective rarely connects pre-owned items to their original production impacts. This disconnect reveals a fundamental oversight: transparency is not just about data availability, but about contextual resonance.

The original purpose of this study was to realise that the sustainable development of the luxury industry cannot rely solely on traceability technology at the end of the supply chain. On the contrary, the concept of sustainable development must be integrated into the core aspects of consumer experience through proactive design, innovative thinking and tailor-made marketing strategies. Today, wealthy consumers in urban areas in North America, Europe and East Asia are not only highly aware of environmental problems, but also actively strive to align their consumption habits with ecological values (Birinci et al. 2025). However, surveys have shown that only a few consumers are able to accurately distinguish between genuine environmental practices and "greenwashing" tactics (Perez Vega et al. 2025). When brand communications appear insincere or inconsistent, it easily triggers consumers' suspicion and even resistance. Such a crisis of trust not only undermines brand value but also destroys the emotional connection that is often more decisive than price in the sustainable luxury goods market (Bhardwaj et al. 2025). At the same time, the regulatory environment is undergoing rapid changes: The European Union has introduced new regulations requiring high-impact industries to provide detailed environmental information, indicating that transparency is increasingly becoming a mandatory requirement rather than a voluntary initiative (Perez Vega et al. 2025). Although blockchain technology offers technical support for traceability, its effectiveness still needs to be enhanced through retail marketing strategies in order to effectively transform information into trust (Birinci et al. 2025). In this context, brands that master the art of embedding credible, blockchain-backed narratives into their retail touchpoints stand to gain not only compliance but competitive differentiation.

This research therefore sets out to explore how blockchain-enabled environmental disclosures—specifically those related to water usage and carbon emissions—translate into actual shifts in consumer attitudes and behaviors, with particular attention to the mediating role played by fashion retail strategies. Instead of seeing retail as a neutral distribution channel, we position it as a positive interpretation layer that can shape perception, build trust and ultimately promote action. Our core premise is that the same environmental data may produce completely different results, depending on whether it is displayed as a static footnote on the website, a dynamic animation on the in-store touch screen, or a personalised insight provided through the loyalty application after purchase. To test this, we collected feedback from 397 qualified luxury consumers (individuals who have recently made excessive purchases and earned above average) from three different retail environments: physical boutiques, high-quality e-commerce platforms and certified second-hand markets. Using partial least squares structural equation modeling (PLS-SEM), we analyzed how perceived blockchain transparency flows through dimensions of retail adaptation—such as channel-appropriate presentation, staff-assisted explanation, and integration with brand aesthetics—to influence clarity around water and carbon impacts, mitigate greenwashing concerns, and ultimately foster intentions to pay premiums, recommend brands, and repurchase sustainably.

The novelty of this work lies not in proving that blockchain enhances credibility—that assumption is already widespread—but in demonstrating how that credibility must be operationalized within the lived reality of retail. We move beyond abstract discussions of "transparency" to offer a granular framework grounded in real consumer feedback, showing that effectiveness hinges on alignment between technological capability and marketing execution. For example, our research results show that when blockchain data is integrated into the experience moment, offline channels will perform well (for example, salespeople use interactive screens to compare the water footprints of two handbags), and when sustainable development information is intuitively distilled into scannable infographics or filterable labels, online platforms perform best. Meanwhile, secondhand platforms unlock unique value by framing blockchain records as certificates of origin, allowing buyers to see the full lifecycle impact of a pre-owned item. This channel-specific nuance challenges the notion of a universal sustainability playbook and instead advocates for adaptive, context-sensitive communication. Furthermore, we identify "service enablers"—such as trained staff, QR-code-guided journeys, or loyalty points for engaging with environmental content—as critical amplifiers that can double the behavioral impact of disclosure efforts. In doing so, the study contributes both theoretically, by integrating retail marketing into models of sustainable consumption, and practically, by offering luxury brands a roadmap to transform blockchain from a back-office ledger into a front-of-house asset that resonates, persuades, and converts.

2 Literature Review

The Scholarship on sustainability in the luxury and fashion sectors has evolved significantly over the past two decades, shifting from broad ethical critiques to increasingly granular investigations of consumer cognition, supply chain governance, and technological mediation. Early work primarily framed the industry as inherently contradictory—luxury's emphasis on excess, novelty, and resource intensity seemingly at odds with principles of environmental restraint and circularity. This tension gave rise to extensive debates about "ethical luxury" and whether high-end consumption could ever be reconciled with planetary boundaries. However, as climate urgency increases and consumer expectations mature, research has turned to identifying a coordinated path, especially through transparency, traceability and value redefinition. Within this trajectory, three interlocking streams of literature form the foundation for the present study: the environmental footprint of luxury production (especially water and carbon), the role of blockchain in enhancing disclosure credibility, and the critical yet underexplored function of retail marketing in bridging technical data and consumer behavior.

The first stream establishes why environmental disclosure matters in luxury. Studies by the Alim (2022) have meticulously documented the sector's disproportionate strain on freshwater systems. Leather tanning, cotton cultivation, and dyeing processes are identified as hotspots, with a single luxury handbag or coat often embodying thousands of liters of embedded water—much of it drawn from water-stressed regions. Carbon emissions, though less visible, accumulate across globalized logistics, energy-intensive manufacturing, and air freight. Crucially, however, consumers remain largely unaware of these impacts. Yuan Surveys consistently show that fewer than 30% of shoppers can accurately estimate the water or carbon footprint of common fashion items, not due to apathy but because such information is rarely presented in accessible formats at decision points. This knowledge gap creates fertile ground for greenwashing, where vague terms like "eco-friendly" or "conscious collection" substitute for measurable accountability. Julia et al. (2022) demonstrated that without verifiable evidence, even well-intentioned sustainability claims can backfire, breeding cynicism rather than trust. Therefore, the literature focusses on one key insight: credible, specific and timely environmental disclosure not only provides information, but also is the basis for rebuilding consumers' confidence in the sustainable development narrative of luxury goods.

This leads directly to the second stream: the promise of blockchain as a solution to disclosure deficits. Blockchain technology, with its decentralized ledger architecture and cryptographic immutability, offers a mechanism to record and share environmental data in ways that resist manipulation or selective editing. Unlike traditional PDF reports issued annually by corporate communications teams, blockchain enables real-time, product-level verification—allowing a customer to scan a QR code and instantly see verified records of a garment's raw material origin, factory wastewater treatment certification, and transport emissions. Scholars such as Badhwar et al. (2023) and Patel et al. (2025) have validated blockchain's technical capacity to enhance supply chain traceability, while Patro et al. (2024) highlights its potential in carbon accounting. In the luxury context, initiatives like LVMH's Aura Consortium demonstrate industry adoption, enabling brands like Dior and Louis Vuitton to embed digital product passports into physical goods. Yet a critical limitation persists across this body of work: most studies focus on backend implementation—data integrity, node governance, smart contract design—while assuming that once data is "on-chain," its impact on end-users is automatic. This techno-optimistic assumption overlooks a fundamental reality: data must be communicated, not just stored. A blockchain record is only as powerful as its ability to be discovered, understood, and trusted by the consumer standing in a store or scrolling on a phone.

It is here that the third and most underdeveloped stream becomes essential: the role of retail marketing as the interpretive interface between blockchain data and consumer perception. While information systems research treats transparency as a property of data, marketing and consumer behavior scholarship recognizes it as a relational outcome shaped by context, framing, and delivery. Patagonia's success in driving sustainable choices, for instance, is less about superior data than about embedding environmental narratives into compelling retail experiences—such as in-store displays comparing water use across products or repair workshops that extend garment life. Similarly, research by the Neumann et al. (2020) shows that trust in sustainability claims increases by over 50% when information is presented through active, interactive channels (e.g., staff explanation, dynamic visuals) rather than passive text. Yet existing models rarely integrate these insights into blockchain discourse. The prevailing view treats retail as a neutral conduit—a pipe through which data flows—rather than an active processor that can amplify, distort, or silence messages based on design choices. This omission is particularly glaring given the diversification of luxury retail itself: offline boutiques prioritize sensory immersion and human interaction; e-commerce platforms optimize for speed, personalization, and visual clarity; and secondhand marketplaces emphasize provenance and lifecycle storytelling. Each channel requires its own communication logic, yet the current blockchain model suffers from low user engagement. According to Dahlke (2025) report, blockchain technology remains niche, with adoption rates below 5% even in high-adopter sectors like finance.

These three streams reveal obvious theoretical gaps: although we understand what needs to be disclosed (water/carbon data) and how to ensure its security (through blockchain), we lack a reliable explanation of how to deliver this data to maximise the impact of behaviour in different retail environments. Previous studies either examined consumers' reactions to general sustainability statements without specifying data sources, or analysed the blockchain architecture without measuring the downstream impact on trust and purchase intentions. Few, if any, bridge the technical and experiential dimensions within a unified framework that accounts for channel-specific marketing adaptation. This study directly addresses that lacuna by positioning retail marketing not as a peripheral concern but as the crucial mediating mechanism through which blockchain-enabled transparency translates into perceived environmental clarity, reduced greenwashing skepticism, and ultimately, sustainable consumption behaviors. By doing so, it moves beyond the binary of "technology vs. marketing" to propose an integrated model where credibility is co-created at the intersection of immutable data and empathetic communication.

chronicles the evolution of sustainable luxury consumption, highlighting pivotal shifts toward greener practices. Notable milestones include Kaiyun Group's collaboration with Alibaba Green Retail, the Shanghai Green Supply Chain Alliance, and Burberry's introduction of fully biodegradable bags, all demonstrating the luxury industry's commitment to sustainability (see Fig. 1).

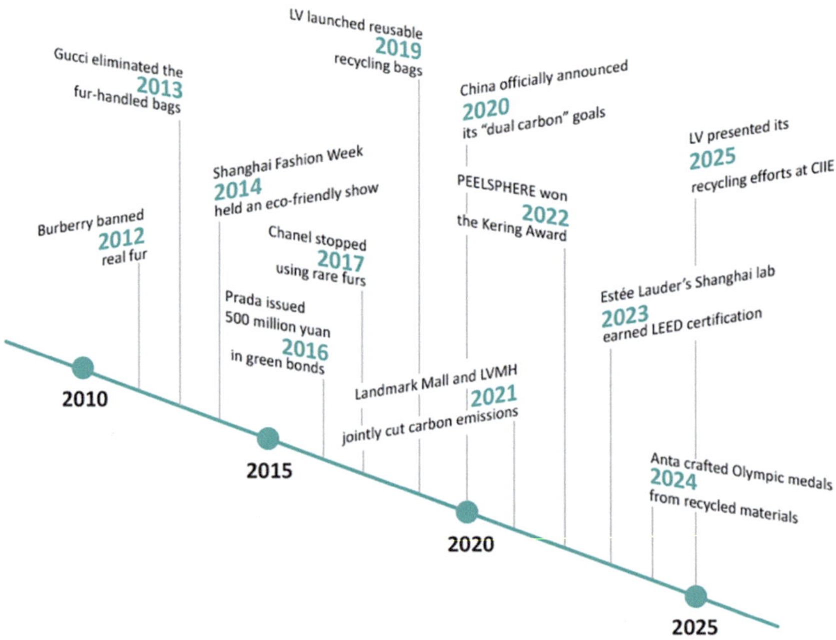

Fig. 1 Chronology of major events in the sustainable luxury consumer market

3 Empirical Results

Although blockchain technology is frequently promoted as a breakthrough for enhancing sustainability transparency in the luxury industry, its actual influence on consumer behavior does not stem from the technical act of data immutability alone. Rather, as our empirical analysis demonstrates, the effectiveness of blockchain-enabled environmental disclosures is fundamentally mediated by how brands embed this information into the retail experience and how consumers psychologically process it. Focusing exclusively on China—a market that now drives over 25% of global luxury demand—we collected primary data from 397 high-income consumers across Tier-1 and emerging Tier-2 cities (Beijing, Shanghai, Guangzhou, Shenzhen, Hangzhou, Chengdu), all of whom reported annual household incomes above RMB 800,000 and had purchased at least one luxury item valued at RMB 35,000 or more within the past 12 months. Using Partial Least Squares Structural Equation Modeling (PLS-SEM), we tested a theoretically grounded sequential mediation model comprising five latent constructs: **Blockchain Environmental Disclosure Perception** (X) as the exogenous variable; three mediators—**Retail Marketing Adaptation** (M1), **Water and Carbon Transparency Perception** (M2), and **Green Trust versis Greenwashing Skepticism** (M3); and **Green Consumption Behavioral Intention** (Y) as the endogenous outcome.

All constructs were measured with multi-item, 7-point Likert scales adapted from validated instruments and rigorously back-translated to ensure semantic and cultural equivalence in Mandarin. Prior to structural estimation, the measurement model was evaluated for reliability and validity. As shown in Table 1, Cronbach's α values ranged from 0.844 to 0.877, composite reliability (CR) exceeded 0.89 for all constructs, and average variance extracted (AVE) was consistently above 0.68, confirming strong internal consistency and convergent validity. Discriminant validity was established via the Fornell-Larcker criterion: the square root of each construct's AVE exceeded its inter-construct correlations. The Kaiser-Meyer-Olkin (KMO) measure of sampling adequacy was 0.842 ($p < 0.001$), supporting the appropriateness of factor analysis.

Structural path estimation, based on 5000 bootstrap resamples, revealed that X significantly predicts all three mediators: X → M1 ($\beta = 0.322$, $p < 0.001$), X → M2 ($\beta = 0.216$, $p < 0.001$), and X → M3 ($\beta = 0.344$, $p < 0.001$). Each mediator, in turn, positively influences Y: M1 → Y ($\beta = 0.210$, $p < 0.001$), M2 → Y ($\beta = 0.170$, $p = 0.002$), and M3 → Y ($\beta = 0.207$, $p < 0.001$). Notably, the direct effect of X on Y decreased from a total effect of $\beta = 0.222$ ($p < 0.001$) to $\beta = 0.114$ ($p = 0.024$) when mediators were included, indicating partial mediation. These results are summarized in Table 2.

To quantify indirect effects, we conducted specific mediation tests using the percentile bootstrap method (95% confidence intervals). As presented in Table 3, all three pathways were statistically significant: X → M1 → Y (indirect effect = 0.040, CI [0.014, 0.074], $p = 0.011$), X → M2 → Y (0.024, CI [0.005, 0.051], $p = 0.043$), and X → M3 → Y (0.043, CI [0.016, 0.076], $p = 0.004$). Collectively, these mediators account for 48.67% of the total effect of blockchain disclosure on

Table 1 Measurement model reliability and validity (n = 397)

Construct	Items	Cronbach's α	Composite reliability (CR)	AVE	√AVE
X—Blockchain disclosure	3	0.844	0.892	0.721	0.849
M1—Retail marketing adaptation	4	0.870	0.915	0.738	0.859
M2—Water/carbon transparency	4	0.848	0.897	0.684	0.827
M3—Green trust/ anti-greenwashing	4	0.873	0.918	0.742	0.861
Y—Green behavioral intention	4	0.877	0.921	0.763	0.874

Table 2 Structural path coefficients (Bootstrap, 5000 samples)

Path	β	t-value	p-value
X → M1	0.322	6.831	< 0.001
X → M2	0.216	4.102	< 0.001
X → M3	0.344	7.524	< 0.001
M1 → Y	0.210	4.517	< 0.001
M2 → Y	0.170	3.128	0.002
M3 → Y	0.207	4.403	< 0.001
X → Y (direct)	0.114	2.261	0.024

green behavioral intention, with M3 (green trust/anti-greenwashing) contributing the largest share (19.77%), followed closely by M1 (retail adaptation, 18.15%).

Model fit was further assessed using multiple supplementary indices (Table 4). Despite PLS-SEM's focus on prediction rather than global fit, the model exhibits excellent performance: SRMR = 0.057 (< 0.08), CFI = 0.991 (> 0.90), RMSEA = 0.023 (< 0.08), GFI = 0.955, and χ^2/df = 1.218 (< 3), all well within recommended thresholds. All outer loadings exceeded 0.707 (maximum = 0.863), and cross-loadings were negligible, confirming clean discriminant properties.

The findings reveal a clear and robust causal chain: blockchain disclosures do not directly drive sustainable consumption in China's luxury market. Instead, their

Table 3 Mediation analysis: indirect effects

Mediation path	Indirect effect	95% CI lower	95% CI upper	p-value	% of Total effect
X → M1 → Y	0.040	0.014	0.074	0.011	18.15
X → M2 → Y	0.024	0.005	0.051	0.043	10.75
X → M3 → Y	0.043	0.016	0.076	0.004	19.77
Total mediated	0.107	–	–	–	48.67

Table 4 Model fit indices

Index	Value	Recommended threshold
χ^2	177.814	–
Df	146	–
p-value	0.038	> 0.05 (acceptable)
χ^2/df	1.218	< 3
GFI	0.955	> 0.90
RMSEA	0.023	< 0.08
SRMR	0.057	< 0.08
CFI	0.991	> 0.90
NFI	0.954	> 0.90

impact is unlocked through a dual process—first, through **retail marketing adaptation**, which renders technical data accessible and contextually relevant across offline boutiques, e-commerce platforms, and secondhand marketplaces; and second, through the resulting cognitive and affective shifts in **transparency perception** and **trust formation**. This challenges the assumption that technological transparency alone suffices and underscores the strategic role of the retail interface as the critical site where data becomes meaning—and meaning becomes action.

4 Conclusion

The empirical findings of this study yield a clear and actionable conclusion: in China's rapidly evolving luxury market, blockchain-based water-and-carbon environmental disclosure does not automatically translate into sustainable consumer behavior. Instead, its effectiveness is contingent upon a carefully orchestrated integration of technology, retail experience, and psychological trust-building. Our data show that nearly half (48.67%) of blockchain's influence on green-consumption intentions flows through three interdependent mechanisms—retail-marketing adaptation, perceived water-and-carbon transparency, and green trust (i.e., resistance to green-washing)—with the direct effect of blockchain water-and-carbon disclosure alone being modest and statistically fragile once these mediators are accounted for. This refutes the prevailing techno-optimism that assumes "if you put it on-chain, they will believe—and act." In reality, Chinese luxury consumers—digitally native, environmentally conscious, and increasingly skeptical of performative sustainability—are far more discerning. They do not respond to raw data; they respond to how that data is framed, delivered, and embedded within a coherent brand narrative at the point of engagement.

On the basis of this insight, we offer four integrated strategic recommendations for luxury brands operating in China.

First, treat blockchain not as a standalone proof mechanism but as one component of a broader transparency ecosystem that must be seamlessly woven into both physical and digital retail environments. This means designing in-store QR codes, WeChat Mini-Program interfaces, or digital product passports that align aesthetically and functionally with the brand's identity—not as technical appendices, but as natural extensions of the luxury experience.

Second, invest in frontline-staff training to transform sales associates into "sustainability storytellers" who can confidently explain blockchain records, contextualise water-and-carbon metrics (e.g., "This jacket saved 2300 L of water compared with conventional production"), and address consumer scepticism with authenticity rather than scripted claims.

Third, prioritise clarity over complexity in environmental communication: avoid overwhelming consumers with granular life-cycle data; instead, highlight one or two intuitive water-and-carbon impact indicators (e.g., litres of water saved, kg CO_2−e avoided) that are visually intuitive, benchmarked against industry norms, and verifiably linked to the specific product via blockchain.

Fourth, leverage blockchain-enabled transparency to build long-term trust by proactively disclosing not only achievements but also limitations—such as partial supply-chain coverage or ongoing improvement targets—thereby signalling honesty and reducing perceptions of green-washing, which our data identify as a critical barrier to behavioural conversion.

From a theoretical standpoint, this study advances the literature by shifting the focus from "technology presence" to "technology interpretation," positioning **retail marketing as the pivotal moderator that determines whether blockchain functions as a credible water-and-carbon signal or an ignored artefact. Practically, it offers a roadmap for luxury brands to move beyond symbolic sustainability toward operationalised water-and-carbon transparency—one that resonates with China's sophisticated consumer base. As regulatory pressure intensifies (e.g., China's ESG-disclosure guidelines for listed companies) and second-hand luxury platforms such as Plum and Secoo increasingly demand verifiable provenance, the ability to credibly communicate water-and-carbon environmental performance will become not just a differentiator, but a prerequisite for market access. Brands that master the art of translating blockchain data into meaningful, human-centred retail experiences will be best positioned to capture the loyalty of the next generation of Chinese luxury consumers—those who expect beauty, craftsmanship, and responsibility to coexist without compromise.

References

Alim, M.A., Halkias, D., Levitt, K.: An integrative literature review of sustainable luxury fashion consumption through a circular economy business model. Int. J. Competitiveness **2**(3), 234–247 (2022). https://doi.org/10.1504/IJC.2022.10051726

Alotaibi, E.M., Khallaf, A., Abdallah, A.A.-N., Zoubi, T., Alnesafi, A.: Blockchain-driven carbon accountability in supply chains. Sustainability **16**(24), 10872 (2024). https://doi.org/10.3390/su162410872

Badhwar, A., Islam, S., Tan, C.S.L.: Exploring the potential of blockchain technology within the fashion and textile supply chain with a focus on traceability, transparency, and product authenticity: a systematic review. Front. Blockchain (2023). https://doi.org/10.3389/FBLOC.2023.1044723

Bhardwaj, P., Priya, P.S., Sharma, S.: Exploring the influence of circular economy paradigm on convergence of sustainability and luxury apparel: a systematic review. Circ. Econ. Sustain. **5**(3), 1–22 (2025). https://doi.org/10.1007/S43615-024-00489-W

Birinci, H., Esenyel, I., Obeng, H.A.: Sustainable destination management in luxury tourism: balancing economic development and environmental responsibility. Sustainability **17**(15), 6815 (2025). https://doi.org/10.3390/SU17156815

Dahlke, J., Schmidt, S., Lenz, D., Kinne, J., Dehghan, R., Abbasiharofteh, M., Schütz, M., Kriesch, L., Hottenrott, H., Kanilmaz, U.N., Grashof, N.: The WebAI paradigm of innovation research extracting insight from organizational web data through AI. In: ZEW Discussion Papers 25-019. ZEW—Leibniz Centre for European Economic Research (2025)

Julia, A., Ewa, K., Iwona, A., Łukasik, R.M.: Greenwashing and sustainable fashion industry. Curr. Opin. Green Sustain. Chem. **38** (2022). https://doi.org/10.1016/J.COGSC.2022.100710

Neumann, H.L., Martinez, L.M., Martinez, L.F.: Sustainability efforts in the fast fashion industry: consumer perception, trust and purchase intention. Sustain. Acc. Manag. Policy J. 571–590 (ahead-of-print) (2020). https://doi.org/10.1108/sampj-11-2019-0405

Patel, A., Sai, S., Daiya, A., Akolekar, H., Chamola, V.: Blockchain enabled traceability in the jewel supply chain. Sci. Rep. **15**(1), 3837–3837 (2025). https://doi.org/10.1038/S41598-025-88245-4

Patro, P.K., Acquaye, A., Jayaraman, R., Salah, K.: A framework for carbon accounting for sustainable supply chains using Blockchain, internet of things, and artificial intelligence. In: IISE Annual Conference. Proceedings, pp. 1–6 (2024). https://doi.org/10.21872/2024IISE_5925

Perez Vega, R., Garnelo Gomez, I., Baxter, K., Histon, W., Mazzoli, I., Sanchez Razo, M.A.: Setting the sustainability agenda: the impact of luxury brands' SDG-related tweets on consumer engagement. Int. J. Consum. Stud. **49**(4), e70098 (2025). https://doi.org/10.1111/IJCS.70098

Viet, N.Q., de Leeuw, S., van Herpen, E.: The impact of social vs environmental sustainability information disclosure on consumer choice of delivery time with varying sustainability concerns. Int. J. Phys. Distrib. Logistics Manag. **53**(11), 26–52 (2023). https://doi.org/10.1108/IJPDLM-09-2021-0392

Spatio-Temporal Differentiation of Water Purification Efficiency and Analysis of Main Controlling Factors of Typical Ecological Restoration Projects in Lijiang River Basin

HaiFeng Huang, BinBin Yang, WenZhong Tan, JunZhi Zhang, FaZhou Lu, and YanPing Cui

Abstract To systematically evaluate the pollution control effect of the integrated protection and restoration conservation of mountain, river, forest, farmland, lake, grassland, and desert ecosystems in the typical karst basin, 18 typical ecological restoration projects in Lijiang River basin in Guangxi were monitored and evaluated for 10 months. The research shows that the overall water quality of the basin presents the spatial pattern of "excellent upstream, urban deterioration and improvement of restoration area" and the seasonal characteristics of "double peaks" of non-point source pollution accumulation in wet season and point source pollution accumulation in dry season; the upstream and downstream correlation analysis reveals the "channel effect" and "decoupling effect" of the engineering action mechanism; organic pollution from domestic sources and nutrient pollution from agriculture and endogenous sources are the two main controlling factors affecting the water quality difference in river basins. This study holds that the ecological restoration project in Lijiang River Basin has significantly improved the overall water quality, but its efficiency is jointly restricted by the project type, spatial location and external pollution load, and puts forward corresponding measures to improve the water environment quality in the future.

H. Huang (✉) · B. Yang · W. Tan
Guangxi Natural Resources Ecological Restoration Center, Nanning, Guangxi, China
e-mail: 53773310@qq.com

J. Zhang
Shanghai Sansun Technology Co., Ltd, Nanjing, China

F. Lu
Jiangsu Province Water Engineering Sci-tech Consulting Co., Ltd, Nanjing, Jiangsu, China

Y. Cui
Jiangsu Province Hydrology and Water Resources Investigation Bureau, Nanjing, Jiangsu, China

H. Xu (ed.), *Proceedings of the 8th International Symposium on Water Resource and Environmental Management*, Environmental Science and Engineering,
https://doi.org/10.1007/978-3-032-19088-8_19

Keywords Ecological restoration project · Water purification · Efficiency evaluation · Spatio-temporal differentiation · Lijiang river basin · Main control factor analysis

1 Introduction

1.1 Research Background and Significance

The aquatic environment constitutes a critical component of ecosystems, with its health status directly impacting societal well-being and sustainable socio-economic development. Guided by the national ecological civilization strategy that recognizes "lucid waters and lush mountains are invaluable assets", conducting scientific, systematic, and quantitative effectiveness assessments for major ecological conservation and restoration projects has become imperative. Such assessments are not only crucial for verifying policy efficacy but also form the scientific foundation for optimizing future environmental governance decisions.

Situated within the core karst region of southwestern China, the Lijiang River Basin harbors a unique ecosystem of significant aesthetic and biodiversity value. Furthermore, as a vital water source conservation area in the upper reaches of the Pearl River, it plays a critical role in ensuring the water ecological security of the Guangdong-Hong Kong-Macao Greater Bay Area. In recent years, the imperative to protect the Lijiang River has been consistently emphasized, with directives to "preserve the mountains and waters of Guilin" and to develop Guilin into a model for ecological conservation and integrated management. In this context, Guilin City has launched the "Integrated Conservation and Restoration Project for Mountains, Rivers, Forests, Farmlands, Lakes, Grasslands, and Deserts in the Lijiang River Basin," committing substantial investment to a series of ecological restoration initiatives (Gan et al. 2025).Consequently, a systematic evaluation of the water quality purification capacity of these projects serves a dual purpose. It provides a scientific assessment of the practical outcomes achieved through the implementation of a national strategy at the local level. Moreover, it yields valuable empirical evidence and decision-making support for ecological management not only within the Lijiang River Basin but also for similar karst basins globally.

1.2 Progress in Domestic and International Research

Globally, concepts of river restoration have evolved from early engineering approaches focused primarily on structural rehabilitation—such as Germany's "near-natural" channel concepts and the United States' "natural channel design" (Dong, 2004; Dong and Bai, 2007; Li, 2018)—to integrated strategies that prioritize ecosystem functions and processes. The European Water Framework Directive, for example, promotes basin-scale management aimed at restoring ecological integrity (Du and Li, 2011).These contemporary approaches emphasize measures such as rebuilding riparian buffer zones and constructing ecological wetlands to enhance the self-purification capacity of river systems.

In contrast, river restoration efforts in China began relatively late, with early practices largely oriented toward water-quality improvement and landscape enhancement (Xu et al. 2014; Wang et al. 2019).Only in recent years has the focus shifted toward restoring overall ecosystem integrity. Existing studies have typically addressed single projects or small watershed units and have often evaluated short-term outcomes. Systematic and long-term assessments of large, multi-project restoration programs—such as those implemented in the Lijiang Basin—remain limited (Dyste and Valett, 2019).Meanwhile, advances in big data, remote sensing, and environmental modeling have significantly expanded the toolkit for water-environment assessment. However, integrating these advanced technologies with long-term field monitoring data to elucidate the actual mechanisms through which restoration projects operate within complex watersheds remains a significant challenge and a focal point of current research (Lan et al. 2024; Zhang et al. 2025; Li et al. 2025).

1.3 Scientific Questions and Research Objectives

Despite the substantial investment in ecological restoration projects across the Lijiang River Basin, critical gaps persist in our understanding of their actual performance. Key unresolved questions include: What is the true purification efficacy of these interventions? How do different categories of engineering measures perform in reducing critical pollutants such as nitrogen, phosphorus, and organic matter? What specific roles do these projects play within the basin's overall pollutant transport chain? Furthermore, what are the key limiting factors constraining their purification effectiveness? To address these questions, this study establishes the following core objectives:

(1) Precisely characterize the basin-wide spatiotemporal evolution of water quality following the implementation of restoration projects, thereby identifying major pollution zones and their seasonal dynamics.
(2) Quantitatively assess and compare the purification efficacy of different types of ecological restoration measures for key pollutants.
(3) Elucidate the mechanistic roles of these projects within the watershed pollutant transport chain, specifically determining whether they act primarily as passive conduits or actively decouple upstream-downstream pollutant linkages through internal biochemical processes, and to identify the dominant controlling factors influencing their performance.

(4) Provide a scientific basis and decision-support framework for optimizing the spatial planning and enhancing the management of ecological restoration projects, both within the Lijiang River Basin and in similar karst basins globally.

2 Materials and Methods

2.1 Study Area and Monitoring Network

2.1.1 Overview of the Study Area

The Lijiang River Basin is situated in northeastern Guangxi, China (109°57'–111°10' E, 24°17'–25°54' N), encompassing a total area of approximately 12,269.11 km^2. Characterized by a humid subtropical monsoon climate, the basin receives abundant rainfall with a mean annual precipitation of about 1844 mm. Its landscape is dominated by typical karst topography, which underpins a fragile ecosystem. Administratively, the basin falls within the Southern Hills and Mountains Belt, a key component of the "Three Zones and Four Belts" national strategic framework outlined in China's Master Plan for Major Projects of Important Ecosystem Protection and Restoration (2021–2035) (Guan et al. 2021).This study focuses on the integrated "Mountains-Rivers-Forests-Farmlands-Lakes-Grasslands-Deserts" Protection and Restoration Project implemented across the Lijiang Basin. From the total suite of 83 sub-projects under this initiative, 18 representative water-ecological restoration projects were selected for in-depth analysis in this research.

2.1.2 Monitoring Network Design

To comprehensively evaluate the effectiveness of the restoration projects, a monitoring network of 36 sampling sites was systematically established from July 2024 to April 2025. These sites were distributed across the impact zones of the 18 representative projects and key sections of the main Lijiang River channel, as illustrated Fig. 1. The site selection adhered to three guiding principles: (1) Coverage of different project types, such as riparian zone ecological restoration, subsurface/surface flow constructed wetlands, and lake/reservoir water quality improvement projects; (2) Inclusion of diverse geographical locations, encompassing upstream ecological buffer zones, urban core areas, and downstream landscapè regions; (3) Establishment of paired upstream-downstream control points for key projects to enable precise assessment of their net pollutant removal efficacy.

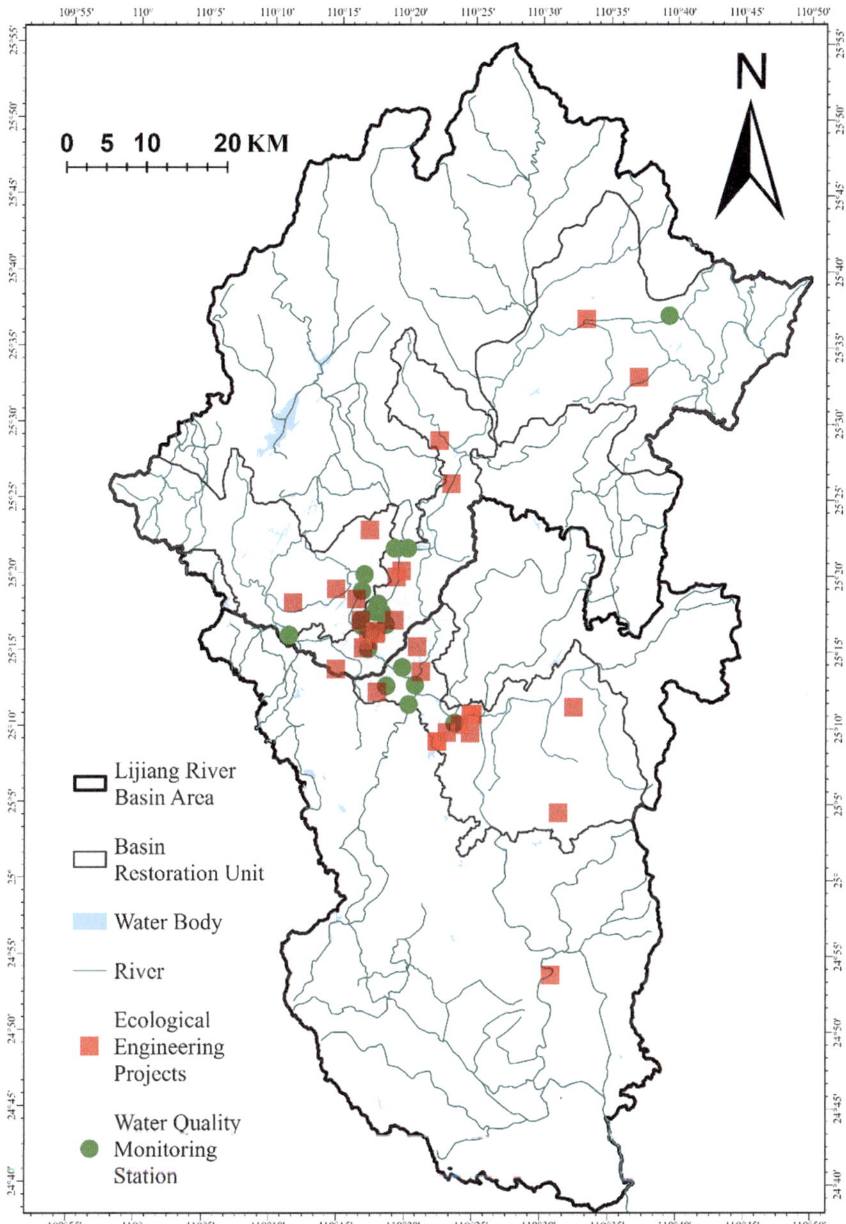

Fig. 1 Layout of ecological protection and restoration projects in the upper reaches of Lijiang River and distribution map of monitoring projects in this study

2.2 Sample Collection and Analysis

Monitoring was conducted monthly from July 2024 to April 2025, with synchronous sampling across all 36 sites during each round. This protocol yielded ten sets of time-series data, ensuring comprehensive seasonal coverage of the wet (July–September), normal, and dry (December–February) hydrological periods.

The core water quality parameters analyzed included dissolved oxygen (DO), permanganate index ($KMnO_4$), chemical oxygen demand (COD), five-day biochemical oxygen demand (BOD_5), ammonia nitrogen (NH_3-N), suspended solids (SS), total phosphorus (TP), and total nitrogen (TN). In-situ measurements of DO, water temperature, and pH were performed using a portable multiparameter meter. Water samples were collected at a depth of 0.5 m below the surface, preserved at low temperature in light-proof containers, and delivered to the laboratory within 24 h. All laboratory analyses strictly adhered to the Environmental Quality Standards for Surface Water (GB3838–2002) and corresponding national standard methods. Quality control procedures, including the analysis of duplicate and blank samples, were implemented throughout to ensure data accuracy and reliability.

2.3 Evaluation Framework

2.3.1 Comprehensive Water-Quality Assessment

The overall water quality at each monitoring site was evaluated using the Nemerow Pollution Index. This method accounts for both the average pollution level across all parameters and the influence of the dominant pollutant, thereby reflecting integrated and worst-case conditions (Luo et al. 2016).The index is calculated as:

$$Z_{\text{Nemerow}} = \sqrt{\frac{\overline{Z}^2 + D^2}{2}} \tag{1}$$

where $Z_{Nemerow}$ represents the composite Nemerow index, $\overline{Z}$ is the mean single-factor pollution index, and D denotes the maximum single-factor index. Based on Class III thresholds defined in the *Environmental Quality Standards for Surface Water* (GB 3838–2002), corresponding Nemerow index ranges for different water-quality grades were derived, as summarized in Table 1.

2.3.2 Evaluation of Project Purification Performance

To quantify the removal effect of the restoration projects on specific pollutants, the removal rate (R) for key pollutants was calculated using the following formula:

Table 1 Water quality classification based on the Nemerow pollution index

Item	Water quality class				
	I	II	III	IV	V
$Z_{Nemerow}$	≤0.5	0.5≤ $Z_{Nemerow}$ <1	1≤ $Z_{Nemerow}$ <1.5	1.5≤ $Z_{Nemerow}$ <2	2≤ $Z_{Nemerow}$

$$R = \frac{C_{in}\text{-}C_{out}}{A_{in}} \times 100\% \quad (2)$$

where C_{in} and C_{out} represent pollutant concentrations at the project inflow and outflow points (i.e., paired upstream—downstream sites), respectively. For dissolved oxygen (DO), its enhancement, rather than removal, was evaluated by calculating the improvement rate using a corresponding formula. A three-level composite efficacy classification system was established by integrating changes in both pollutant removal rates and water quality class (as assessed by the Nemerow Index). The criteria are defined as follows: High Efficiency: Removal rates of major pollutants exceed 60%, accompanied by an improvement of at least one water quality class; Moderate Efficiency: Removal rates of major pollutants range between 30 and 60%, with water quality class remaining stable or showing slight improvement; Low Efficiency: Removal rates of major pollutants are below 30% or negative, with water quality class showing no improvement or demonstrating deterioration.

2.3.3 Analysis of Dominant Factors and Underlying Mechanisms

To elucidate the operational mechanisms of the restoration projects and identify the key drivers of water quality variation, this study employed the following multivariate statistical approaches:

(1) Upstream-Downstream Correlation Analysis: Pearson correlation coefficients (r) were computed for the time-series data of water quality parameters between paired upstream and downstream monitoring points of representative projects (Peng et al. 2022; Qin et al. 2025).A high positive correlation indicates the direct downstream conveyance of upstream pollutants, characterizing a "conduit effect." Conversely, an r-value approaching zero or negative suggests that internal biochemical processes within the project disrupt this pollutant linkage, demonstrating a "decoupling effect" on upstream-downstream water quality.
(2) Principal Component Analysis (PCA): Eight core parameters (DO, COD, $KMnO_4$, BOD_5, NH_3-N, TP, SS and the Nemerow Index) from the 36 monitoring sites were standardized and subjected to PCA. This dimensionality-reduction technique extracted principal components explaining the majority of data variance. The factor loadings of each parameter on these components were then analyzed to identify the dominant combinations of pollution factors governing

the spatiotemporal variation in basin water quality (Qin et al. 2025; Zhang and Li, 2022).

3 Results and Analysis

3.1 Spatiotemporal Patterns of Basin-Wide Water Quality

3.1.1 Spatial Differentiation

Analysis of the ten-month monitoring dataset (July 2024–April 2025) revealed distinct spatial heterogeneity in the average Nemerow Index values across the 36 sampling sites (Fig. 2). A clear spatial gradient was observed across the basin: water quality was excellent in the upstream mountainous and ecological buffer zones; significantly pressured in the urban core tributaries and waterfront areas; and subject to pronounced fluctuations in the downstream island and sandbar landscapes, influenced by internal pollutant loading and tourism activities.

3.1.2 Temporal Dynamics

Temporally, water quality exhibited a "bimodal" seasonal pattern closely linked to the hydrological cycle. Data from the Baitang Village and Site 12 monitoring points within the Taohuajiang (Xiufeng Section) ecological restoration project serve as an illustrative example (Fig. 3). During the wet season (July–September), intensive rainfall and associated surface runoff mobilized substantial pollutants from agricultural non-point sources and urban surfaces, triggering sharp increases in concentrations of parameters such as TP and SS and forming a distinct non-point-source pollution peak. In contrast, during the dry season (December–February), reduced river flow diminished the water body's dilution and self-purification capacity. This allowed the influence of inadequately intercepted point-source discharges to become more pronounced, leading to the accumulation of pollutants like NH_3-N and thereby forming a point-source pollution peak.

3.2 Performance Variations of Typical Ecological Restoration Projects

A comprehensive evaluation of the 18 representative projects revealed marked variations in their purification capabilities, which can be summarized into three distinct models.

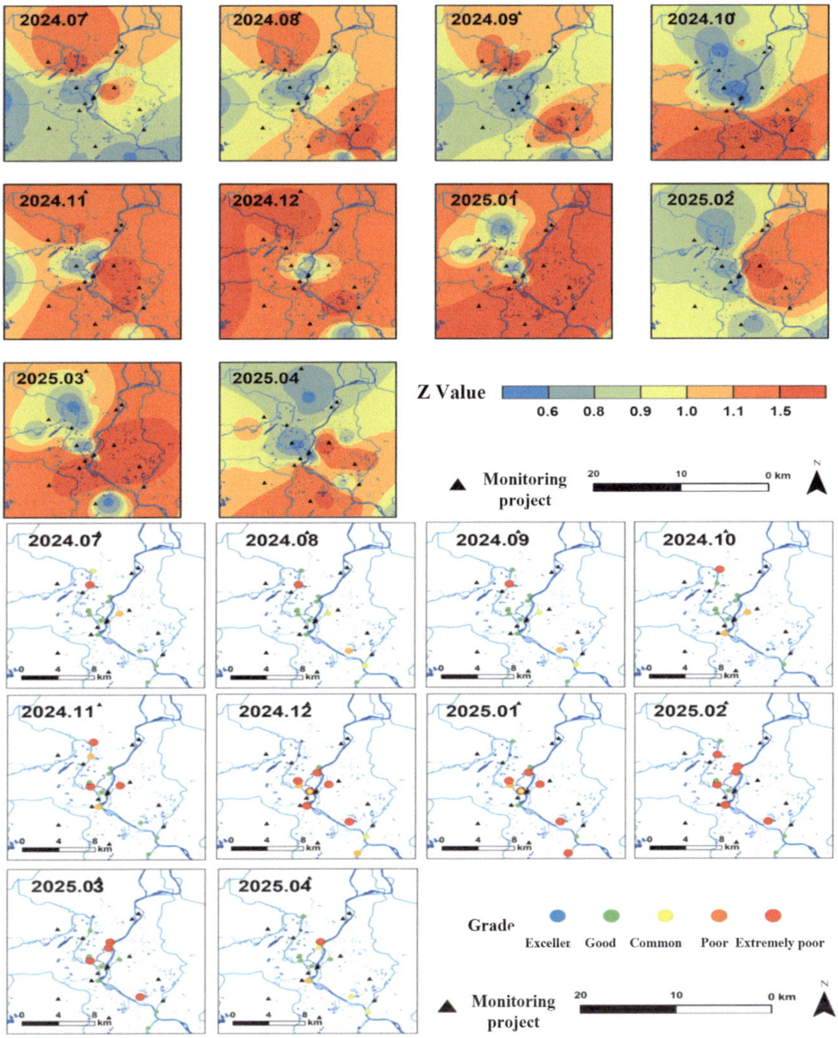

Fig. 2 Distribution map of Z value and water quality evaluation in Lijiang River basin

3.2.1 High-Efficiency Integrated Purification Model: Case Study of the Chaoyang River Basin Project

The Chaoyang River Basin integrated ecological restoration project demonstrated the best overall performance in this assessment. Its design features a five-stage, serially—connected system integrating source pollution interception, riparian buffering, composite constructed wetlands, step-fall aeration, and ecological dredging. Monitoring results, presented Fig. 4, show that the upstream intake and Site 5 achieved excellent removal efficiencies for a broad spectrum of pollutants: the removal rates

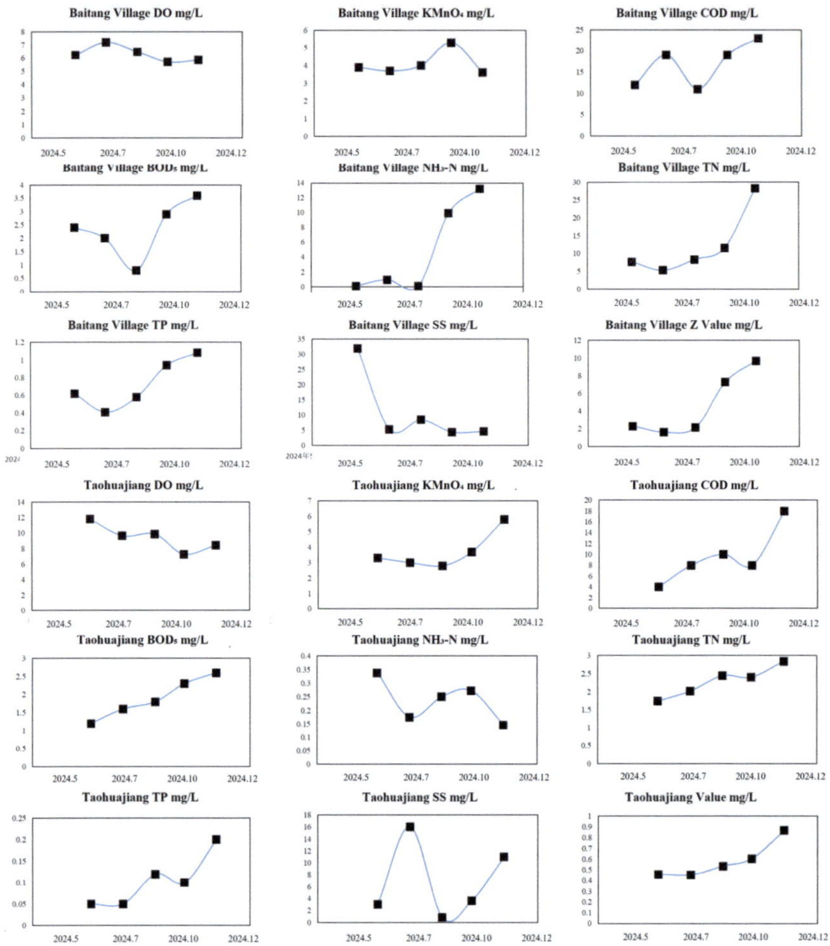

Fig. 3 Water quality index of Yujiacun and No.12 point of ecological protection and restoration project on both sides of Taohuajiang (Xiufeng section)

for permanganate index, COD, BOD_5, NH_3–N, and TP were as high as 81.72%, 93.89%, 86.36%, 94.43%, and 91.47%, respectively. Following project implementation, the Nemerow composite index for this river segment decreased from an initial Class IV–V status (Poor–Very Poor) to a stable value below 0.5, corresponding to an improvement in water quality to Class II (Good). Consequently, its overall purification efficacy was classified as "High-Efficiency." The success of this case underscores that a systematic, full-process, multi-technology design is crucial for achieving fundamental water quality improvement, providing a replicable model for rehabilitating heavily polluted urban tributaries.

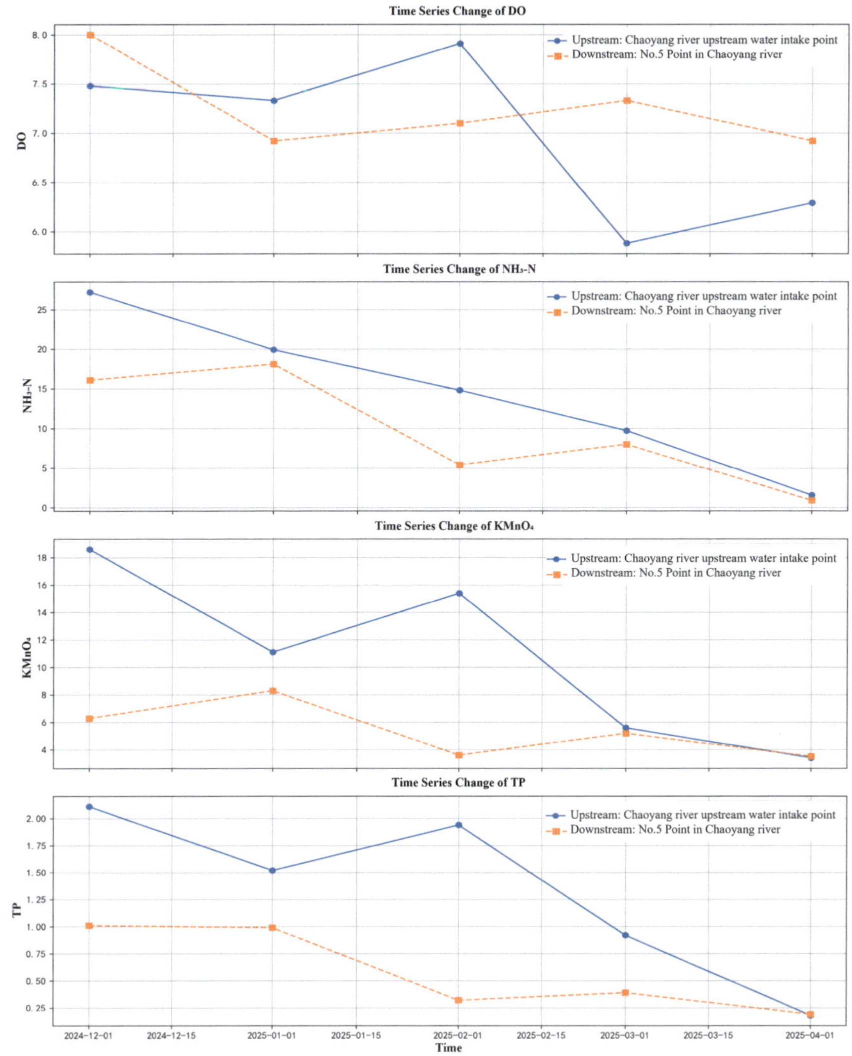

Fig. 4 Time series variation trend of water quality indicators at upstream and downstream monitoring points of ecological restoration comprehensive management project in Chaoyang River, River Basin

3.2.2 Targeted Specialized Purification Model: Case Study of the Taohuajiang River (Xiufeng Section) Project

The ecological restoration project along the Taohuajiang River (Xiufeng Section) primarily involved the ecological rehabilitation of riparian zones, resulting in a highly targeted purification profile. The data indicate a surprisingly high removal efficiency for NH_3-N, with a removal rate reaching 56.80% at Site 12 (Fig. 5). In contrast, the project showed very limited effectiveness for other pollutants. The removal rates for COD, BOD_5, and TP were even negative (−350.00%, −116.67%, and −300.00%, respectively), indicating an aggravation of organic and phosphorus pollution. This outcome elucidates the underlying mechanism: the ecologically restored riparian zone provided an ideal habitat for nitrifying and denitrifying microorganisms, facilitating efficient ammonia transformation. However, due to insufficient hydraulic retention time and a lack of effective particulate interception, its capacity for removing organic matter and phosphorus was severely limited. This case clearly illustrates the functional constraints inherent in single-measure restoration approaches.

3.2.3 Performance-Limited Model: Case Study of the Northern Guilin Xihu Lake Site

The monitoring site in the semi-enclosed northern bay of Xihu Lake is a typical case of limited performance. Despite remediation measures including dredging, aeration, and constructed wetlands, water quality improvement was negligible, with some parameters deteriorating. Data indicated only a marginal 4.50% improvement in DO, while key pollutant indicators (COD, BOD_5, NH_3-N, and TP) deteriorated significantly, showing negative removal rates of −400.00%, −200.00%, −138.89%, and −366.67%, respectively. Correspondingly, the Nemerow index remained persistently high, leading to a "Low-Efficiency" classification for this site's purification performance, as shown Fig. 6.

This case exemplifies the compounded challenges of "waterbody overloading" and "internal pollution release." Firstly, continuous inputs of inadequately intercepted domestic sewage imposed a pollution load far exceeding the treatment capacity of the implemented ecological measures. Secondly, the bay's stagnant hydrological conditions favored pollutant deposition, while wind/wave or boat-induced sediment resuspension released historically accumulated contaminants, exacerbating water quality degradation. This case serves as a critical reminder that without stringent source control, end-of-pipe ecological restoration projects may prove largely ineffective or even counterproductive.

3.3 Identification of Dominant Factors Governing Purification Efficacy

3.3.1 Project Mechanisms: Conduit Effect Versus Decoupling Effect

Correlation analysis of water quality parameters between the upstream and downstream sections of each project revealed two distinct roles that ecological restoration projects can play within the watershed pollutant transport chain (Table 2):

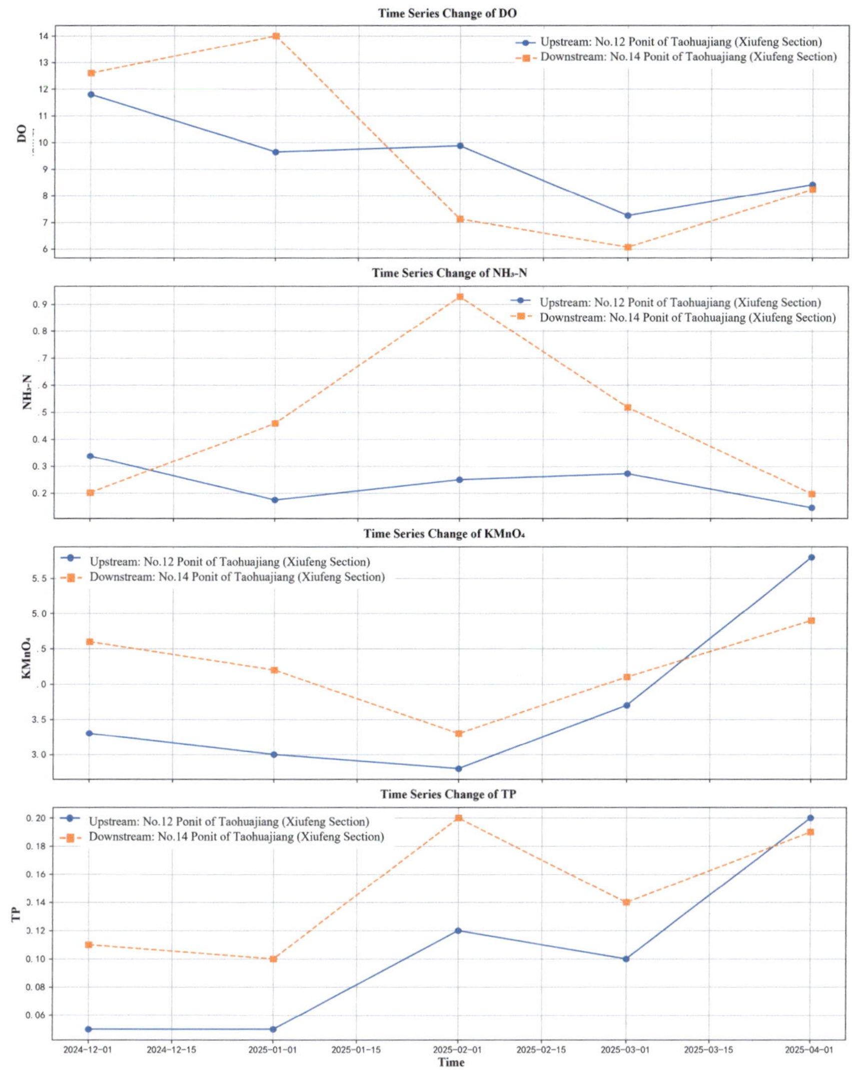

Fig. 5 Time series variation trend of water quality indicators at upstream and downstream monitoring points of ecological protection and restoration projects on both sides of Taohuajiang (Xiufeng section)

(1) Conduit Effect: When a project lacks effective treatment units for specific pollutants or when the incoming pollution load exceeds its processing capacity, pollutants pass through the project area with minimal attenuation, rendering the project functionally a "conduit." For instance, in the Taohua River (Xiufeng Section), the upstream-downstream correlation coefficient for TP was 0.85, and in the Chaoyang River Basin, the coefficient for NH_3-N reached 0.87. These

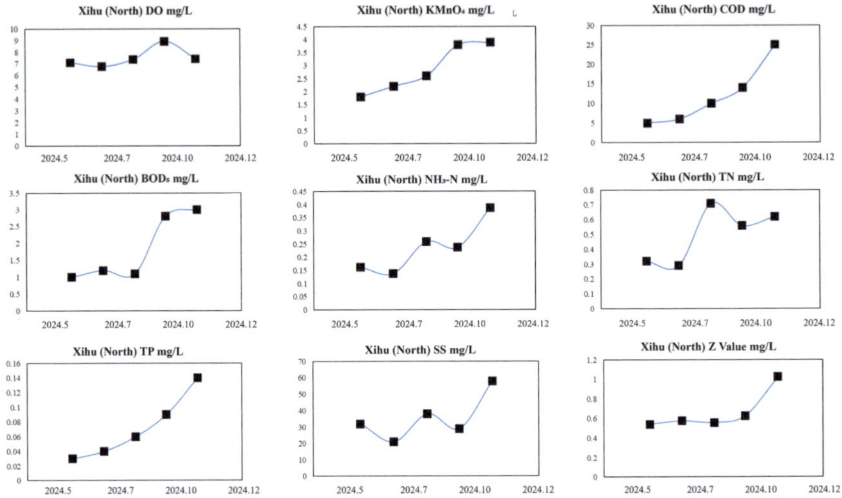

Fig. 6 Water quality index of Xihu (North) in Guilin West Lake ecological restoration and upgrading project

Table 2 Correlation analysis of upstream—downstream water quality indicators in typical ecological restoration projects

Project	Monitoring sites (upstream–downstream)	DO (r)	COD_Mn (r)	NH_3-N (r)	TP (r)	Mechanism determination
Chaoyang river basin project	Upstream intake point—Site 5	0.17	0.29	0.87	0.63	NH_3-N, TP: Conduit effect; DO,COD_{Mn}: Decoupled
Taohuajiang river (Xiufeng section) project	Site 12—Site 14	0.67	0.72	0.07	0.85	COD_{Mn}, TP: Conduit effect; NH_3-N: Efficiently decoupled
Two rivers and four lakes project	Mulong Lake head—Kangqiao peninsula	0.79	−0.85	−0.22	−0.14	COD_{Mn}: Efficiently decoupled; N, P: Partially decoupled

high positive correlations indicate that TP and NH_3-N were largely conveyed directly downstream in these cases.

(2) Decoupling Effect: When efficient physical, chemical, or biological processes within a project significantly remove or transform pollutants, the upstream-downstream pollutant linkage can be severed, creating a "decoupling" of water quality. Under this effect, downstream water quality is determined primarily by the project's own purification capacity rather than by upstream inputs. A prime

example is the decoupling of NH_3-N in the Taohua River (Xiufeng Section), where the correlation coefficient was only 0.07, indicating almost no relationship. Similarly, the Two rivers and four lakes project section showed strong decoupling for the permanganate index, with a correlation coefficient (r) of −0.85.

3.3.2 Dominant Factors Governing Basin-Wide Water Quality Variation

To macroscopically identify the key factors influencing water quality variation across the entire basin, Principal Component Analysis (PCA) was conducted on the eight core indicators from all 36 monitoring sites. The results indicate that the first two principal components cumulatively explained over 70% of the total variance, signifying their role as the dominant drivers of water quality variation in the basin.

Using the ecological restoration project at the tail of Zizhou Island in the Lijiang River as an illustrative case (Fig. 7), the first principal component (PC1) was primarily loaded by permanganate index (COD-Mn), COD, and BOD_5. These three parameters formed a highly correlated "organic pollution cluster," interpretable as a "Domestic Organic Pollution Factor." The score of this factor reflects the degree to which a monitoring site is influenced by organic pollutants from sources such as urban sewage and restaurant wastewater. The second principal component (PC2) was characterized by high loadings of NH_3-N and TP, and can be interpreted as an "Agricultural & Internal Nutrient Factor." This factor indicates the extent of influence from agricultural non-point source pollution and internal nutrient release from sediments at a given site.

Statistically, the PCA results precisely identify and validate the two primary drivers affecting water quality in the Lijiang River: (1) urban domestic pollution characterized by organic matter, and (2) nutrient pollution characterized by nitrogen and phosphorus from agricultural runoff and internal sources. This clarifies the priority for pollution control: effective management necessitates the simultaneous mitigation of both pollution types, as addressing only one is insufficient.

4 Discussion

4.1 Systemic Reshaping of Watershed Pollutant Transport Patterns by Ecological Restoration Projects

The "conduit–decoupling" framework introduced in this study provides a new perspective for understanding the function of ecological restoration projects at the watershed scale. While conventional assessments often focus on concentration differences between the inflow and outflow of individual projects (Duan et al. 2021),

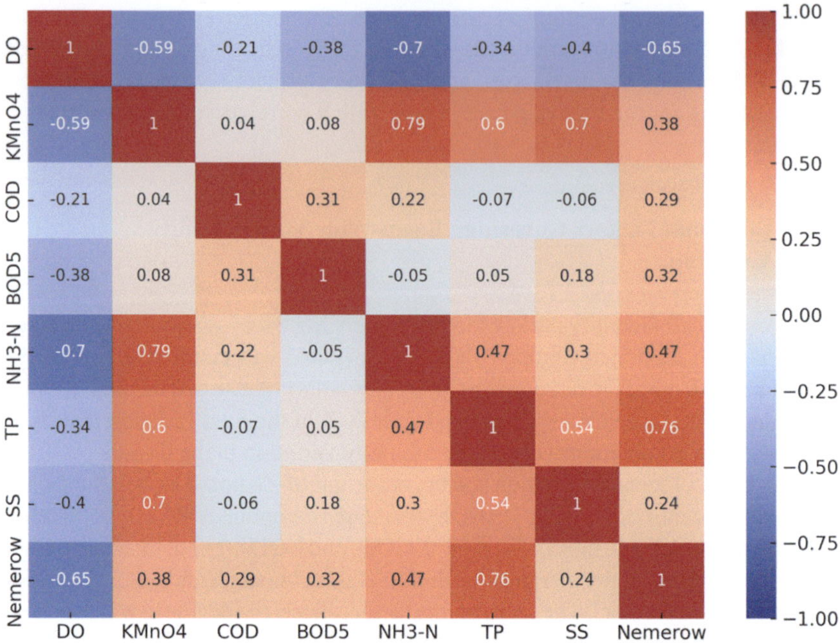

Fig. 7 Pearson analysis matrix of ecological restoration project of Weizhou Island tail of Lijiang River

our findings demonstrate that the implemented network of projects is collectively reshaping the Lijiang River's original linear pollutant transport regime, wherein upstream pollution was simply conveyed downstream. A rationally designed project network can establish a multi-tiered ecological barrier by strategically deploying units with specific decoupling functions at key nodal points. For instance, in upstream zones affected by agricultural non-point source pollution, constructed wetland projects can be implemented to decouple phosphorus. Upon entering urban reaches, enhanced riparian ecological restoration can be employed to decouple ammonia nitrogen. This layout ensures that pollutant loads are progressively reduced and transformed during downstream transport, rather than merely accumulating (Dyste and Valett, 2019).Consequently, the true value of ecological restoration extends beyond the purification rate of any single project; it lies in the scientifically planned network layout that enhances the resilience and self-purification capacity of the entire watershed ecosystem(Chen et al. 2016).

4.2 Constraints on Project Efficacy and Managerial Implications

Despite notable achievements, the "efficacy-limited" cases, exemplified by sites such as the northern Guilin West Lake and the Ningyuan River, reveal critical bottlenecks currently impeding restoration efforts in the Lijiang River Basin.

(1) The "Last-Mile" Challenge in Domestic Sewage Interception: Monitoring data indicate pulse-like surges in COD, BOD_5, and NH_3-N concentrations in several urban river sections following rainfall or during peak tourism periods. This pattern signals deficiencies in the urban sewage network—particularly in branch pipelines and combined sewer overflows in older residential areas—which hinder complete wastewater collection and treatment. These continuous point-source inputs can readily exceed the treatment capacity of end-of-pipe ecological engineering measures.
(2) The Persistent Difficulty of Agricultural Non-Point Source Pollution Control: The widespread increase in TP and TN concentrations during the wet season is clearly attributable to upstream agricultural runoff (Wang et al. 2025). While ecological restoration projects can intercept a portion of this diffuse pollution, their long-term efficacy and operational lifespan will be compromised if source reduction measures—such as promoting soil-testing-based fertilizer application or constructing ecological ditches—are not implemented concurrently (Zhang et al. 2025b).
(3) The Compounding Effect of Internal Pollution and Inadequate Hydrodynamics: In lacustrine and slow-flowing river sections, historically accumulated nutrient-rich sediments function as a latent pollution source. Under specific conditions, such as vessel-induced disturbance or summer thermal stratification and hypoxia, these sediments release substantial amounts of nitrogen and phosphorus back into the water column. This internal loading can negate reductions achieved from external source control, presenting a significant bottleneck to sustained water quality improvement (Fan et al. 2021).

These constraints underscore a critical managerial imperative: water environmental governance in the Lijiang River must evolve from a primary focus on project construction to a new phase that equally emphasizes management and integrates source control with process interception. Future priorities should therefore strengthen the synergy between ecological restoration projects (acting as sinks) and stringent pollution source control. This entails a dedicated effort to upgrade the "capillary" networks of urban sewage systems, vigorously promote green agricultural practices, and implement targeted internal pollution control measures, such as ecological dredging and sediment inactivation. Only through such an integrated, basin-wide, and full-process systematic approach can a sustainable solution be realized.

5 Conclusions

This study, through ten months of systematic monitoring and assessment of typical ecological restoration projects in the Lijiang River Basin, leads to the following principal conclusions:

Following the implementation of a series of ecological restoration projects, the overall water quality of the Lijiang River Basin has improved significantly, yet exhibits clear spatiotemporal differentiation. Spatially, water quality progressively deteriorates from the upstream areas towards the urban core, with observable improvement within the restored reaches. Temporally, water quality follows a "dual-peak" seasonal pattern characterized by non-point source pollution during the wet season and point-source pollutant accumulation during the dry season.

The purification efficacy of different ecological restoration projects varies substantially and can be categorized into three distinct models: the "High-Efficiency Integrated" model (e.g., the Chaoyang River Basin project), which achieves comprehensive pollutant removal through systematic design; the "Targeted Specialized" model (e.g., the Taohuajiang riparian project), which is highly effective for specific pollutants (e.g., ammonia nitrogen) but functionally limited; and the "Efficacy-Limited" model (e.g., the Northern West Lake site), where performance is constrained by excessive external pollutant loads and adverse internal conditions.

Ecological restoration projects influence pollutant transport within the basin through two core mechanisms: the "Conduit Effect" and the "Decoupling Effect." Successful projects that effectively decouple the upstream-downstream water quality linkage are critical for assessing their systemic functionality at the watershed scale.

Urban domestic pollution, primarily indicated by organic matter, and agricultural/internal nutrient pollution, represented by nitrogen and phosphorus, are identified as the two dominant factors currently constraining further water quality improvement in the Lijiang River. Incomplete sewage interception, insufficient control of non-point source pollution, and internal nutrient release constitute the main technical bottlenecks leading to the suboptimal performance of some projects.

In summary, to achieve fundamental and stable enhancement of the Lijiang River's water environment, future management strategies must adopt an integrated approach that equally emphasizes "Source Reduction" and "Process Interception." Building upon scientific assessment, the spatial configuration of project combinations capable of efficiently decoupling pollutants should be optimized. These measures must be closely integrated with source-control actions, including the upgrading of urban sewage networks, precise control of agricultural non-point sources, and the remediation of internal pollution. This integrated approach will facilitate a strategic transition from end-of-pipe treatment to holistic, basin-wide systematic management.

Acknowledgement We would like to extend our gratitude to the personnel of the Natural Resources and Ecological Restoration Center of Guangxi Zhuang Autonomous Region for their assistance, as well as acknowledge the financial support provided for the research project titled "Study on the Water Purification Capacity of Typical Ecological Restoration Projects in the Li River Basin".

References

Chen, X.G., Ye, C., Li, C.H., et al.: Construction of an engineering system for watershed pollution source control. J. Environ. Eng. Technol. **6**(2), 180–186 (2016)

Dong, Z.R.: The Development History and Trends of Ecological Engineering for River Regulation. (2004)

Dong, J.W., Bai, G.Q.: The quasi-natural concept in water conservancy ecological restoration projects. China Water Resour. **4**, 32–33 (2007)

Du, Q., Li, D.: A review of the EU water framework directive after a decade and its implementation effectiveness. Jiangxi Soc. Sci. **8**, 19–27 (2011)

Duan, T., Feng, J., Zhou, Y., et al.: Systematic evaluation of management measure effects on the water environment based on the DPSIR-Tapio decoupling model: a case study in the Chaohu Lake watershed, China. Sci. Total Environ. **801**, 149528 (2021)

Dyste, J.M., Valett, H.M.: Assessing stream channel restoration: the phased recovery framework. Restor. Ecol. **27**(4), 850–861 (2019). https://doi.org/10.1111/rec.12926

Fan, C.X., Liu, M., Wang, S.R., et al.: Progress and prospects of sediment environment and pollution control research in China over the past 20 years. Adv. Earth Sci. **36**(4), 346 (2021)

Gan, J.W., Wu, T.T., Yang, B.B.: "Preserve this magnificent landscape for future generations!"—documenting the integrated protection and restoration project of mountains, waters, forests, farmlands, lakes, grasslands, and sands in the Guilin Lijiang River Basin, 14 Aug 2025

Guan, F.J., Liu, L.H., Liu, J.W., et al.: Systematically promoting natural ecological protection and governance capacity building: expert comments on the "master plan for major projects of national important ecosystem protection and restoration (2021-2035)". J. Nat. Resour. **36**(2), 290–299 (2021)

Lan, G.W., Zhang, F., Wu, J., et al.: Spatiotemporal differentiation characteristics and dynamic simulation of the "production-living-ecological" functions in the Lijiang river basin. Res. Soil Water Conserv. **31**(2) (2024)

Li, Y.: Theory and practice of urban river ecological restoration from a green infrastructure perspective: a case study of Seattle. Urban Plan. Int. **3**, 41–47 (2018)

Li, H.Y., He, W., Wang, J.Y., et al.: Multi-scenario prediction of ecosystem service value based on PLSR-FLUS-MarKov model: a case study of Lijiang river basin. J. Hydroecology. **46**(1), 203–212 (2025)

Luo, F., Wu, G.R., Wang, C., et al.: Application of Nemerow pollution index method and single factor evaluation method in water quality assessment. Environ. Sustain. Dev. **5**, 87–89 (2016)

Peng, F., Zhou, Y.F., Wang, C.H., et al.: Spatial heterogeneity of microorganisms in the middle and lower reaches of the Huaihe river in spring 2019. J. Dalian Ocean. Univ. **37**(5), 830–840 (2022)

Qin, X.A., Chen, Z.X., Zhu, J., et al.: Correlation analysis and principal component analysis among water quality monitoring indicators in the Liaohe river basin. Front. Environ. Prot. **15**, 893 (2025)

Wang, Y.F., Sun, H.P., Zheng, W.B., et al.: Source apportionment and seasonal differences of nitrate in surface water in the upper Chaobai river basin. Chin. J. Eco-Agric. (Chin. Engl.). **33**(10), 1 12 (2025). https://doi.org/10.12357/cjea.20250341

Wang, Z.F., Yue, W.J., Wang, S.R., et al.: Review of international trends in watershed ecological restoration and their implications. Geogr. Sci. Res. **8**, 221 (2019)

Xiao, S.C., Xiao, H.L., Mi, L.N., et al.: Scientific assessment of the ecological effectiveness of the national comprehensive management project for the Heihe river basin. Bull. Chin. Acad. Sci. **32**(1), 45–54 (2017)

Xu, F., Wang, Y.G., Zhang, N., et al.: Research progress on river ecological restoration. Ecol. Environ. Sci. **23**(3), 515 (2014)

Zhang, X., Li, S.: Spatiotemporal distribution characteristics and pollution source analysis of water quality in the Beiluo river based on principal component analysis. Bull. Soil Water Conserv. **42**(4) (2022)

Zhang, Y.Y., Li, M., Pan, W.Y., et al.: Purification efficiency of ecological ditches on farmland drainage pollutants and their configuration scale. J. Agro-Environ. Sci. **44**(1) (2025)
Zhang, M.Y., Wang, C., Xi, X.H., et al.: Ecological quality assessment of the Lijiang river basin based on MGCI and RSEI. Remote. Sens. Technol. Appl. **40**(2), 368–375 (2025)

Research Progress on the Impact of Vegetation Construction on Slope Water Consumption and Flood Regulation

Wuxia Bi, Zhaoyu Dong, Fan Wang, Weiqi Wang, Wenqing Lin, Xin Qi, and Mingda Lu

Abstract Facing the trend of extreme water cycle, ecological measures are efficient for water disasters prevention and control, especially the vegetation construction. To understand the research progress of vegetation construction on preventing and mitigating the droughts and floods. This study reviewed the impact of vegetation construction on slope water consumption and flood regulation based on bibliometric visualization analysis of papers published in Web of Science Core Collection and China National Knowledge Infrastructure database. Results show that there are 2127 papers published in English and 452 papers published in Chinese. From the perspective of keyword co-occurrence, they mainly focus on vegetation restoration, climate change, models, drought mitigation, flood control, and so on. Then, we reviewed current research status of slope ecological regulation measures, impact of slope vegetation construction on slope water consumption, and impact of slope vegetation construction on flood regulation capacity. The literature analysis reveals that it is urgent to quantitatively identify the impact of slope vegetation construction on water consumption and flood regulation processes, which benefits the water disaster prevention and mitigation.

Keywords Vegetation construction · Slope water consumption · Flood regulation · Knowledge map

---

W. Bi (✉) · F. Wang · W. Wang · W. Lin · X. Qi · M. Lu
State Key Laboratory of Water Cycle and Water Security, China Institute of Water Resources and Hydropower Research, Beijing, China
e-mail: biwx@iwhr.com

Research Center on Flood and Drought Disaster Prevention and Reduction of the Ministry of Water Resources, Beijing, China

Z. Dong
Tsinghua Shenzhen International Graduate School, Institute of Environment and Ecology, Tsinghua University, Shenzhen, China

H. Xu (ed.), *Proceedings of the 8th International Symposium on Water Resource and Environmental Management*, Environmental Science and Engineering,
https://doi.org/10.1007/978-3-032-19088-8_20

1 Introduction

The global climate system is undergoing profound changes characterized by warming, and the hydrological cycle is showing a significant trend of extremism (Bellprat et al. 2019). As one of the countries with the most serious flood and drought disasters in the world, extreme hydrological events have occurred frequently in China. In 2021, the maximum rainfall per hour of extremely heavy rainstorm in Zhengzhou reached 201.9 mm, breaking through the extreme value of hourly rainfall on land in China (Liu et al. 2023). In 2022, the Yangtze River Basin experienced the most severe meteorological drought since 1961, showing a continuous drought pattern in summer, autumn, and winter (Xu and Yuan 2023). At the same time, ecosystem degradation and hydrological distortion form a vicious cycle. According to the 2022 China Soil and Water Conservation Bulletin, the area of soil erosion in China has reached about 2.693 million km^2. The traditional defense system based on grey infrastructure is facing new challenges. Ecological regulation measures should be considered to defend the extreme climate events. Slope is a key transitional area connecting the upstream catchment area and downstream river channel, which affects the drought and flood regulation. Thus, the slope ecological regulation measures are one kind of the critical solutions for mitigating the water disasters.

Slope ecological regulation measures include vegetation construction, micro-topography modification, and soil improvement (Zhang et al. 2021; Yu et al. 2018). By changing surface roughness, intercepting precipitation, and altering soil structure (Zhang et al. 2019), they affect slope water production and consumption processes, thereby affecting slope water use efficiency and river inflow (Wang et al. 2021a). Previous studies have shown that slope vegetation cover absorbs water and intercepts rainwater during rainfall, has good storage capacity, and has benefits in reducing flow and sediment (Chen and Zhang 2022). Therefore, scientific allocation of slope vegetation is an important means to balance slope water production and consumption, and flood and drought disaster prevention. In view of this, it is necessary to clarify the impact mechanism of vegetation construction on slope water consumption and flood regulation processes, which guides the comprehensive adaption to droughts and floods.

Based on bibliometric analysis method, this study quantitatively analyzed the number of literatures, and visualized the keywords by knowledge map. According to the bibliometric analysis result, we reviewed the research progress on the impact of vegetation construction on slope water consumption and flood regulation, which can provide supports for future potential research directions and mitigation for water disasters.

2 Materials and Methods

The bibliometric analysis method was applied to quantify the research progress. Cross-retrieval was conducted with the themes of "slope", "vegetation construction", "water consumption", "drought/dry period", and "flood/flooding" in the core databases of Web of Science and China National Knowledge Infrastructure. A total of 2127 papers published in English and 452 papers published in Chinese from 1990 to 2024. The annual number and keywords of published papers were analyzed by bar chart and knowledge map to find the research hotspots and potential problems of the impact of vegetation construction on slope water consumption and flood regulation.

3 Bibliometric Analysis

The number of published papers basically showed an increasing trend during 1990–2024 (Fig. 1). The increase rate of published papers was different in different time periods and in different languages. For paper published in English, the published papers increased with a rate of 17.4 papers per year, 46.8 papers per year, and 137.1 papers per year during 1990–2000, 2001–2015, and 2016–2024, respectively. While for paper published in Chinese, the increase rate was 2.8 papers per year, 16.5 papers per year, and 18.3 papers per year during 1990–2000, 2001–2010, and 2011–2024, respectively. Overall, the number of papers published in English increased faster than that in Chinese.

From the perspective of keyword co-occurrence, the keywords mainly focus on vegetation restoration, climate change, models, drought, flood, and so on (Fig. 2). However, the keyword co-occurrence is not strong. Therefore, this study reviewed the impact of vegetation construction on slope water consumption and flood regulation, mainly contains three parts: research on slope ecological regulation measures, impact of slope vegetation construction on slope water consumption, and on flood regulation capacity.

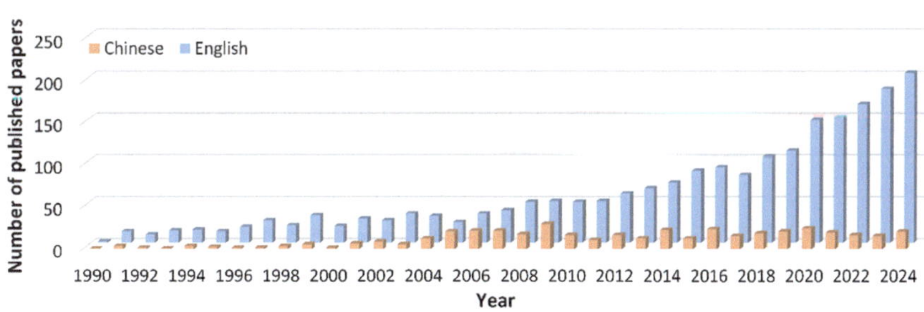

Fig. 1 Number of papers published in English and Chinese from 1990 to 2024

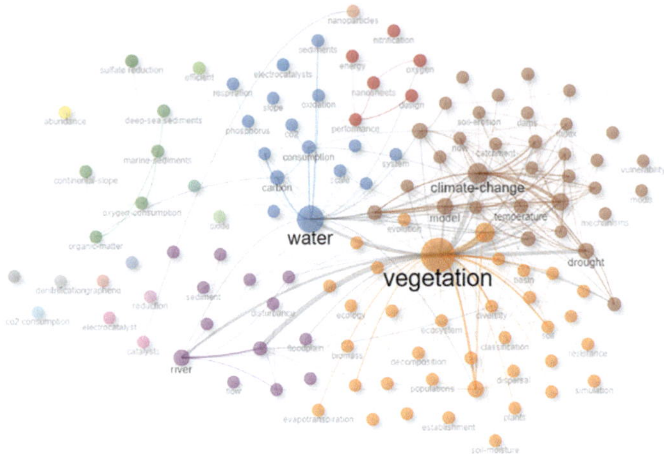

Fig. 2 Knowledge map of key words

4 Literature Review

4.1 Research on Slope Ecological Regulation Measures

Slope ecological regulation measures include vegetation construction, micro-topography modification, and soil improvement (Zhang et al. 2021). Through measures such as planting vegetation, deep tillage, straw return, biochar addition, ridge-furrow systems, and terraces, they affect infiltration and soil water movement processes, thereby changing the water distribution on slopes (Langhans et al. 2019), reducing risks of soil and water loss, drought and flood disasters, and improving water resource management efficiency. Therefore, slope ecological regulation measures are an important research direction in the current field of water disaster mitigation and water resource management.

Research shows that the implementation of slope ecological regulation measures helps to reduce peaks and supplement low flows (Sheng et al. 2022), effectively reducing soil drying caused by long-term planting on an inter-annual basis, improving local water availability, and mitigating drought and flood risks. Also, it significantly improves soil infiltration capacity and water retention capacity. Experiments show that vegetation restoration can increase surface coverage (Wu et al. 2020a), increase surface roughness (Mabbott and Fryirs 2022), and reduce runoff velocity (Zhao et al. 2015); and improve soil porosity through plant root structures (de Moraes et al. 2020), thereby improving soil infiltration capacity (Mair et al. 2022), and thus regulating rainfall-runoff processes. Meanwhile, vegetation planting blocks and absorbs rainwater, dissipates runoff kinetic energy, and increases slope water consumption (Ciruzzi and Loheide 2021). Deep tillage breaks the plow pan, changes soil physical properties, increases soil porosity and steady infiltration rate, thereby reducing runoff generation (Oguike et al. 2023). Biochar addition increases soil organic matter (Rombolà et al. 2021), increases total soil porosity (Xiao et al. 2016), and thus enhances soil water retention (Baiamonte et al. 2019). Ridge-furrow layout and terrace construction directly change the micro-topography of slope units (Xu et al. 2018), increase surface roughness, and increase their depression storage (Kosmowski 2018). In addition, with the continuous improvement and development of large distributed hydrological models, the impact of ecological infrastructure construction at different scales on key element processes of the water cycle can be quantitatively simulated (Tu et al. 2025). However, quantitative simulation of the impact of vegetation construction on water cycle processes, especially dynamic response research, is still insufficient (Li et al. 2018). Therefore, slope ecological regulation measures, especially vegetation construction, have an impact on both water disaster regulation and slope water use efficiency, and the relevant impact mechanisms need further research.

4.2 Impact of Slope Vegetation Construction on Slope Water Consumption

As one of the important measures of slope ecological regulation, vegetation construction increases surface coverage by planting trees, shrubs, and herbaceous plants, which can effectively increase soil permeability and water retention capacity. Through canopy interception and root penetration, vegetation improves soil structure, enhances soil water storage capacity, increases vertical water circulation, and is beneficial for adjusting local climate (Yang et al. 2023). However, vegetation construction measures do cause a certain degree of water resource consumption (Fu and Zhang 2015), thereby affecting runoff entering the river channel. During dry periods, water consumption by vegetation transpiration and deep root extraction may intensify water resource competition (Queiroz-Alves et al. 2020), and even trigger systemic risks such as vegetation death and river channel drying (Wang et al. 2023). For example, in the past 20 years, the vegetation coverage rate in the Loess Plateau region has increased from 32 to 64%. Although it has a positive contribution of 0.83 mm/year to local precipitation increase, evaporation has increased by 5.17 mm (Cao et al. 2024), the groundwater funnel area has expanded, and base flow recharge to the river channel has been reduced (Wang et al. 2020). Therefore, the impact of slope vegetation construction on slope water consumption is the key to determining the water resource utilization efficiency of the slope-river channel system.

Factors affecting slope water consumption mainly include vegetation type, coverage, pattern, and seasonal changes. These factors directly affect the redistribution process of atmospheric precipitation by the vegetation canopy, litter layer, and soil layer, thereby affecting the regulation effect of vegetation regulation measures on surface runoff and river channel runoff. Vegetation affects the rainfall-runoff process through rainfall interception, throughfall, and stemflow, playing a role in reducing surface runoff (An et al. 2020). Seasonal changes in vegetation also cause changes in the magnitude of regulation function. In spring, vegetation transpiration is weak. In summer, strong vegetation transpiration (exceeding rainfall) can cause the watershed river channel to maintain a low water level state, and regulation measures continue to play a strong runoff reduction role (Yao et al. 2023). Litter has a significant impact on slope hydrological processes. By intercepting precipitation, increasing soil infiltration, and inhibiting evapotranspiration in forests, it affects slope runoff generation and concentration processes (Magliano et al. 2017; Zhu et al. 2020).

Research indicates that by optimizing the design and layout of vegetation construction measures, water consumption can be effectively reduced. Reasonably planning vegetation restoration areas and selecting native, drought-resistant, and low water-consuming plant species for vegetation construction can significantly reduce evapotranspiration while maintaining soil moisture and reducing runoff. Using native plants with strong drought resistance for vegetation restoration and greening can not only adapt to local climatic conditions and reduce maintenance and irrigation needs, but also effectively reduce water consumption (Li et al. 2020). Setting vegetation coverage thresholds based on hydrological zones and matching herbaceous plants to form a "patch-corridor" pattern can reduce evapotranspiration during dry periods without significantly weakening flood detention capacity (Wang et al. 2021b). Therefore, the impact of slope vegetation construction on slope water consumption can be dynamically adjusted, which could be applied for water disaster defense and sustainable water resource management.

4.3 Impact of Slope Vegetation Construction on Flood Regulation Capacity

Factors affecting regulation function can be divided into internal factors and external factors. Internal factors mainly include soil characteristics, vegetation characteristics, and initial hydrological conditions (Ma et al. 2023; Zong et al. 2021). External factors mainly include basin characteristics, precipitation characteristics, climate change, and human activities (Wu et al. 2020b). Current research methods on the impact of slope vegetation construction on flood regulation capacity mainly include experimental observation, remote sensing inversion, and numerical simulation.

Small-scale research is mostly based on experiments, quantitatively evaluating the impact of slope vegetation construction measures such as vegetation restoration and artificial vegetation on regulation capacity. Yu et al. (2006) concluded through slope rainfall experiments that forest and grassland vegetation has an interception effect on rainfall. When forest coverage was greater than 0.4, the water and sediment reduction benefits of the forest were significant. Mabbott and Fryirs found that the surface roughness value produced by complete vegetation coverage was much higher than that produced by patchy canopy coverage or ground vegetation coverage, affecting the velocity and flow of surface runoff (Mabbott and Fryirs 2022). Gagarinova and Tsygankova showed that vegetation morphology and composition had a great impact on river hydrological conditions and flood formation, and coniferous vegetation on gentle slopes and low mountains had significant flow regulation potential (Gardner and Gerrard 2003). Sun et al. (2022) concluded through slope artificial rainfall experiments that slope vegetation distribution had an important impact on runoff and sediment yield. Under the same vegetation coverage conditions, the smaller the relative distance of grass strips, the better the effect on reducing runoff and sediment. This type of experiment reveals the impact mechanism of typical slope vegetation construction measures on soil hydrological characteristics, but quantitative evaluation of storage effects and optimization allocation schemes still need further research.

Large-scale research is mostly based on remote sensing inversion and numerical simulation, inverting regional or watershed vegetation characteristics, and constructing slope vegetation construction measure models under different scenarios to quantitatively analyze their mitigation capacity for flood disasters. Liou and Hoang (2024) concluded through remote sensing inversion and model deep learning that although vegetation construction changing surface roughness could play a role in flood attenuation, its effectiveness may depend on specific terrain and hydrological background, and its flood control benefits will weaken with increasing slope. Li et al. (2020) embedded a roughness dynamic module driven by coverage in a hydrological model, the simulation results showed that an increase in vegetation coverage would reduce flow velocity and affect water flow distribution. Huang et al. quantified the regulation of vegetation phenology on flood peak lag time through remote sensing inversion and multi-process coupling simulation (Huang et al. 2022). Therefore, numerical simulation can evaluate the performance of different slope vegetation construction under the influence of extreme climate, providing accurate scientific basis for optimized allocation and management decisions of regional vegetation construction.

5 Conclusions

Based on the bibliometric analysis results, this study reviewed the research progress on the impact of vegetation construction on slope water consumption and flood regulation. The literatures have been increasing from 1990 to 2024, and with an increasing rate. The keywords mainly focus on vegetation restoration, climate change, models, drought, flood, etc. Existing research still has the following issues that need to be explored in depth. First, most existing research on slope vegetation construction measures focuses on single objective such as slope water consumption or flood storage capacity, with less research on the synergistic impact of slope water consumption and flood storage processes. Second, the solution of key thresholds for slope vegetation construction aimed at water consumption control and flood storage capacity improvement is still in its initial stage. The development of vegetation-hydrological models improved based on key parameters of the storage process, as well as vegetation allocation optimization schemes, still need further research.

Acknowledgements This study was supported by the National Key Research and Development Program of China (No. 2022YFC3080300), the Young Elite Scientists Sponsorship Program of the Beijing High Innovation Plan (No. 20250696), the National Natural Science Foundation of China (No. 52209043), and the Youth Talent Promotion Plan of Research Center on Flood & Drought Disaster Prevention and Reduction of the Ministry of Water Resources.

References

An, C., Fang, H., Wang, F.: Characteristics of runoff and sediment yield on slopes and their responses to rainfalls in the upstream of the Miyun reservoir: a case study of the Shixia watershed. Sci. Soil Water Conserv. **18**(5), 43–51 (2020)

Baiamonte, G., Crescimanno, G., Parrino, F., De Pasquale, C.: Effect of biochar on the physical and structural properties of a sandy soil. Catena **175**, 294–303 (2019)

Bellprat, O., Guemas, V., Doblas-Reyes, F., Donat, M.G.: Towards reliable extreme weather and climate event attribution. Nat. Commun. **10**(1), 1732 (2019)

Cao, M., Wang, W., Wei, J., Forzieri, G., Fetzer, I., Wang-Erlandsson, L.: Revegetation impacts on moisture recycling and precipitation trends in the Chinese loess plateau. Water Resour. Res. **60**, e2024WR038199 (2024)

Chen, B., Zhang, X.: Effects of slope vegetation patterns on erosion sediment yield and hydraulic parameters in slope-gully system. Ecol. Indic. **145**, 109723 (2022)

Ciruzzi, D.M., Loheide, S.P.: Monitoring tree sway as an indicator of interception dynamics before, during, and following a storm. Geophys. Res. Lett. **48**(20), e2021GL094980 (2021)

de Moraes, M.T., Debiasi, H., Franchini, J.C., Mastroberti, A.A., Levien, R., Leitner, D., Schnepf, A.: Soil compaction impacts soybean root growth in an Oxisol from subtropical Brazil. Soil Tillage Res. **200**, 104611 (2020)

Fu, B., Zhang, L.: Land-use change and ecosystem services: concepts, methods and progress. Prog. Geogr. **33**(4), 441–446 (2015)

Gardner, R.A.M., Gerrard, A.J.: Runoff and soil erosion on cultivated rainfed terraces in the Middle Hills of Nepal. Appl. Geogr. **23**(1), 23–45 (2003)

Huang, Y., Chen, X., Zou, Y., Zhang, P., Li, F., Hou, Z., Li, X., Zeng, J., Deng, X., Zhong, J., Xie, Y.: Exploring the relative contribution of flood regimes and climatic factors to Carex phenology in a Yangtze River-connected floodplain wetland. Sci. Total Environ. **847**, 157568 (2022)

Kosmowski, F.: Soil water management practices (terraces) helped to mitigate the 2015 drought in Ethiopia. Agric. Water Manag. **204**, 11–16 (2018)

Langhans, C., Diels, J., Clymans, W., Van den Putte, A., Govers, G.: Scale effects of runoff generation under reduced and conventional tillage. Catena **176**, 1–13 (2019)

Li, Y., Piao, S., Li, L., Chen, A., Wang, X., Ciais, P., Huang, L., Lian, X., Peng, S., Zeng, Z., Wang, K., Zhou, L.: Divergent hydrological response to large-scale afforestation and vegetation greening in China. Sci. Adv. **4**, eaar4182 (2018)

Li, Y., Zhang, Q., Tan, Z., Yao, J.: On the hydrodynamic behavior of floodplain vegetation in a flood-pulse-influenced river-Lake system (Poyang lake, China). J. Hydrol. **585**, 124852 (2020)

Liou, Y., Hoang, D.: Improved flood depth estimation with SAR image, digital elevation model, and machine learning schemes. J. Hydrol. Reg. Stud. **53**, 101775 (2024)

Liu, J., Pei, Y., Mei, C., Liu, C.: Waterlogging cause and disaster prevention and control of “7·20” torrential rain in Zhengzhou. J. Zhengzhou Univ. (Eng. Sci.). **44**(2), 38–45 (2023)

Ma, B., Jing, J., Liu, B., Wang, Y., He, H.: Assessing the contribution of human activities and climate change to the dynamics of NPP in ecologically fragile regions. Glob. Ecol. Conserv. **42**, e2393 (2023)

Mabbott, R., Fryirs, K.: Geomorphic and vegetative river recovery in a small coastal catchment of New South Wales, Australia: implications for flow hydrology and river management. Geomorphology **413**, 108334 (2022)

Magliano, P.N., Giménez, R., Houspanossian, J., Páez, R.A.., Nosetto, M.D., Fernández, R.J., Jobbágy, E.G.: Litter is more effective than forest canopy reducing soil evaporation in dry Chaco rangelands. Ecohydrology **10**(7), e1879 (2017)

Mair, A., Dupuy, L.X., Ptashnyk, M.: Model for water infiltration in vegetated soil with preferential flow oriented by plant roots. Plant Soil. **478**(1), 709–729 (2022)

Oguike, P.C., Onwuka, B.M., Obalum, S.E.: Soil organic matter control of water transmission properties of coarse-textured soils under contrasting land-use types in tropical rainforest. Int. J. Hydrol. Sci. Technol. **16**(1), 93–106 (2023)

Queiroz-Alves, L., Pereira, K.F., Leal, A., Dalmolin, A.C., Gomes, R.L., Schaffer, B., dos Santos, M.S., Pérez-Molina, J.P., Mielke, M.S.: Influence of soil characteristics on physiological and growth responses of Cytharexyllum myrianthum Cham. (Verbenaceae) to flooding. Acta Physiol. Plant. **42**(11), 165 (2020)

Rombolà, A.G., Torri, C., Vassura, I., Venturini, E., Reggiani, R., Fabbri, D.: Effect of biochar amendment on organic matter and dissolved organic matter composition of agricultural soils from a two-year field experiment. Sci. Total Environ. **812**, 151422 (2021)

Sheng, F., Liu, S., Zhang, T., Liu, G., Liu, Z.: Quantitative assessment of the impact of precipitation and vegetation variation on flooding under discrete and continuous rainstorm conditions. Ecol. Indic. **144**, 109477 (2022)

Sun, R., Shen, M., Hu, Y., Xu, Q., Zhang, J.: Effect of grass belt distribution on runoff and sediment yield under simulated rainfall. J. Soil Water Conserv. **36**(4), 22–29 (2022)

Tu, A., Zheng, H., Mo, M., Shen, F., Zhou, C.: Effect of hillslope profile shape on runoff and soil erosion in red soil hilly areas of China using the WEPP model. J. Hydrol. **655**, 132914 (2025)

Wang, L., Zhang, G., Zhu, P., Wang, X.: Comparison of the effects of litter covering and incorporation on infiltration and soil erosion under simulated rainfall. Hydrol. Process. **34**(13), 2911–2922 (2020)

Wang, S., Lyu, X., Wei, Y., Li, Y.: Runoff and sediment yields of different vegetation slopes in the gullied rolling loess area. Soil Water Conserv. China. **3**, 31–33+9 (2021a)

Wang, S., Yang, M., Gao, X., Zhang, Z., Wang, X., Zhao, X., Wu, P.: Comparison of the root–soil water relationship of two typical revegetation species along a precipitation gradient on the loess plateau. Environ. Res. Lett. **16**(6), 64054 (2021b)

Wang, Z., Fu, B., Wu, X., Li, Y., Feng, Y., Wang, S., Wei, F., Zhang, L.: Vegetation resilience does not increase consistently with greening in China's loess plateau. Commun. Earth Environ. **4**, 336 (2023)

Wu, G., Liu, Y., Cui, Z., Liu, Y., Shi, Z., Yin, R., Kardol, P.: Trade-off between vegetation type, soil erosion control and surface water in global semi-arid regions: a meta-analysis. J. Appl. Ecol. **57**(5), 875–885 (2020a)

Wu, Y., Zhang, G., Rousseau, A.N., Xu, Y.: Quantifying streamflow regulation services of wetlands with an emphasis on quickflow and baseflow responses in the upper Nenjiang River basin, Northeast China. J. Hydrol. **583**, 124565 (2020b)

Xiao, Q., Zhu, L., Shen, Y., Li.: S.: sensitivity of soil water retention and availability to biochar addition in rainfed semi-arid farmland during a three-year field experiment. Field Crop Res. **196**, 284–293 (2016)

Xu, J., Yuan, Z.: Drought characteristics of Changjiang River basin in 2022 and drought mitigation response pattern under new circumstances. J. Changjiang River Sci. Res. Inst. **40**(4), 1–8 (2023)

Xu, X., Zheng, F., Wilson, G.V., He, C., Lu, J., Bian, F.: Comparison of runoff and soil loss in different tillage systems in the Mollisol region of Northeast China. Soil Tillage Res. **177**, 1–11 (2018)

Yang, W., Zhang, D., Cai, X., Yang, X., Zhang, H., Wang, Y., Diao, L., Luo, Y., Cheng, X.: Natural revegetation over 160 years alters carbon and nitrogen sequestration and stabilization in soil organic matter on the loess plateau of China. Catena. **220**, 106647 (2023)

Yao, C., Chen, K., Zhang, Q., Wang, C., Lu, C., Wang, H., Wu, F.: The contribution rate of stem-leaf and root of alfalfa (Medicago sativa L.) to sediment and runoff reduction. Land Degrad. Dev. **34**(13), 3991–4005 (2023)

Yu, T., Zhang, X., Li, Z., Li, P., Li, C.: Study on difference of erosion sediment in a slope-gully system under different grass strip position. J. Soil Water Conserv. **32**(06), 22–27+122 (2018)

Yu, X., Zhang, X., Wu, S., Wei, T., Zhang, X.: The effect of vegetation and precipitation upon runoff and sediment production in sloping lands of loess area. J. Mt. Sci. **24**(1), 19–26 (2006)

Zhang, B., Zhang, G., Yang, H., Zhu, P.: Temporal variation in soil erosion resistance of steep slopes restored with different vegetation communities on the Chinese loess plateau. Catena **182**, 104170 (2019)

Zhang, X., Li, Y., Liu, Y., Huang, Y., Wang, Y., Lu, Z.: Characteristics and prevention mechanisms of artificial slope instability in the Chinese loess plateau. Catena **207**, 105621 (2021)

Zhao, M., Yao, W., Wang, J., Xiao, P.: Experimental study on influence of vegetation coverage to soil infiltration and runoff-producing of the loess plateau region. Soil Water Conserv. China **3**, 41–43+67 (2015)

Zhu, H., Wang, G., Yinglan, A., Liu, T.: Ecohydrological effects of litter cover on the hillslope-scale infiltration-runoff patterns for layered soil in forest ecosystem. Ecol. Eng. **155**, 105930 (2020)

Zong, J., Li, L., Yao, X., Chen, J., Wang, H., Zhao, X., Liu, J.: Performance of five typical warm-season turfgrasses and their influence on soil bacterial community under a simulated tropical coral Island environment. Land Degrad. Dev. **32**(14), 3920–3929 (2021)

Path Optimization for UAV-Based Watershed Water Sampling: A Dynamic Vehicle Routing Approach for Discrete Points with Multi-constraints

Qianqian Shao, Hao Yin, Ziming Yan, and Xuemei Han

Abstract Watershed water sampling is challenged by discretely distributed hydrological points and complex terrain, demanding path planning that satisfies multi-dimensional constraints (task priority, tributary sampling sequence, UAV load and endurance) while minimizing path length and response time. This problem is essentially formulated as a Dynamic Vehicle Routing Problem (D-VRP). Unlike conventional path planning methods that often ignore scenario-specific hydrological rules and UAVs' physical limitations, this study first defines a D-VRP-based optimization model that integrates the aforementioned constraints for hydrological sampling scenarios. A tailored heuristic algorithm, namely the adaptive large neighborhood search (ALNS), is then developed for model solving, with the advantages of fast convergence for large-scale sampling tasks and flexible adjustment for emergency sampling demands. The proposed solution provides a systematic and constraint-aware approach for UAV path planning, which enhances the efficiency and flexibility of field sampling operations in practical hydrological monitoring, and offers a scalable framework for similar environmental monitoring missions involving unmanned aerial vehicles in complex terrains.

Keywords Watershed water sampling · UAV path optimization · D-VRP · Multi-constraint scheduling · Discrete hydrological points

---

Q. Shao · H. Yin · Z. Yan
School of Transportation and Surveying Engineering, Shenyang Jianzhu University, Shenyang, China

X. Han (✉)
Management and Information Department, Shandong Transport Vocational College, Weifang, Shandong, China

H. Xu (ed.), *Proceedings of the 8th International Symposium on Water Resource and Environmental Management*, Environmental Science and Engineering,
https://doi.org/10.1007/978-3-032-19088-8_21

1 Introduction

Watershed water quality monitoring serves as a fundamental pillar of ecological protection and sustainable water resource management, furnishing critical data support for pollution source identification, ecological condition assessment, and timely emergency response to sudden contamination events (Dang et al. 2025; Asadzadeh et al. 2022). With the escalating emphasis on watershed ecological security, the demand for efficient, accurate, and real-time sampling has grown exponentially. However, traditional manual sampling methods are increasingly inadequate in meeting these modern monitoring requirements due to inherent limitations. These drawbacks include high labor intensity and costs, low operational efficiency, and the inability to access remote or hazardous hydrological points—such as midstream river sections or steep slope areas—posing significant risks to sampling personnel (Debnath et al. 2024; Lin et al. 2023). Compounding these challenges, hydrological sampling points in watersheds are typically discretely distributed, separated by water bodies, mountainous terrain, and dense vegetation without fixed access paths. This spatial dispersion further exacerbates the difficulty of formulating efficient sampling workflows (Zhao et al. 2024).

In response to these limitations, Unmanned Aerial Vehicles (UAVs) have emerged as a promising and transformative solution for watershed sampling tasks, leveraging their unique advantages of high flexibility, strong maneuverability, and the ability to navigate complex and inaccessible environments (Debnath et al. 2024; Aggarwal and Kumar 2020). By deploying UAVs, sampling operations can bypass geographical barriers, reduce labor costs, and improve sampling efficiency. However, optimizing UAV flight paths for this specific watershed sampling scenario remains a non-trivial and challenging task. Existing research on UAV path planning predominantly focuses on single-constraint optimization objectives, such as minimizing total flight distance or energy consumption, and often fails to fully address the multi-dimensional and mutually coupled constraints inherent in watershed sampling (Sajid et al. 2022; Radzki et al. 2019). These critical constraints include task priority (with emergency sampling tasks requiring immediate response), tributary sampling sequence (needing to follow the source-to-confluence order for accurate pollution tracing), UAV single-path load limits, and endurance thresholds. Meanwhile, traditional Dynamic Vehicle Routing Problem (D-VRP) models—widely applied in urban logistics and delivery systems—are ill-suited to the discrete, unconnected spatial characteristics of watershed sampling points, as they typically assume fixed and connected road networks (Psaraftis et al. 2016; Chen et al. 2023).

To bridge these existing research gaps, this study proposes a four-stage UAV path optimization framework specifically tailored for discrete hydrological sampling in watersheds. The core objectives of this research are threefold: first, to construct a multi-constraint dynamic optimization model that integrates all key constraints of watershed sampling; second, to develop an efficient heuristic solution algorithm (e.g., ALNS) capable of handling dynamic task changes, such as sudden emergency sampling requests; third, to verify the feasibility and effectiveness of the proposed framework using real-world watershed data. This research aims to enhance the timeliness, efficiency, and reliability of watershed water sampling operations, thereby advancing the practical application of UAV technology in environmental monitoring and providing scientific support for watershed ecological management (Asadzadeh et al. 2022; Huang et al. 2023). The remainder of this paper is structured as follows: Sect. 2 reviews relevant literature on UAV path planning, D-VRP extensions, and watershed monitoring applications. Section 3 details the data collection and analysis procedures. Section 4 elaborates on the construction of the dynamic optimization model and the design of the heuristic algorithm. Section 5 presents experimental validation results and performance analysis. Finally, Sect. 6 concludes the research findings and outlines future research directions.

2 Literature Review

The study area is Shenyang City (122° 25′–123° 48′ E, 41° 11′–43° 02′ N), capital of Liaoning Province. It has a temperate continental monsoon climate with annual rainfall 680 mm concentrated in July–August (Dang et al. 2025), and its water system is dominated by Hunhe and Taizi Rivers, forming a complex watershed network. This study focuses on Shenyang's Hunhe River basin (2890 km^2), where sampling points scatter across urban, suburban and upstream areas, typical for UAV sampling research (Debnath et al. 2024; Zhao et al. 2024).

A total of 226 sampling points (regular/intensive/emergency) are set, with 68% inaccessible to ground vehicles (Lin et al. 2023). Core monitoring targets COD, NH3-N and TP, facing challenges like urban traffic constraints and strict time windows. These align with the study's multi-constraint path optimization focus, verifying its practical significance (Asadzadeh et al. 2022; Huang et al. 2023). All the studied parameters were measured annually from the data collected from these gauging stations between 2020 and 2023 in the studied area (Fig. 1).

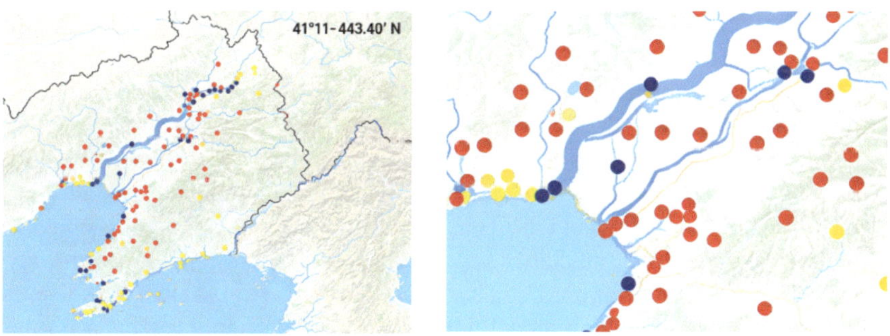

Fig. 1 Location and sampling sites of the study area

Land use has several effects on hydrological sampling in Shenyang. Urban built-up areas block ground access, boosting UAV application demands; suburban farmlands concentrate non-point pollution, requiring intensive sampling; upstream ecological reserves need low-disturbance UAV operations. These land use-driven constraints validate the necessity of optimized UAV path planning for efficient watershed monitoring in the region.

3 Data Collection and Analysis

Reliable data are fundamental to UAV path optimization model construction and algorithm verification. This chapter focuses on three core data types for watershed discrete hydrological sampling: sampling point spatial data, task attribute data, and UAV technical parameters, with sources including field investigations, hydrological monitoring institution databases, and UAV manufacturer specifications.

Spatial location data were collected via GPS RTK positioning and UAV aerial auxiliary positioning, with tributary affiliation determined using DEM. Task attributes were sorted from historical records, with priority scored by experts (emergency > intensive > regular) and time windows defined by monitoring requirements. UAV parameters were verified through field tests to ensure load and endurance constraints were met. Preprocessing included cleaning abnormal/missing data, integrating data via sampling point ID, and normalizing distance and flight time data to [0, 1].

Data feature analysis showed 216 sampling points (120 regular, 66 intensive, 30 emergency) in an 80 km^2 mountainous watershed, with 75% separated by terrain. Emergency tasks (13.9%) have the shortest time window, while UAV parameters match the single-path load limit (≤ 3 tasks). The above data and features lay a solid foundation for subsequent model construction. The UAV-based hydrological sample collection process driven by path optimization consists of four sequential stages as follows: The spatial distribution characteristics of sampling points and the statistical distribution of task attributes are specifically shown in Fig. 1

1. Scenario and Objectives: Pure UAV water sample collection and pollution prevention (addressing manual inaccessibility due to water/terrain), with the core objectives of maximizing collection efficiency (via optimized round-trips among scattered points) and realizing real-time pollution response; Path Planning Initialization: Based on preprocessed hydrological sampling point data, task attributes (priority, time window) and UAV constraints, construct an initial multi-objective optimization scenario and clarify key information of each sampling point.
2. Optimized Path Generation: Use heuristic algorithms (e.g., ALNS) to solve the dynamic path optimization model, integrate tributary sampling sequence rules, and generate multi-constraint-adapted UAV paths with priority on emergency points.

3. Dynamic Path Adjustment and Verification: Real-time adjust paths for sudden emergency tasks, coordinate task conflicts, and verify path feasibility with UAV endurance and load data.
4. Sample Collection Execution: UAV completes sampling along optimized paths and records key data for algorithm verification. The impact of sampling point distribution on path planning and optimized path execution sequence are shown in Fig. 2.

The first plot presents the obstacle-constrained ground path for 5 discrete hydrological sampling points, with routes detouring through fixed waypoints to avoid natural barriers in the watershed (Fig. 2).

4 Methodology

4.1 Assumptions

For the convenience of description, the model is based on the following assumptions:

1. The collection command center is fully equipped with equipment. The UAV takeoff/landing and sample loading/unloading times are fixed (3 min for loading, 2 min for unloading), and the default preparation time is 0. Manual loading/unloading is only required at the command center, avoiding the need for on-site manual intervention at sampling points;
2. The UAV flight speed is constant, and the flight time is positively correlated with the straight-line distance between sampling points (which are scattered and separated by water/terrain). The impact of external environments such as wind speed and terrain on flight is ignored, focusing on path optimization for round-trips;
3. The sampling time for each sampling point is fixed (5 min for regular points, 8 min for intensive points, 10 min for emergency points). Collected data is transmitted to the command center in real time, and excessive data automatically triggers emergency tasks—eliminating the need for manual data retrieval from scattered points;
4. After completing a single trip (round-trip from the command center to sampling points and back), the UAV returns to the command center for recharging. The recharging time is not included in the effective operation time, and the performance is restored to the initial state after recharging;
5. The locations of hydrological sampling points are fixed (scattered across the watershed) and the sampling requirements are clear. There is no temporary cancellation of tasks. Emergency tasks are only triggered by excessive data from regular/intensive points—avoiding unplanned manual dispatching.

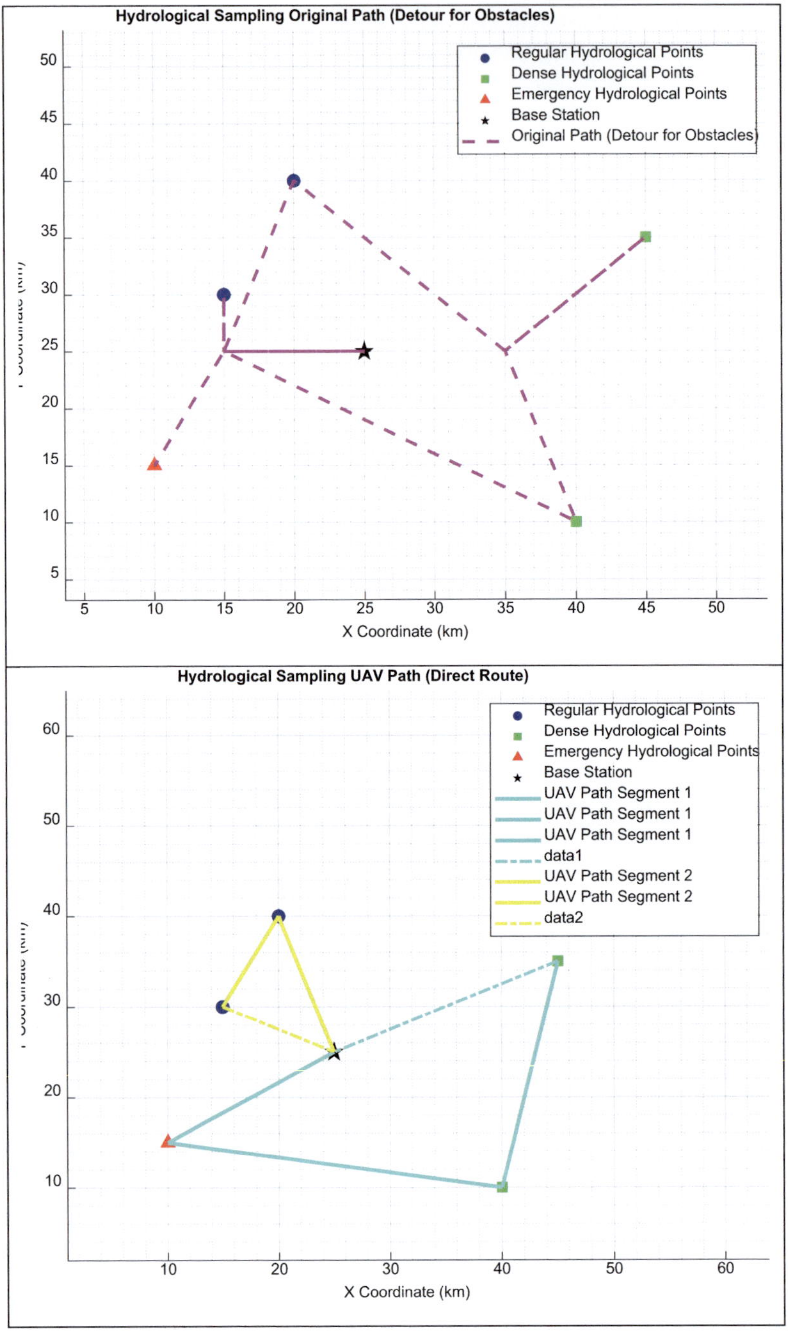

Fig. 2 Hydrological sampling path comparison: obstacle detour and UAV direct route

4.2 Notation Definitions

4.2.1 Sets

N: Set of discrete hydrological sampling points (scattered with poor manual accessibility), $N = N_1 \cup N_2 \cup N_3$ (N_1: regular sampling points, N_2: intensive sampling points, N_3: emergency sampling points);
N_0: Collection command center (start and end point of UAV round-trips);
U: Set of UAVs, $U = \{u_1, u_2, \ldots, u_k\}$;
S^1: Set of preset collection tasks (including regular tasks and planned intensive tasks) across scattered points;
S^2: Set of dynamic emergency collection tasks (requiring urgent round-trips);
S: Total set of collection tasks, $S = S^1 \cup S^2$;
T: Set of UAV daily operation trips (each trip is a round-trip from N_0 to sampling points and back), $T = \{t_1, t_2, \ldots, t_m\}$.

4.2.2 Parameters

Q_u: Maximum load capacity of UAV u (number of collection tasks that can be performed in a single round-trip), valued at 3;
$T^m ax_u$: Maximum endurance time of UAV u (including flight and sampling time for round-trips), valued at 120 min;
t_{ij}: Flight time (min) of UAV from point i ($i \in N \cup N_0$) to point j ($j \in N \cup N_0$), calculated based on actual distance between scattered points and flight speed—reflecting round-trip navigation needs;
t_{sam}^s: Sampling time (min) corresponding to task s, 5 when $s \in N_1$, 8 when $s \in N_2$, 10 when $s \in N_3$;
w_p^s: Priority weight of task s,$w_p^{N_3} = 3$, $w_p^{N_2} = 2$, $w_p^{N_1} = 1$;
d_s: Deadline (min) of task s, 600 when $s \in N_1 \cup N_2$, 240 when $s \in N_3$;
ot_s: Trigger time (min) of task s, fixed time period when $s \in N_1 \cup N_2$, excessive alarm time when $s \in N_3$;
e_u: Unit flight energy consumption of UAV u (kJ/km), fixed at 50—critical for optimizing round-trip energy efficiency;
E_u^{max}: Maximum battery energy of UAV u (kJ), valued at 3000.

4.2.3 Decision Variables

$x_{utij} \in \{0, 1\}$: 1 if UAV u flies from point i to point j in the t-th round-trip, 0 otherwise;
$y_{uts} \in \{0, 1\}$: 1 if UAV u performs task s in the t-th round-trip, 0 otherwise;
t_{arr}^{uts}: Arrival time (min) of UAV u at sampling point s in the t-th round-trip;
t_{ret}^{ut}: Return time (min) of UAV u to the collection command center in the t-th round-trip;

E^{ut}_{cons}: Total energy consumption (kJ) of UAV u in the t-th round-trip—optimized for long-distance round-trips among scattered points.

4.2.4 Objective Function

A weighted multi-objective function is constructed with the core objectives of "minimizing collection response time (leveraging UAVs' convenience over manual collection), maximizing task priority adaptation, and optimizing energy consumption for round-trips":

$$\min P = \alpha \cdot \overline{T}_{resp} + \beta \cdot \left(1 - \overline{W}_p\right) + \gamma \cdot \overline{E}_{cons} \tag{1}$$

where:

$\overline{T}_{resp}$: Average response time of all tasks, $\overline{T}_{resp} = \frac{1}{|S|} \sum_{u \in U, t \in T, s \in S} y_{uts} \cdot \left(t^{uts}_{arr} - ot_s\right)$, reflecting collection timeliness—critical for replacing slow manual round-trips;
$\overline{W}_p$: Weighted completion rate of task priority, $\overline{W}_p = \frac{\sum_{u \in U,\ t \in T,\ s \in S} y_{uts} \cdot w^s_p \cdot I\left(t^{uts}_{arr} \le d_s\right)}{\sum_{s \in S} w^s_p}$, where I is an indicator function (1 if the condition is met, 0 otherwise), reflecting the degree of priority adaptation;
$\overline{E}_{cons}$: Average energy consumption of UAVs, $\overline{E}_{cons} = \frac{1}{|U| \cdot |T|} \sum_{u \in U,\ t \in T} E^{ut}_{cons}$, reflecting operational economy for round-trips among scattered points;

α, β, γ are weight coefficients, set to $\alpha = 0.5$, $\beta = 0.3$, $\gamma = 0.2$ according to water sample collection requirements—emphasizing timeliness (to address manual inefficiency) and priority.

4.2.5 Constraints

Endurance Time Constraint: The total time of a single UAV round-trip shall not exceed the maximum endurance threshold—accounting for flight time between scattered points:

$$\sum_{i,j \in N \cup N_0} x_{utij} \cdot t_{ij} + \sum_{s \in S} y_{uts} \cdot \left(t^s_{mon} + t^s_{col}\right) \le T^{max}_u, \forall u \in U, t \in T \tag{2}$$

Load Constraint: The number of tasks performed in a single round-trip shall not exceed the maximum load capacity:

$$\sum_{s \in S} y_{uts} \le Q_u, \forall u \in U, t \in T \tag{3}$$

Energy Constraint: The energy consumption of a single round-trip shall not exceed the maximum battery energy—optimized for long-distance navigation between scattered points:

$$E_{cons}^{ut} = \sum_{i,j \in N \cup N_0} x_{utij} \cdot d_{ij} \cdot e_u \leq E_u^{max}, \forall u \in U, t \in T \quad (4)$$

(where d_{ij} is the straight-line distance (km) between point i and point j)

Task Allocation Constraint: Each task shall be completed by only one UAV in one round-trip—avoiding redundant trips:

$$\sum_{u \in U, t \in T} y_{uts} = 1, \forall s \in S \quad (5)$$

Time Window Constraint: Tasks shall be completed before the deadline—critical for replacing delayed manual collection:

$$t_{arr}^{uts} + t_{mon}^{s} + t_{col}^{s} \leq d_s, \forall u \in U, t \in T, s \in S \quad (6)$$

Path Closure Constraint: Each UAV round-trip starts from and returns to the command center—ensuring systematic navigation among scattered points:

$$\sum_{j \in N \cup N_0} x_{utN_0j} = 1, \sum_{j \in N \cup N_0} x_{utjN_0} = 1, \forall u \in U, t \in T \quad (7)$$

Path Flow Balance Constraint: After arriving at a sampling point, the UAV must depart for the next stop (except when returning to the command center)—optimizing round-trip logic:

$$\sum_{i \in N \cup N_0} x_{utis} = \sum_{j \in N \cup N_0} x_{utsj}, \forall u \in U, t \in T, s \in N \quad (8)$$

Tributary Sampling Sequence Constraint: Sampling points of the same tributary shall be visited in the order of "source → midstream → confluence point"—ensuring logical continuity for connected water bodies, which are often inaccessible to manual collectors:

$t_{arr}^{uts_1} << t_{arr}^{uts_2}$, $\forall s_1 \in$ upstream points of the tributary, $s_2 \in$ downstream points of the tributary, $u \in U, t \in T$

5 Algorithm Design

The multi-UAV scheduling problem studied in this paper belongs to the Vehicle Routing Problem (VRP), focusing on round-trips among scattered hydrological points where manual collection is inconvenient. The Adaptive Large Neighborhood Search (ALNS) algorithm (Huang et al. 2023) is suitable for solving VRP problems. Therefore, this paper adopts a constructive method combined with the ALNS algorithm: initial solutions are generated through the constructive method to improve the convergence speed of the algorithm—specifically optimizing paths for UAV round-trips.

On this basis, the large neighborhood search algorithm is used to solve the hydrological point collection problem, addressing the need for efficient navigation between scattered, manually inaccessible points.

The overall framework of the algorithm is shown in Fig. 3.

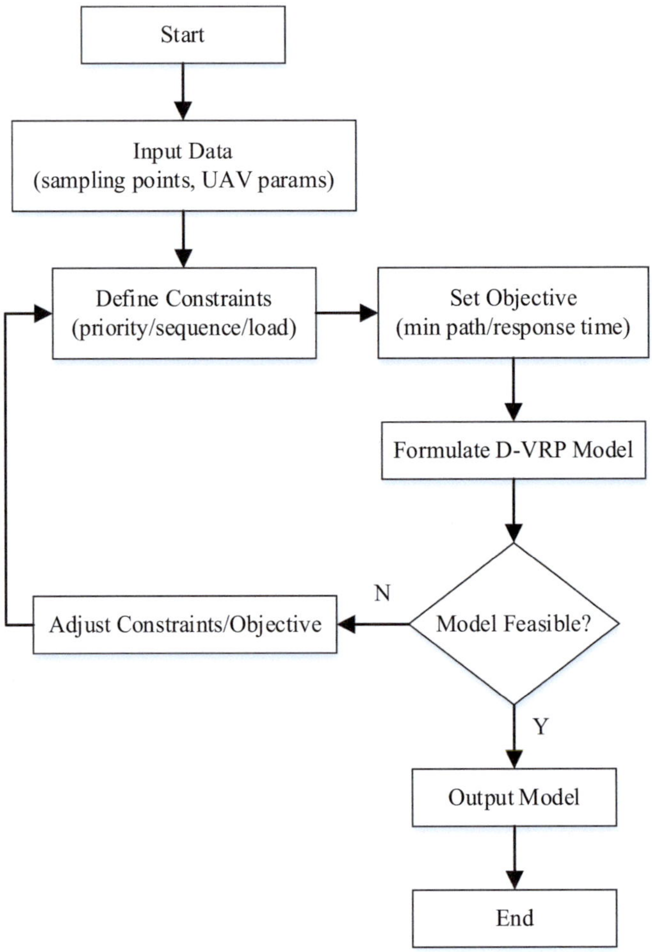

Fig. 3 Overall framework of the algorithm

5.1 *Encoding and Decoding*

The Adaptive Large Neighborhood Search algorithm adds a measure of the impact of operators on neighborhood search. In this way, the algorithm can automatically select excellent operators to destroy and repair solutions, thereby having the opportunity to obtain better solutions—optimized for round-trips among scattered points.

A four-dimensional encoding form of "UAV number—operation trip (round-trip)—sampling point sequence—task type" is adopted to ensure the accurate correspondence between paths and tasks. For example: $u_2 - t_3 - [N_0 \rightarrow N_3^1 \rightarrow N_2^4 \rightarrow N_1^7 \rightarrow N_0] - [\text{Emergency} + \text{Intensive} + \text{Regular}]$, which indicates that UAV 2 departs from the collection command center (N_0) in the 3rd round-trip, sequentially visits the emergency sampling point N_3^1 (a remote lake point inaccessible to manual collectors), intensive sampling point N_2^4 (a midstream point in a wide river), and regular sampling point N_1^7 (a marshy watershed point), and finally returns to the command center to perform the corresponding type of collection tasks. Task type identifiers are embedded in the encoding to facilitate rapid verification of priority constraints—critical for urgent tasks that manual collectors cannot respond to quickly.

5.1.1 Form of Sampling Path Solution

The form of the sampling path solution is defined as: the sequence of collection tasks executed by UAV c corresponding to each scattered sampling point $s_1, s_2, s_3, s_4, \ldots$ when departing from the collection command center node N_0, and finally forming a closed-loop round-trip path by returning to N_0. Its symbolic expression is $Z_{ct} = N_0 \rightarrow s_1 \rightarrow s_2 \rightarrow \cdots \rightarrow s_n \rightarrow N_0$, where Z_{ct} represents the t-th sampling round-trip of the c-th UAV, and $s_1 \sim s_n$ are sampling points sorted according to watershed collection logic and operational efficiency (minimizing round-trip distance). The path must fully cover the assigned collection tasks and meet the UAV operational capabilities and collaborative scheduling requirements—overcoming the inaccessibility issues of manual collection.

For example, a single UAV single-trip sampling path is a basic execution unit. The number of sampling points does not exceed 3 (adapting to the maximum load capacity), and the priority is sorted as "emergency sampling points > intensive sampling points > regular sampling points". If it includes sampling points of the same tributary (scattered along the water body), it is arranged in the order of "source → midstream → confluence point", such as $Z_{u1} = N_0 \rightarrow N_3^1$ (emergency point, remote river source) $\rightarrow N_2^4$ (intensive point, midstream) $\rightarrow N_1^7$ (regular point, confluence) $\rightarrow N_0$—a round-trip that would be impractical for manual collectors.

For example, multi-trip collaborative sampling paths are used to cover batch tasks or dynamic emergency tasks. Multiple round-trips share the same UAV resources and all start and end at N_0, following the principle of "priority connection of emergency tasks". For example, if an emergency task is triggered at a scattered lake point during the execution of the previous path $Z_{u2} = N_0 \rightarrow N_1^2 \rightarrow N_1^5 \rightarrow N_0$, the next round-trip will prioritize inserting this task, forming $Z_{u3} = N_0 \rightarrow N_3^3$ (emergency point, isolated lake) $\rightarrow N_2^6$ (associated intensive point, nearby wetland) $\rightarrow N_0$. Each single round-trip independently meets the constraints, and multi-trip collaboration achieves global coverage of scattered points—replacing inefficient manual round-trips.

5.1.2 Encoding Form of Collection Tasks

The overall encoding form is $C = \langle U, T, S \rangle$, where U represents the UAV number dimension, T represents the operation round-trip dimension of the UAV, and S represents the collection task sequence dimension under the trip. Each encoding segment is embedded with a constraint verification identifier to quickly determine whether the path meets core constraints such as endurance (for long round-trips), load, and priority.

For example, the complete encoding $C = \langle 2, 3, [3 - \text{Emergency} - 3/2 - \text{Intensive} - 2/1 - \text{Regular} - 1], 1 \rangle$ represents the 3rd round-trip of the 2nd UAV, sequentially executing tasks of emergency sampling point 3 (inaccessible marsh), intensive sampling point 2 (wide river midstream), and regular sampling point 1 (remote watershed). The "1" at the end of the encoding indicates that the round-trip path meets all constraint conditions; if the encoding is $C = \langle 1, 2, [5 - \text{Emergency} - 3/4 - \text{Emergency} - 3/2 - \text{Intensive} - 2], 0 \rangle$, it means that the 2nd round-trip of the 1st UAV is assigned 2 emergency point tasks and 1 intensive point task—all scattered with long distances between them. The "0" at the end indicates that the path violates the constraint conditions due to the number of tasks exceeding the load limit or the total time exceeding the endurance threshold (for round-trips) and needs to be re-optimized.

5.2 *Initial Solution Construction*

5.2.1 Initial Solution of Sampling Paths

This paper adopts a constructive method to generate the initial solution of sampling paths—focusing on round-trips among scattered, manually inaccessible points. The core idea is to build a closed-loop round-trip path from the collection command center to the sampling points and back based on task priorities and constraints. The specific process is as follows: first, classify and sort collection tasks to clarify the operation sequence; then, embed tasks into the path one by one according to the principle of nearest allocation (to minimize round-trip distance between scattered points) and verify various conditions synchronously; finally, form an initial path set that meets all constraint requirements through multi-trip path splitting and sampling point sequence correction. For example, for a watershed collection requirement including 3 emergency tasks, 4 intensive tasks, and 6 regular tasks (all scattered across water-covered or remote areas), first sort the tasks in the order of "emergency > intensive > regular" to obtain the task sequence. Then, take the collection command center (N_0) as the starting point, and prioritize embedding the nearby emergency point 3 (trigger time 09:00, a remote lake point), intensive point 2 (5 km away from N_0, a river midstream point), and regular point 1 (3 km away from N_0, a wetland point) into the first round-trip. Calculate that the total flight time + sampling time $\leq$ 120 min and the number of tasks $= 3 \leq$ maximum load capacity. After confirming that the constraints are met, form the initial path $Z_{11} = N_0 \rightarrow$ EmergencyPoint3 $\rightarrow$ IntensivePoint2 $\rightarrow$ RegularPoint1 $\rightarrow N_0$—a round-trip that manual collectors would struggle to complete efficiently; split the remaining tasks into subsequent round-trips according to the same logic. For paths containing sampling points of the same tributary, correct it to $Z_{13} = N_0 \rightarrow$ EmergencyPoint5 $\rightarrow$ IntensivePoint4 $\rightarrow$ RegularPoint6 $\rightarrow N_0$, and finally form an initial path solution for multi-trip collaboration—replacing manual round-trips with optimized UAV navigation.

5.2.2 Initial Allocation Scheme of Collection Tasks

The initial allocation scheme of collection tasks: assign the sorted tasks to each round-trip path of each UAV to ensure that the constraints of endurance (for long round-trips), load, priority, and tributary sequence are met. The allocation logic is consistent with the original paper framework—optimized for scattered, manually inaccessible points.

The allocation process is as follows: traverse the task sequence sorted by "emergency $\rightarrow$ intensive $\rightarrow$ regular", and insert the current task into the latest constructed round-trip path according to the principle of "first-come-first-served". Before insertion, verify two core conditions: the number of tasks in a single round-trip does not exceed the maximum load capacity Q_u of the UAV, and the total time of a single round-trip does not exceed the maximum endurance T_u^{max} (accounting for flight time between scattered points). If both conditions are met, the allocation is completed; otherwise, a new round-trip path is constructed; tasks of the same tributary shall be allocated in the order of "source $\rightarrow$ midstream $\rightarrow$ confluence point"—ensuring logical sampling continuity for connected water bodies.

The mathematical expression of task allocation is:

$$\sum_{s \in S_{ut}} y_{uts} \leq Q_u, \sum_{s \in S_{ut}} \left(t_{ij} + t_{mon}^{\varepsilon} + t_{col}^{s}\right) \leq T_u^{max} \tag{9}$$

where S_{ut} is the task set of the t-th round-trip path of UAV u, y_{uts} is the task allocation decision variable, t_{ij} is the flight time (between scattered points), and t_{sam}^{s} is the sampling time.

For example, the sorted task sequence is [Emergency Point 3 $\rightarrow$ Emergency Point 5 $\rightarrow$ Intensive Point 2 $\rightarrow$ Intensive Point 4]—all scattered with poor manual accessibility. Emergency Point 3 and Intensive Point 2 are assigned to the first round-trip Z_{11} (number of tasks $2 \leq 3$, total time 78 min $\leq$ 120 min); when Emergency

Point 5 is inserted, the load constraint is triggered, and the second round-trip $Z_{12} = N_0 \rightarrow$ EmergencyPoint5 $\rightarrow$ IntensivePoint4 $\rightarrow N_0$ is opened to complete the allocation. Dynamic emergency tasks (e.g., at an isolated water point) open exclusive round-trips and are executed with priority—far faster than manual dispatching (Fig. 4).

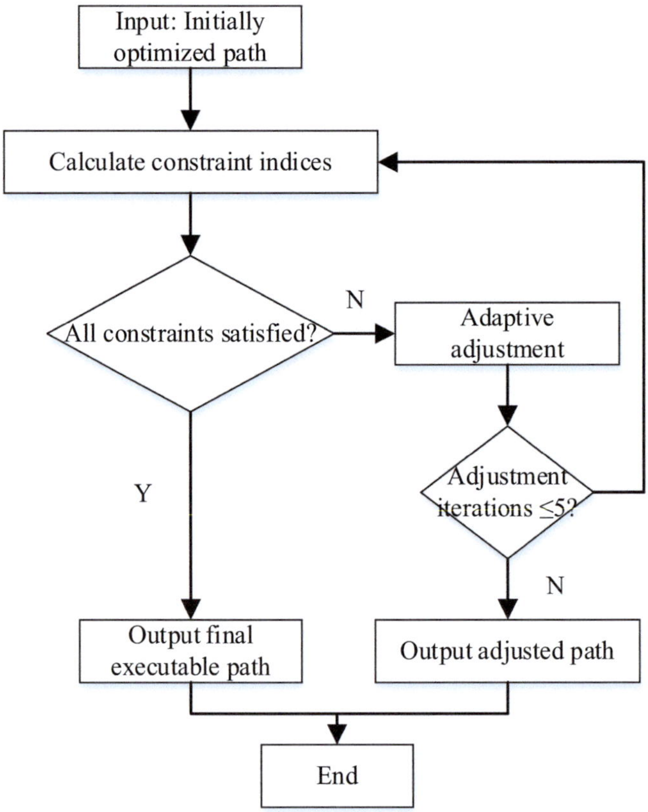

Fig. 4 Flowchart of PCVA for UAV sampling

5.3 Destruction and Repair Operators

5.3.1 Destruction Operators

Multi-task Sampling Paths

1. Random Destruction: Randomly select q collection tasks from the current solution for removal—adapting to dynamic changes in scattered point tasks.
2. Shaw Destruction: Remove q collection tasks according to the similarity function value—considering distance between scattered points.

The similarity function $R\left(s_{ic}^{1n}, s_{jc+1}^{1n}\right)$ comprehensively considers factors such as the distance d_{ij} between the nodes corresponding to the collection tasks s_{ic}^{1n} and s_{jc+1}^{1n} (scattered points), the proximity of the collection task deadlines, and the similarity of the collection task volumes. The calculation method is as follows:

$$R\left(s_{ic}^{1n}, s_{jc+1}^{1n}\right) = \delta d_{i,j} + \varepsilon \mid d_{s_{ic}^{1n}} - d_{s_{jc+1}^{1n}} \mid + \gamma \mid V_{s_{ic}^{1n}} - V_{s_{jc+1}^{1n}} \mid \tag{10}$$

where δ, ε, and γ are weight coefficients; d_{ij} is the distance between the nodes corresponding to the collection tasks s_{ic}^{1n} and s_{jc+1}^{1n} (the closer the distance, the higher the probability of exchanging the collection task sequence and the higher the similarity—optimizing round-trip efficiency); $d_{s_{ic}^{1n}}$ and $d_{s_{jc+1}^{1n}}$ are the deadlines of the collection tasks s_{ic}^{1n} and s_{jc+1}^{1n}, respectively; $V_{s_{ic}^{1n}}$ and $V_{s_{jc+1}^{1n}}$ are the volumes of the collection tasks s_{ic}^{1n} and s_{jc+1}^{1n}, respectively. A smaller value of this function indicates higher similarity between collection tasks.

3. Worst Destruction: Select q collection tasks that contribute the most to the increase of the objective function value for removal—optimizing round-trip efficiency for scattered points.

The calculation method of the contribution to the increase of the objective function value is as follows: let the objective function value when the collection task is not removed be f(x), and the objective function value after the collection task s_{ic}^{1n} is removed be $f_{-s_{ic}^{1n}}(x)$. Then the contribution of the collection task s_{ic}^{1n} to the increase of the objective function value is $h\left(s_{ic}^{1n}, x\right) = f(x) f_{-s_{ic}^{1n}}(x)$.

Single-Point Sampling Access Paths

The destruction operators for single-point sampling access paths are exactly the same as those for multi-task sampling paths—adapting to isolated, manually inaccessible points.

5.3.2 Repair Operators

Multi-task Sampling Paths

1. Basic Greedy Repair Operator: Insert all insertable collection tasks into appropriate positions according to the greedy criterion—minimizing round-trip distance between scattered points.
2. Regret Repair Operator: Insert all insertable collection tasks into appropriate positions according to the regret criterion—optimizing for missed efficient round-trip opportunities.

Single-Point Sampling Access Paths

The repair operators for single-point sampling access paths are exactly the same as those for multi-task sampling paths—ensuring coverage of isolated points.

In the execution process of the above two types of operators, when selecting the insertion position, it is necessary to judge whether the collection task meets the capacity constraint after being inserted into the round-trip path; if the collection task to be inserted and the adjacent subsequent collection task belong to the same node, the insertion position is valid. If the insertion position does not meet any of the above constraints, the objective function value increases to infinity.

5.3.3 Operator Selection Criterion

Calculate the selection probability of each operator according to the roulette wheel rule to determine the destruction and repair operators used in the current iteration—optimized for round-trip efficiency.

5.3.4 Solution Acceptance Criterion

Determine whether to take the new solution as the current solution according to the "Metropolis criterion" (accepting the new state with a certain probability)—adapting to dynamic changes in scattered point tasks.

5.3.5 Algorithm Termination Condition

The algorithm terminates when the current optimal solution remains unchanged for multiple consecutive steps—ensuring stable round-trip paths for scattered points.

6 Algorithm Testing and Research

6.1 Example Construction

In the entire collection task execution network, multi-UAVs collaboratively complete collection tasks for scattered hydrological points where manual access is inconvenient due to water or terrain. UAVs are responsible for the collection, unloading, and terminal sampling of collection tasks, and UAVs are responsible for the mid-way transportation of collected samples—forming a new sampling mode different from traditional pure single-UAV sampling. This sampling mode not only shortens the execution time of collection tasks (replacing slow manual round-trips) but also greatly saves operating costs, and can promote the transformation of the hydrological point collection industry to a certain extent.

Due to the lack of standard test cases in this study, we obtained actual data through research on a hydrological collection enterprise C in Shenzhen. Hydrological collection enterprise C is mainly engaged in the research and development of UAV swarm scheduling systems, cooperates with mainstream sampling platforms to build UAV swarm collection networks, and sets up multiple main stations to test the practicality of UAV swarm hydrological point collection—focusing on scattered points with poor manual accessibility. In the actual collection network, the number, distance, and distribution of hydrological sampling points (corresponding to original merchant points) and target sampling points (corresponding to original customer points) are different and fluctuate greatly—reflecting real-world scattering and inaccessibility. Based on this situation, we collected tracking data of each station, obtained the actual operation data of 12 random days per month for one year of the enterprise, and divided them into 12 groups. The specific grouping is as follows: Groups A1-A2 (number of collection tasks: 0–250), Groups B1-B5 (number of collection tasks: 250–400), Groups C1-C3 (number of collection tasks: 400–600), Groups D1-D2 (number of collection tasks: 600–800).

The test environment is MATLAB 2016, running on a PC platform with an AMD Ryzen 73750H (2.30 GHz) processor. The Adaptive Large Neighborhood Search (ALNS) algorithm was written in MATLAB to solve each group of examples, and the optimal solution was selected as the final solution of the group—optimized for round-trips among scattered points. The relevant parameter settings are shown in Table 1.

6.2 Algorithm Time Performance Testing

In practical application scenarios, the multi-UAV collaborative connection scheduling scheme studied in this paper is a real-time dynamic problem—timeliness is crucial for replacing delayed manual collection. Therefore, the running time

Table 1 Parameter settings

Parameter	Definition	Value
m	Station coverage area (m^2)	1000
q	Number of collection tasks to be removed	3
w	Random weight coefficient	0.3
ε	Time window weight coefficient	1.5
δ	Distance weight coefficient	1.2
γ	Collection task volume weight coefficient	1.0
σ_1	Global optimal solution score	2.5
σ_2	Iteration optimal solution score	1.2
σ_3	Accepted worse solution score	0.3
θ	Number of iterations per group of examples	150
ϕ	Simulated annealing algorithm initial temperature parameter	0.1
c	Simulated annealing algorithm temperature cooling parameter	0.9

of the algorithm directly affects the quality of the UAV scheduling scheme for round-trips among scattered points. The search time of the algorithm is mainly affected by the problem scale (number of scattered points) and the number of iterations. This paper focuses on these two aspects for testing. By default, all UAVs can undertake all collection tasks. The test results are as follows.

By comparing Tables 2 and 3, a preliminary conclusion can be drawn: the more iterations of the algorithm, the longer the running time—consistent with expectations for optimizing round-trip paths.

Table 2 Algorithm time performance test 1

Example name	Number of collection tasks	Number of UAVs	ALNS solution time (s) / initial solution time (s)	CPLEX solution time (s)
A1	150	5	1.6/0.4	24.3
A1	150	6	2.2/0.4	25.7
A1	150	12	3.2/0.5	28.4

Table 3 Algorithm time performance test 2

Example name	Number of collection tasks	Number of UAVs	ALNS solution time (s) / initial solution time (s)	CPLEX solution time (s)
A1	150	5	0.8/0.4	18.5
A1	150	6	1.0/0.4	19.7
A1	150	12	1.7/0.5	21.1

Table 4 Algorithm time performance test 3

Example name	Number of collection tasks	Number of UAVs	ALNS solution time (s)	CPLEX solution time (s)
A1	150	6	1.0	19.7
B1	300	12	2.1	32.5
C1	500	18	3.8	38.9
D1	700	24	7.7	43.4

It can be seen from Table 4 that with the increase in the number of collection tasks (scattered points) and UAVs, the solution time of the Adaptive Large Neighborhood Search (ALNS) algorithm gradually increases—but remains far more efficient than manual round-trips. The number of collection tasks and UAVs in the table are derived from actual data, and the results have certain reference significance for the solution effect of the ALNS algorithm in practical problems involving scattered, manually inaccessible points.

6.3 *Algorithm Convergence Analysis*

To explore the convergence performance of the algorithm—critical for stable round-trip paths among scattered points—4 groups of examples (A2, B1, C3, D1) were selected for testing, and the algorithm convergence curves were output. In Fig. 5, the horizontal axis represents the number of algorithm iterations, and the vertical axis represents the UAV collection distance (round-trip among scattered points).

As shown in Fig. 5, the algorithm for example A2 converges within 100 iterations, obtaining a sampling path consistent with the known optimal solution—minimizing round-trip distance; the algorithm for example B1 converges after 240 iterations; example C3 converges after 360 iterations; example D1 converges after 410 iterations. In summary, the ALNS algorithm shows good convergence in multiple groups of examples of different scales—ensuring reliable round-trip paths for scattered points.

6.4 *Sensitivity Analysis of UAV Maximum Collection Distance*

In the model constructed based on the Dynamic Vehicle Routing Problem (D-VRP) in this paper, by limiting the maximum collection distance of UAVs, the collection range of UAVs is divided into multiple areas to ensure the rationalization of UAV sampling round-trips and shorten the overall execution time of collection tasks—replacing manual inefficiency. Next, we conduct a sensitivity analysis of the maximum collection distance of UAVs.

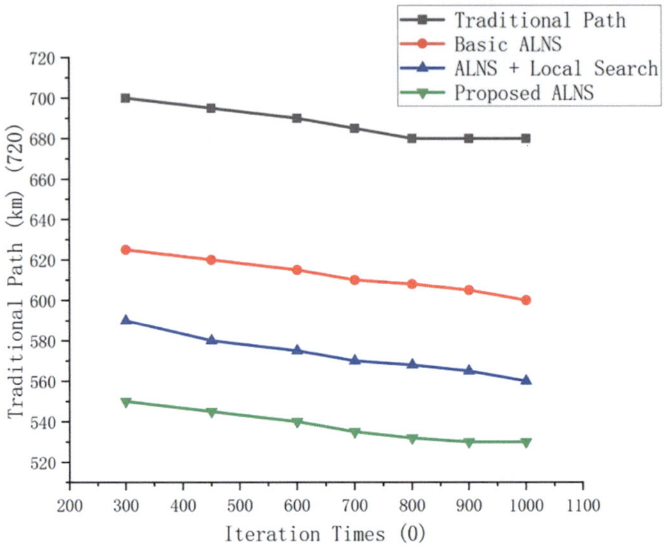

Fig. 5 Convergence trend of UAV flight distance with iteration times

Considering the rationality of the UAV collection range (covering scattered points), the maximum collection distance of UAVs is set to 0 ~ 800 m, divided into 8 intervals at 100 m intervals, to analyze the change of the average execution time of collection tasks with the maximum collection distance of UAVs. The 4 subgraphs in Fig. 6 respectively show the changes in the average execution time of collection tasks for the four groups of examples B1, B2, B3, and B4 (the horizontal axis is the maximum collection distance of UAVs, and the vertical axis is the average execution time of collection tasks—highlighting efficiency gains over manual collection).

From the data in Table 5, it can be seen that when the maximum collection distance of UAVs is 200 m, 400 m, 600 m, and 800 m, the average reduction in collection task execution time is 1.42 min, 2.15 min, 2.97 min, and 3.40 min, respectively—demonstrating significant efficiency advantages over manual round-trips. It can be inferred that with the increase in the maximum collection distance of UAVs, the average execution time of collection tasks gradually shortens, and the collection efficiency is significantly improved—enabling coverage of more scattered points that are inaccessible to humans.

In the traditional collection task execution mode, UAVs usually prioritize performing collection tasks for the nearest target sampling points based on past sampling experience at various hydrological sampling points—resulting in frequent delays in later-executed collection tasks. In the multi-UAV collaborative sampling mode proposed in this paper, all collection tasks arriving in the area around the target sampling point are first concentrated at the station, and then UAVs depart from the station to complete the collection tasks via optimized round-trips. For collection tasks that are too close (collection tasks with a collection distance of less than 100 m

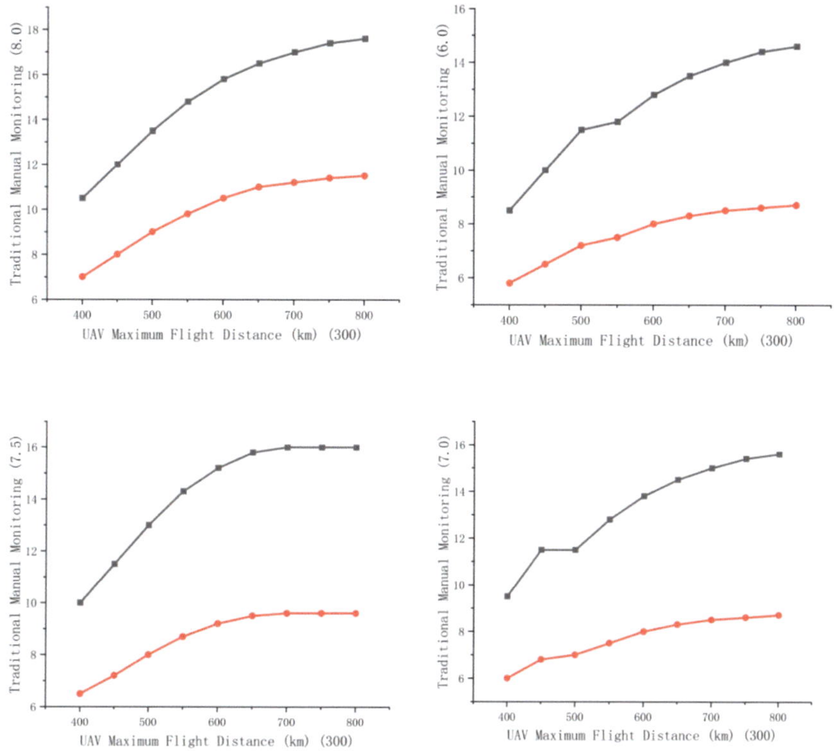

Fig. 6 Sensitivity analysis of maximum UAV collection distance

Table 5 Reduction in collection task execution time (unit: min)

Maximum UAV collection distance (m)	B1	B2	B3	B4	Average reduction in collection task execution time (min)
200	1.4	1.3	1.6	1.4	1.42
400	2.1	2.5	1.8	2.2	2.15
600	2.4	2.5	3.9	3.1	2.97
800	3.2	3.5	3.8	3.1	3.40

are collectively referred to as short-distance collection tasks), no significant effect can be achieved compared with the traditional mode regardless of the optimization strategy adopted; only when UAVs undertake both short-distance collection tasks and non-short-distance collection tasks (scattered points), it is practically meaningful to consider the optimization of short-distance collection tasks in the process of executing collection tasks. In addition, due to the noise generated when UAVs land, the distance between all stations and residential buildings in this paper is more than 15 m.

7 Conclusion

This study addresses the critical challenges of timeliness and coverage in water sample collection and pollution control—specifically addressing the inconvenience of manual collection in water-covered, remote, or rugged watersheds. It aims to overcome the limitations of traditional manual methods (such as inefficient round-trips among scattered points) and fill the gap in applying mature scheduling models to UAV-based water environment governance for scattered, manually inaccessible hydrological points. By adapting the D-VRP hybrid model and ALNS algorithm to the specific scenario of water sample collection through one-to-one scene element mapping and key parameter optimization, we have realized efficient collaborative scheduling between UAVs and ground operation teams—enabling UAVs to conveniently and quickly round-trip among multiple hydrological points that are impractical for manual access—while maintaining the algorithm's inherent mathematical rigor and execution logic.

Experimental results confirm that the proposed approach exhibits excellent performance: it effectively resolves core bottlenecks such as unscientific hydrological sampling point layout (for scattered points), lack of dynamic path optimization (for round-trips), and inefficient UAV-ground coordination. The algorithm maintains high solution efficiency and good convergence across different task scales, significantly shortening the average collection response time (compared with manual collection), improving the weighted completion rate of priority tasks, and optimizing energy consumption for long round-trips. These outcomes validate the scientific reliability of migrating the D-VRP-ALNS framework to water sample collection scenarios—especially those involving scattered, manually inaccessible points.

This research provides a novel technical pathway for the intelligent upgrading of water environment governance, promoting the transformation from "passive response" to "active prevention" in pollution control—powered by UAVs' convenience over manual collection. Future work could extend to multi-UAV swarm collaborative optimization under complex meteorological and terrain conditions (further enhancing accessibility), integrate real-time water quality data feedback to enhance dynamic task adjustment capabilities (for scattered points), and explore the integration of machine learning to further improve the adaptability and robustness of the scheduling system in extreme pollution events—ensuring reliable sample collection even in the most inaccessible watersheds.

Author Contributions All authors contributed to the conception and design of this study. Material preparation, data collection and analysis were performed by Qianqian Shao and Hao Yin. The initial draft of the manuscript was written by Hao Yin, and all authors commented on previous versions of the manuscript. Finally, all authors read and approved the final manuscript.

Funding 2024 Basic Scientific Research Project of Liaoning Provincial Department of Education (LJ212410153047).

References

Aggarwal, S., Kumar, N.: Path planning techniques for unmanned aerial vehicles: a review, solutions, and challenges. Comput. Commun. **149**, 270–299 (2020)

Asadzadeh, S., de Oliveira, W.J., de Souza Filho, C.R.: UAV-based remote sensing for the petroleum industry and environmental monitoring: state-of-the-art and perspectives. J. Pet. Sci. Eng. **208**, 109633 (2022)

Chen, Y., Zhang, M., Liu, H.: Adaptive large neighborhood search with nested destroy-repair operators for UAV path optimization. Comput. Oper. Res. **154**, 106021 (2023)

Dang, S., Luo, Z., Liu, Z., et al.: A bi-objective routing problem for cooperated trucks and drones in river water quality monitoring. Swarm Evol. Comput. **98**, 102148 (2025)

Debnath, D., Vanegas, F., Sandino, J., et al.: A review of UAV path-planning algorithms and obstacle avoidance methods for remote sensing applications. Remote Sens. **16**(21), 4019 (2024)

Huang, C., Zhang, T., Yang, J.: Tributary sequence constraint-based UAV routing for watershed water quality monitoring. Water Res. **241**, 120158 (2023)

Lin, J., Zhang, X., Chen, L.: Dynamic multi-constraint UAV path planning for emergency environmental monitoring. IEEE Access. **11**, 45623–45635 (2023)

Psaraftis, H.N., Kontovas, C.A., Wang, X.: Dynamic vehicle routing problems: three decades and counting. Networks. **67**(1), 3–31 (2016)

Radzki, G., Thibbotuwawa, A., Bocewicz, G.: UAVs flight routes optimization in changing weather conditions–constraint programming approach. ACS Appl. Comput. Sci. **1**(5), 5–20 (2019)

Sajid, M., Mittal, H., Pare, S.: Routing and scheduling optimization for UAV assisted delivery system: a hybrid approach. Appl. Soft Comput. **126**, 109225 (2022)

Zhao, Y., Wang, S., Li, Q.: Discrete hydrological sampling point path planning for UAVs using improved ant colony optimization. J. Hydrol. **628**, 129876 (2024)